Integated IT Project Management

A Model-Centric Approach

For a listing of recent titles in the *Artech House Effective Project Management Library,* turn to the back of this book.

Integrated IT Project Management

A Model-Centric Approach

Kenneth R. Bainey

Artech House
Boston • London
www.artechhouse.com

Library of Congress Cataloging-in-Publication Data
Bainey, Kenneth R.
Integrated IT project management: a model-centric approach / Kenneth R. Bainey.
p. cm.—(Artech House project management library)
ISBN 1 58053-828-2 (alk. paper)
1. Information technology—Management. 2. Project management. I. Title. II. Series.
T58.64.B33 2004
658.4'038—dc22 CIP
2004041026

British Library Cataloguing in Publication Data
Bainey, Kenneth R.
Integrated IT project management : a model-centric approach.—(Artech House project management library)
1. High technology industries—Marketing 2. New products—Management 3. Project management
I. Title
620'.00688
ISBN 1-58053-828-2

Cover design by Igor Valdman

685 Canton Street
Norwood, MA 02062

International Standard Book Number: 1-58053-828-2
Library of Congress Catalog Card Number: 2004041026

10 9 8 7 6 5 4 3 2 1

To my mom Lutchmin Bainey and my dad Ramnanan Bainey,
a retired education principal who has a gift for the Shakespeare-style poetic language—
My admiration for his forceful command of the English language
was an inspiration to my writing aspirations.

Contents

Preface

I keep six honest serving men,
(They taught me all I know);
Their names are What and Why and When
And How and Where and Who.
—Rudyard Kipling

This book is a compilation of my extensive project management, enterprise architecture, and applications development knowledge, skills, and industry experiences gained during my 28 years as an information technology (IT) professional in Canada and the United States. It is not intended to be another theoretical book on project management. There are many excellent books on project management concepts and theory, some of which are mentioned in the selected bibliographies at the end of each chapter. If you are in search of a book on Project Management 101, Elementary Project Management, then this is not the book for you. This book assumes a certain level of understanding of basic project management concepts and theory.

However, if you are in search of real-world practical applications of project management practices and desire to obtain examples of policies, roles and responsibilities, processes and procedures, templates, and checklists for managing one or more IT projects, and if you are interested in how project management integrates with business and IT processes, then this book may be the magical solution to your project management knowledge needs. The objective of this book is to provide an integrated IT project management (IPM-IT) framework that integrates business, IT, and project management components and to demonstrate the applicability of this framework to the management of one or many projects, using a model-centric or deliverables-based approach.

During my 28 years as an IT professional, I have had the opportunity to experience both high excellence and less desirable practices in project management, according to the ESI International Project Management Maturity

Model, as defined in their *Project Framework* publication. Certainly, project management requires human resource management skills, but these skills alone are definitely not sufficient for project success. The Project Management Body of Knowledge (PMBOK) from the Project Management Institute (PMI) identifies human resource management as one of the nine knowledge areas required for the management of projects. However, some senior managers strongly believe that human resource management talent is the only solution to project management excellence, to resolving the communication problems internal and external to projects. This view that human resource management talent is the only skills requirement of project managers is often expressed by executive and senior management staff, who may have limited knowledge, skills, and experiences of project management processes. Project managers who do not demonstrate confidence and the ability to apply or objectively challenge executive management directives or to provide valuable suggestions or recommendations usually lack the required project management knowledge, skills, and processes as defined in PMBOK or similar project management processes.

The key to solving communication problems internal and external to projects is the establishment and deployment of consistent project management processes that can be effectively applied and integrated into the business and IT processes.

Effective application of modern project management principles in this technologically advancing world requires professional IT project managers with generalized business conceptualization skills; specialized process, people, and technology integration skills; and excellent risk management skills.

As a result of my real-world experiences, I decided to compile in a single volume the project management problems and challenges I experienced in both the highly excellent and less desirable project management environments and to provide a set of real-world solutions with an emphasis on the integration of business, IT, and project management processes.

The real-world solutions presented in this book demonstrate a model-centric or deliverables-based approach to managing projects that focus on the three major modern project management skills requirements of conceptualization, integration, and risk management, as discussed in Harold Kerzner's book, *Project Management: A Systems Approach to Planning, Scheduling, and Controlling, Seventh Edition.*

In order to survive in this dynamically changing information industry, we need to be progressing towards the management of projects using conceptualization, integration, and risk management skills, rather than the traditional project management skills based on schedule and cost manipulations that are sometimes applied by human resource managers, auditors, sales executives, or glorified secretaries disguised as project managers. To accomplish this transition effectively, we need to:

- Establish and implement an IPM-IT framework to guide the management and delivery of projects, optimize resource utilization efficiency, and ensure that IT projects are aligned with the goals and objectives of the corporation while meeting stakeholders' expectations of effectiveness;

- Employ professional project managers with conceptualization, integration, and risk management skills who can balance the four project management constraints—scope, quality, cost, and schedule—with the objective to optimize the utilization of IT resources—people, process, and technology—to adequately meet stakeholders' expectations;
- Educate senior and executive management in project management integration principles to assist them in making effective project management decisions;
- Demonstrate to senior and executive management the value of project management to the corporation by showing them how the integration of business management, IT management, and project management is essential to effectively managing projects to provide economic, technical, and operational value to the business.

I hope that this book will provide some valuable insights to those project and program managers, functional managers (business and IT), and executive managers (business and IT) in their search for the magic, or silver bullet, solution. To those readers who may view the contents as very advanced and may not gain any immediate value—which is unlikely—I do hope that you appreciate the what, why, when, where, who, and how of project management presented in this book and that the readings at least created some revealing, thought-provoking ideas. In this book, I attempt to make clear or create objectivity from subjective or confusing project management themes through the use of the English language interrogatives: what, why, when, how, where, and who.

Purpose of this book—why?

Widely published IT statistics from John Hopkins University reveal that:

- 30% of IT projects never reach fruitful conclusion.
- $75 billion was invested in IT in 2001.
- 51% exceed budgets by 189% and delivered only 75% functionality.
- IT investments may equate to as much as 50% of a firm's entire capital budget.

These real-world statistics confirm the need to:

- Optimize the allocation and utilization of IT resources;
- Select the best projects with the highest probability of success.

The executives at 2002 CIO Panel of Experts see integration as a business imperative—not as one possible strategy among several options, but as the only strategy.

The purpose of this book is to provide my view from an IT industry perspective as to how to address these issues and the integration imperative by providing an integrated framework of processes, practices, and real-world scenarios based my 28 years as an IT professional in Canada and United States. I compiled these practical experiences into a single comprehensive volume—this book—in the hope that IT professionals involved in project management practices will understand and use the components of this integrated project management framework as a valuable reference guide during the management and delivery of IT projects.

In this book, you will:

- Learn the key components of an IPM-IT framework and how to integrate these components vertically and horizontally;
- Learn how to improve communications internal and external to projects by applying simple, flexible, and consistent processes based on a model, which integrate with business, IT, and project management practices;
- Develop an appreciation for project management best practices that I have provided by means of real-world examples of policies, roles and responsibilities, deliverables templates, process flow templates, and checklists templates;
- Understand real-world issues involved in project management problems that have caused stressful situations for IT project managers and senior IT managers and develop an appreciation for the recommended solutions based on real-world scenarios;
- Understand the roles and responsibilities involved in managing and delivering multiple IT projects, which focus on delivering integrated solutions; this understanding will provide the needed guidance in determining specific roles and responsibilities for IT projects;
- Learn the key components of the Rational Unified Process (RUP) from IBM Rational Corporation and PMBOK from PMI and how these components align during the management and delivery of IT projects, based on these processing standards.

Who should read this book?

First and foremost, this book is intended as a reference source for the practicing IT project managers whose interests are focused toward the integration of business management, IT management, and project management. It is also intended to serve as a project management reference for functional managers (IT and business), enterprise architects (IT and business), and executive managers (IT and business) involved in making project management decisions. Finally, this text can serve as a reference for project management stakeholders, such as software application developers, business analysts, and systems architects, who are involved during project execution and wish to expand their knowledge of IPM-IT processes.

How to use this book

IT professionals who are involved in project management practices at organizations that have adopted the project management integration philosophy, in whole or part, should read this book sequentially. The chapters have been organized logically, based on the components of the IPM-IT model.

Executive managers and senior managers can limit their readings to Chapters 1 and 2, which provide an introduction to the deliverables (what), processes (how), and people resources (who) involved in implementing this IPM-IT model. These chapters provide some real-world scenarios on the value of integration that executive and senior managers will appreciate; they will perceive the information as being very informative and insightful.

Project managers and program managers should read this book linearly in order to develop a complete understanding of the what, why, how, and who of implementing integrated solutions to IT development projects. They may have to tailor the IPM-IT framework model to fit within the context of their existing organizations.

IT architects, methodologists, business analysts, and developers can limit their readings to Chapters 2, 6, and 7. Chapter 2 provides an introduction to the IPM-IT model, Chapter 6 shows how this model can be applied during the software development process or phase of the IT project delivery life cycle (PDLC), and Chapter 7 provides guidelines on how this IPM-IT framework can be used to align RUP with PMBOK processes.

Organization of this book

Integrated IT Project Management Framework: A Model-Centric Apporach provides a model-centric framework solution that is based on a specific set of business management, IT management, and project management components that formed the basis for the integrated (horizontal and vertical) solution. At the end of each chapter, there is a summary that highlights the key model-centric concepts discussed in that chapter; footnotes further explain the meaning and context of for some of these concepts; and a questions section at the end of each chapter provides some questions for the readers to objectively determine the level of project management maturity at their company. The reader will find texts or diagrams reproduced throughout this book in order to provide the necessary continuous and logical flow of information to enable a more comprehensive understanding of the foundation principles and real-world applicability.

- Chapter 1—Introduction to Integrated IT Project Management—introduces the components of IPM-IT, discusses real-world problems, and provides real-world solutions during the management and delivery of a major IT project. Guidelines on gaining executive management support that challenge project management myths are also presented to provide some helpful hints and useful insights for the practicing project managers.

- Chapter 2—Integrated IT Project Management Model (Framework) —introduces the IPM-IT model and discusses the business management, project management, and IT management components of the model. The key responsibilities of the program delivery manager are provided to demonstrate the integration skills required during the management and delivery of a program or multiple IT projects.
- Chapter 3—Business Management Model—expands on the business management model introduced in Chapter 2, discusses the components of the model, and provides the purpose, policies, or guiding principles, roles and responsibilities, and deliverable, process flow, and checklist (measurement criteria) templates for each of the components. The responsibilities of the program delivery manager discussed in Chapter 2 are further expanded to include the roles and responsibilities of executive sponsors—business, project management and IT, program managers—business, project management and IT, and the project team. These roles and responsibilities demonstrate the nature of business management integration with executive sponsors, program managers, and the project team during the management and delivery of a program or multiple IT projects.
- Chapter 4—Project Management Model—expands on the project management model introduced in Chapter 2, discusses the components of the model, and provides the purpose, policies, or guiding principles, roles and responsibilities, and deliverable, process flow, and checklist (measurement criteria) templates for each of the components. The responsibilities of the program delivery manager discussed in Chapter 2 are further expanded to include the roles and responsibilities of executive sponsors—business, project management and IT, program managers—business, project management and IT, and the project team. These roles and responsibilities demonstrate the nature of project management integration with executive sponsors, program managers, and the project team during the management and delivery of a program or multiple IT projects.
- Chapter 5—IT Management Model—expands on the IT management model introduced in Chapter 2, discusses the components of the model, and provides the purpose, policies or guiding principles, roles and responsibilities, and deliverable, process flow, and checklist (measurement criteria) templates for each of the components. The responsibilities of the program delivery manager discussed in Chapter 2 are further expanded to include the roles and responsibilities of executive sponsors—business, project management and IT, program managers—

business, project management and IT, and the project team. These roles and responsibilities demonstrate the nature of IT management integration with executive sponsors, program managers, and the project team during the management and delivery of a program or multiple IT projects.

- Chapter 6—Integrated IT Project Delivery Life-Cycle Model—expands on the IT PDLC model introduced in Chapter 2, discusses the components and phases of the model, and provides the alignment or horizontal integration with business management, project management, and IT management components, based on the IT PDLC phases.
- Chapter 7—Aligning PMI-PMBOK Processes with IBM Rational Corporation RUP—shows the integrative nature of these processes and their alignment with the IPM-IT framework presented in this book. The IPM-IT framework is modified to include RUP components—business management, project management, and RUP management components. This framework forms the basis for understanding the alignment or horizontal integration with RUP, based on IBM Rational Corporation RUP life-cycle phases.
- Appendixes are presented to show the practical automated implementation of the integrated work breakdown structure (WBS).
- The Internet Web sites http://www.ICCP.org and http://www.kenbainey.ca contain Microsoft PowerPoint presentations and templates presented in this book. These can be easily modified, adapted, and applied to delivering integrated IT project management (IPM-IT) solutions.

Acknowledgments

I wish to acknowledge the editorial assistance provided by Ron Powley, CMC senior project manager at AGTI Consulting Services (West), Calgary, Alberta, Canada; Michael Frenette, ISP vice president PMI chapter, Nova Scotia, Canada; Eldon Wig, M.Sc., PMP principal consultant, Soft-Coach IT project management consulting, Canada; Kewal Dhariwal, CCP, ISP lecturer, Athabasca University, Canada; and Myles Diamond, senior project manager, Edgeware Inc., Edmonton, Alberta, Canada. I wish to express my thanks and gratefully appreciate their valuable suggestions.

Finally, and most importantly, I wish to acknowledge my wife, Carol Bainey, my older son Kevin Bainey, M.D., and my younger son Kristian Bainey, computing science student, for all their encouragement; without such support, this book would never have been written.

CHAPTER

Contents

Introduction to Integrated IT Project Management

If we built houses the way we build software, the first woodpecker to come along would destroy civilization.
—U.S. Deputy Defense Secretary John J. Hamre, in testimony before the U.S. Senate Armed Services Committee, June 1998.

1.1 Introduction

There is no silver bullet.
—Fred Brooks

Integrated information technology (IT) project management, as defined in this book, is the process of integrating business management, IT management, and project management components during the software development process with the objective of effectively and efficiently optimize the utilization of IT resources—people, processes, and technology, while adequately meeting the expectations of stakeholders.[1] Figure 1.1 is a conceptual representation of these three major components that will be discussed in further detail throughout this book. The main focus is on the urgent need for the optimization of IT resources while meeting stakeholders' expectations, by improving time, effort, and costs utilization, and enhancing quality and scope objectives, through the application of an integrated IT project management framework (IPM-IT). This framework is a set of practical and simple policies, roles and responsibilities, procedures-deliverables, process flows, and

1. Stakeholders: As stated in PMBOK, "Individuals and organizations that are actively involved in the project, or whose interests may be positively or negatively affected as a result of project execution or project completion. They may also exert influence over the project and its results."

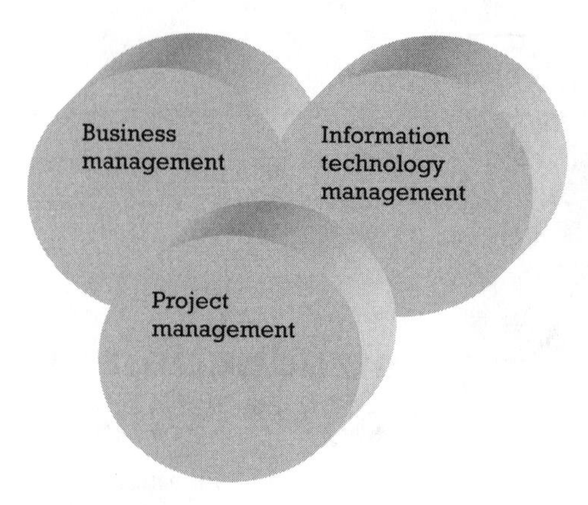

Figure 1.1 IPM-IT: Conceptual view.

checklists that demonstrate the integrated nature of these three components. The uniqueness of this framework is its model-centric[2] approach, based on deliverables or "nouns," rather than the traditional process-, or "verb-," oriented method. The model-centric approach discussed in this book is a unique and simple solution that can be readily applied during the management and delivery of one or more IT projects to effectively address the root causes of project management major challenges—communications problems internal and external to projects.

During my 28 years of IT industry experiences at various large companies in Canada and the United States, I have had the opportunity to work with senior project managers who, rightly so, stress the importance of the four famous project management objectives:

1. On schedule (time);
2. Within budget (cost);
3. According to requirements (scope);
4. Meeting acceptable criteria (quality).

However, some of these senior project managers seem to manage projects based on these four famous project management objectives without any knowledge or consideration for business, IT, and project management integration. In the majority of cases, these project managers have announced success, using the criteria of successfully meeting approved schedule (time) and budget (cost) objectives, with little regard for scope and quality objectives and without any consideration for integration, maintainability, reusability, consistency, and completeness.

Senior project management focus on the human resources aspects of project management with little or no concern for what, why, or how these deliverables are produced to support the business is sometimes a common

2. Model-centric: An approach to implementing IPM-IT using a deliverables-oriented or noun-based method to represent information and a process-oriented or verb-based method to represent processes.

directive by those senior managers who may have limited exposure to the fundamental principles of project management integration. Project managers, when confronted by executive management about business, IT, and project management integration requirements, manage to recommend and convince executive management that the solution is a project management methodology. After closer assessments of this solution, it is evident that certain project managers' perceptions of a given methodology are based on unstructured documentation, which can be classified as a "Victorian-novel-style" set of processes that support some ancient traditional verb-oriented method. In my view, we have just compounded the problem rather than solving the root causes of project management's major challenges—communication problems internal and external to projects.

In a majority of such cases, rework is the acceptable mode of operations, through a new set of "disguised" projects with similar functionality as previous projects, but with different labels, renamed as phase 2 of the original project name. The end result is additional people, processes, and technology resources (time, effort, and costs) being actively applied to these disguised projects to cover up the failures of the original projects. This is a typical case of redundant and nonproductive time, effort, and costs being applied to projects that keep adding to the company's valuable IT resources. The following problems continue to resurface, resulting in the search for the magical silver bullet solution:

- Complex scope;
- High costs;
- Schedule slippage;
- Inconsistent and incomplete quality;
- Dissatisfied stakeholders;
- Redundant and duplicated efforts.

Senior management's perception of the causes of such project failures blames the methodology solution—there is too much emphasis on methodology and too little emphasis on people. The new directive from senior management to focus on people rather than processes normally results in a chaotic project management environment with inconsistent, redundant, and incomplete processes and with project managers having no documented project accountability—the consultant project manager's favorite working environment. The cycle in the search for the silver bullet solution, presentations on business and IT value to executive management for more budget, and promises for more effective management of projects continues to reoccur with IT senior management carefully crafting excuses to business management in search of the nonexistent silver bullet solution.

There is no silver bullet solution. These problems continue to resurface because of the lack of a business, IT, and project management integration framework that is necessary and essential to resolving the root causes of project management challenges—communication problems internal and external to projects.

1.2 Problems during management of IT projects

Projects fail because of ineffective processes, tools, and techniques. They always fail because of project managers' execution of these processes, tools, and techniques.

The common problems that project managers experience while managing the delivery of one or more IT projects, as identified in the introductory section above, are further elaborated with the objective of demonstrate why an integration framework is necessary and essential to solving the root causes of project management major challenges—communication problems internal and external to projects.

- Complex scope:
  - No linkage to business processes and objectives;
  - No integrated view of multiple projects requirements;
  - Ineffective use of Project Management Office (PMO)[3] processes.
- High costs:
  - Poor estimates;
  - Lack of effective funding approval processes;
  - Lack of understanding of data conversion and applications interface issues.
- Schedule slippage:
  - Use of activity-based, not deliverables-based, schedule;
  - Lack of integrated master schedule;
  - Lack of consistent work breakdown structure (WBS)[4].
- Inconsistent and incomplete quality:
  - No measurement criteria for deliverables completeness;
  - Lack of consistent and complete project plans and deliverables;
  - Inconsistent understanding of completed deliverables.
- Dissatisfied stakeholders:
  - No project completion criteria communicated;
  - Poor communication of how technical solutions solve business requirements;
  - Lack of adequate program reporting processes.
- Redundant and duplicated efforts:
  - No reuse of deliverables and processes;

3. PMO: The project management group responsible for providing project infrastructure support for the project team. Project infrastructure support includes project management methodology and process deployments—training, metrics reporting, measurement criteria guidelines, tools support and project reporting support.
4. Work breakdown structure: As stated in PMBOK, "A deliverables-oriented group of project elements that organizes and defines the total work-scope. Each descending level represents an increasing detailed definition of the project work."

- Unclear roles and responsibilities;
- Inconsistent use of tools, techniques, and processes among project managers.

The root causes of most common project management problems presented above seem to focus on the need for consistent and integrated use of process, tools, and techniques as the foundation guiding principles to improve communication with all external and internal stakeholders. To be effective and efficient, project managers, preferably professional project managers with conceptualization, integration, and risk management skills,[5] must effectively apply these guiding principles during the management and delivery of projects.

1.3 The increasing demand for integration

Integration requires consistency, and consistency demands standardization.

Every area of modern technology requires some form of integration, yet project managers manage the implementation of software development projects without any consideration for integration and consistency. No wonder we have so many IT projects that are over budget, behind schedule, and poor in quality, mainly because of the lack of integration, consistency, and standardization. In certain situations, the innocent takes the brunt of the blame and the indecisive decision makers, with secrecy-style management, get praises and awards for excellence—the reward of excellence for the incompetent. Severe communication challenges, with no sound reasoning for the root causes of these problems, continue to manifest themselves within the company, and this nonproductive project management cycle continues without any real corrective action.

Project Management Institute (PMI) has published project management integration guidelines in the publication *A Guide to the Project Management Body of Knowledge (PMBOK)* that is highly recognized internationally within the project management discipline. ESI International has also published excellent models on project management integration guidelines in the publication *PROJECTFRAMEWORK*. Carnegie Mellon University, Software Engineering Institute (SEI)[6], has published the internationally recognized Capability Maturity Model Integration® (CMMI®). International Business

5. Conceptualization, integration, and risk management skills: The skills required for modern project management that require generalized business knowledge; specialized people, process, and technology integration skills; and excellent risk management skills.

6. SEI CMMI® project management maturity levels: 1-Initial—Use of inconsistent processes, tools, and techniques; project management process is informal, ad hoc, and unpredictable. 2-Consistent—Implementing a structured approach to project management; project management information system (PMIS) is established. 3-Integrated—Implementing IPM processes throughout the organization. 4-Comprehensive—Product and processes are controlled with commitment to a project management culture. 5-Optimizing—Process improvements are made to project management practices.

Machine (IBM), Computer Sciences Corporation (CSC), and various universities and IT standards committees have published IT integration guidelines that are well established within the IT discipline. BPMI.org has published integration guidelines in the business process modeling language (BPML),[7] which is slowly gaining acceptance within the business management discipline. However, there are no widely accepted published guidelines that I am aware of that provide any real-world practical solutions on what, why, and how these three major components can be integrated in managing the delivery of IT projects.

Managing IT projects cannot be accomplished in isolation. It requires an understanding of business management, IT, and project management components to ensure that the projects being developed support the business processes and provide value and benefits to the business. The requirements for integration in this modern, technologically advancing world cannot be met with the crude integration methods of the early days of IT. New integration methods, such as the framework presented in this book, must be established and effectively applied and communicated to significantly improve the way business management is aligned with project management and IT management to ensure the successful delivery of a program or multiple projects that represent a major business initiative.

1.4 Role of executive management

If you have no executive staff involvement and approvals, you have no support and commitment by definition.

Effective IPM-IT requires many executive-level management talents and skills. IPM-IT is defined in this book as the process of integrating business management (BM), information technology management (ITM), and project management (PM) components during the software development processes with the objective of efficiently and effectively optimizing the utilization of people, processes, and technology resources to adequately meet stakeholders' expectations. The roles and responsibilities of executive management from the business, IT, and project management disciplines must be established and communicated to ensure optimum utilization of vital people resources.

- Executive business management (support integrated business and IT initiatives):
  - Define company's business objectives and strategies.
  - Establish company's corporate business priorities and direction.
  - Approve recommended programs.
  - Allocate program budgets based on approved programs.

7. Business process modeling language (BPML): A metalanguage for the modeling of business processes, just as XML is a metalanguage for the modeling of business data.

  - Release funds based on business objectives, strategies, and program progress.
  - Perform roles of program executive sponsor and executive business sponsor, including taking overall accountability for program success.
- Executive IT management (support integrated business and IT initiatives):
  - Define company's IT objectives and strategies.
  - Establish company's IT priorities and direction.
  - Approve IT policies, procedures, and standards for program delivery.
  - Recommend appropriate budget levels and program budget allocations for IT investments to executive business management.
  - Obtain approval from executive business management on recommended IT budgets based on approved programs.
  - Release funds based on business objectives, strategies, and program alignment with IT objectives, strategies, priorities, and direction.
  - Perform role of executive IT sponsor, including retaining accountability for optimizing the allocation of IT resources (people, processes, and technology) to align with business strategies.
- Executive project management (support integrated business, IT, and PM initiatives):
  - Define company's PM objectives and strategies.
  - Establish company's PM priorities and direction.
  - Approve PM policies, procedures, and standards for program delivery.
  - Recommend appropriate budget levels and program budget allocations for PM investments to IT and executive business management.
  - Obtain approval from IT and executive business management on recommended PM budgets based on approved programs.
  - Release funds based on business objectives, strategies, and program alignment with IT and PM objectives, strategies, priorities, and direction.
  - Perform role of executive PM sponsor to champion the project management methodology, including accountability for project management integration.

1.5 A real-world problem scenario: managing a major IT project

There is no silver bullet.
—Fred Brooks

You can't solve a problem with the same thinking that created it in the first place.
—Albert Einstein

Some consulting firms turn clients' problems into their personal goldmines.

A senior business manager at a large company mentioned:

> We seemed to be experiencing several major challenges in managing a major customer billing project. We had expended extensive amounts of effort documenting business, IT, and project management requirements, but could not justify the value of such extensive documentation. We had project plans and schedules with extensive amounts of activities/tasks and budget forecasting and tracking documents that were gradually becoming unmanageable. An enormous amount of documentation was being produced, which looked impressive, but lacked integration and consistency, resulting in the demand for additional IT, business, and project management resources. We seemed to be going through some sort of information crisis with unclear roles and responsibilities, with business and IT specifications being delivered and redelivered as different system specifications with new labels, and with project plans and schedules being managed by several enthusiastic but nonproductive project management teams. The headless chicken environment seemed to be the project team's mode of operations, with reoccurring requests for approval appearing on my desk for additional systems architects, business analysts, systems and data analysts, developers, and project managers. Executive management needed proper justification for the escalating costs, and my frustration level seemed to have reached that limiting point.

My heightened frustration resulted in a decision to sign a contract with an "expert" project management consulting firm to assist in managing this project, with the objective of optimizing the utilization of business and IT resources and to improve the project management communications reporting, both internally and externally to the project. My search for the silver bullet solution had started, with the hope that this IT consulting company will magically deliver the solution to this information jungle problem.

This project management consulting firm, overexcited by this opportunity, produced more extensive documentation based on its self-proclaimed unique project management methodology, resulting in additional consulting resources, regular daily nonproductive status meetings, and more detailed, disintegrated documentation. The project team, overexcited about the new project management methodology, adopted the consulting firm's fresh-start approach with great respect and enthusiasm. The result was another set of similar business, IT, and project management documents, repackaged with the consulting company's logo and confidentiality and copyright terms appearing on the title page of each document.

The original problem of the lack of consistency and integration prevailed with unclear linkages among project management, IT management, and business management deliverables. I was now confronted with additional legal contracting issues as a result of the consulting firm's confidentiality and copyrights terms appearing on the title page of each document. To add to this existing information jungle, the project schedule now contained an additional 100 activities, mainly to justify the need for five more senior consulting resources. The project was now way over budget and behind schedule, with scope and quality requirements being addressed with Victorian-novel-style documentation.

A polished report from the consulting firm recommended a solution to the problems, which required an organizational and cultural change. This change recommended the establishment of two separate project management groups, one to support the business management functions and the other to support the IT management functions. This recommendation suggested the need for additional business and IT resources that escalated the project budget to a level that was difficult to justify to executive management.

The search for the silver bullet solution continued, and my position seemed to be in more severe trouble, with valid excuses to executive management becoming nonexistent. I was no further ahead in answering the original questions of: What should I do to get this major IT project back to an acceptable schedule, budget, scope, and quality constraints level? and What approach or technique should I apply? These questions still remain unsolved, but now I was confronted with additional resource and legal complexities, as a result of this desperate search for that silver bullet solution.

I finally realized that there is no silver bullet solution, so I decided to approach the information jungle, inconsistency, and disintegration problems with the original project team members and to search for a senior project manager. This individual needed to have experience with business management, IT management, and project management integration skills, as well as customer billing systems architecture and project management expertise. Now, I was in search of an individual solution after considering certain risk factors. I believed that the risk impact in this case was small and eventually hoped to address these project management problems with a senior project manager who could demonstrate extensive practical integration, conceptualization, and risk management skills, resulting from successful implementation of customer billing systems.

This problem scenario, experienced at one of my consulting assignments, shows some of the major project management errors that occur on projects:

- Committing to project schedule and budget, whether or not scope and quality objectives can be met effectively.
- Establishing separate project management groups, one for business and one for IT, each establishing different and unique project management policies, processes, and procedures.
- Holding business responsible for describing the business requirements with little or no IT involvement and IT responsible for development based on the business requirements developed with little or no IT involvement. Business then gets involved during the integration and deployment/production phase. Integration only exists during the integration phase. This approach may work well using the traditional waterfall approach to software development where business requirements are well known in advance. However, in most cases, business requirements are rarely known and, as such, require a more

iterative/prototyping method that demands the need for integration, conceptualization, and risk management.

- Assigning a project manager who lacks integration, conceptualization, and risk management skills.
- Addressing the communications issue with excessive status meetings.
- Ignoring the three major requirements of modern project management—conceptualization, integration, and risk management.
- Disregarding the root causes of communication problems—lack of consistency, completeness, and integration.
- Project manager's disregarding clear and articulated documentation of project team roles and responsibilities in a project structure environment to show how these roles and responsibilities are reflected in the project schedule.
- Project schedule's containing hundreds of activities without regard for WBS.
- Ineffective use of project management tools, even by the so-called expert project management consulting firm.
- Separating business management, IT management, and project management functions, with each group happily performing its individual responsibilities without any overall direction or vision for integration—a recipe for chaos.
- Searching for that silver bullet magical solution without a clear understanding of the root causes of the problems.
- Placing the emphasis on human resources management with little or no consideration for integration of business, IT, and project management components.

The solutions discussed in Section 1.6 demonstrate the value of business management, IT management, and project management integration to solve the root-cause communication problems resulting from the lack of consistency, completeness, and integration of these major components. This real-world practical solution addresses most of the critical problems presented above. It should provide some revealing and useful observations and lessons learned for those practicing project managers and senior and executive management staff involved in managing and delivering IT projects whose interest is towards integration.

1.6 A real-world solution scenario: managing a major IT project

This section highlights the project management solution that was implemented, with a high degree of success, at this major corporation in solving the real problem scenario discussed in Section 1.5. This client-server application was managed and delivered by a newly established business systems group (BSG) consisting of both IT and business resources. IT was responsible for technology infrastructure support (hardware, systems software, and network), and the BSG was responsible for the business and software application support. The key project management principles that guided this deliverables-oriented solution are presented in this section as:

- Project organizational structure;
- Project roles and responsibilities;
- Project integrated WBS model and hierarchy;
- Microsoft Project schedule that support the WBS.

The objective of this section is to provide the real-world solution that was successfully delivered to solve the information jungle, inconsistency, and disintegration problems discussed in the problem scenario. This real-world practical solution is based on the model-centric or deliverables-based WBS principle. The tool-based solution to integrating the business, IT, and project management components using Microsoft's project management tool is provided to demonstrate a real-world practical and effective utilization of a project management tool. Establishing order and structure to this chaotic project environment required the use of a project management tool by project management professionals with conceptualization, integration, and risk management skills.

- The project organizational structure solution presented in Figure 1.2 demonstrates integration of business management, project management, and IT management from a project organizational or people perspective. At this corporation, the newly formed BSG was responsible for project management—management, delivery and support—of the billing system. IT management was responsible for technology infrastructure direction and development of and support for hardware and system software, including the database management system (DBMS) and network configuration. Business management was mainly responsible for providing business strategies, billing requirements—internal and external to project—and performing acceptance testing and end-user training.
- Project roles and responsibilities show the integration of business management, project management, and IT management, from process and people perspectives.
- The project integrated WBS model and hierarchy demonstrates integration of business management, project management, and IT management from project management process and technology perspectives.
- Microsoft Project schedule supports the WBS and shows the automated implementation of the integrated WBS for a specific phase. Appendix A contains the detailed Microsoft Project schedule that was executed
- The project organizational structure in Table 1.1 expands on the roles identified in Figure 1.2.

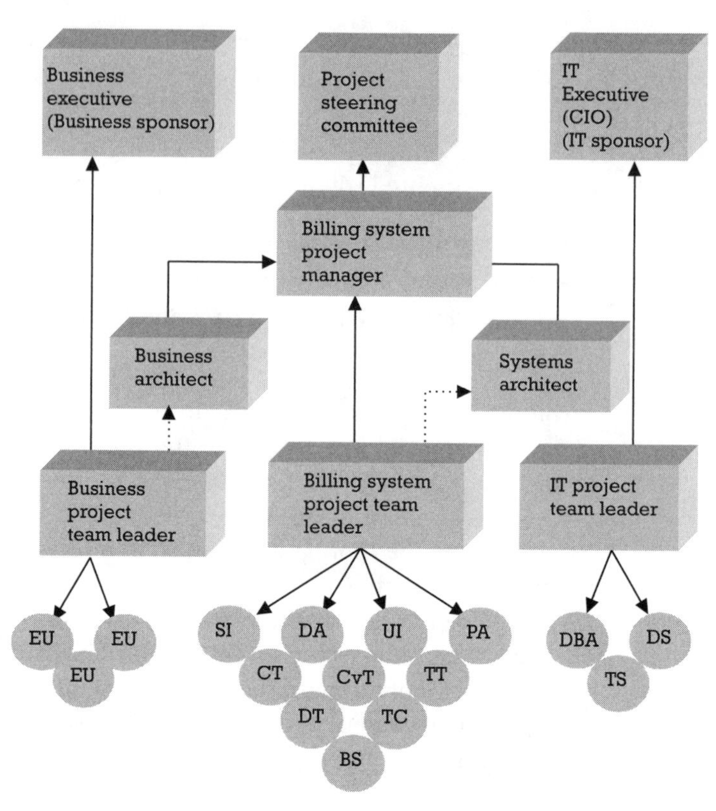

Figure 1.2 Project organizational structure solution.

Table 1.1 Project Organizational Structure

EU: end user	SI: systems integrator; DA: data analyst; UI: user interface analyst	DBA: database administrator
	PA: process analyst; CT: construction team; CvT: conversion team	DS: development support staff
	TT: testing team; DT: deployment team; TC: technical consultant; BS: business systems support analyst	TS: technical support staff

1.6.1 Project team roles and responsibilities

Business executive

Coordination/leadership:

- Take accountability for preparation of the business case and overall project success.
- Ensure that the business analysts commit and deliver the business benefits.
- Allocate the necessary business budgets for the project.
- Determine the membership of the steering committee and perform the role of the steering committee chairperson.

- Approve the project charter.
- Champion the project and its potential benefits/risks to the corporation.
- Role: Business sponsor

IT executive

Coordination/leadership:

- Take accountability for preparation of the IT architecture (data, applications, and technology), IT strategic alignment with the business, risk strategies, and optimization of IT resources.
- Ensure that the IT analysts commit and deliver the business benefits.
- Allocate the necessary IT budgets for the project.
- Determine the membership of the steering committee and perform the role of the IT steering committee advisor.
- Approve the IT architecture and the project charter.
- Ensure that the project delivers a solution that supports IT strategic direction.
- Role: IT sponsor.

Project steering committee

Coordination/leadership:

- Provide high-level business and technical guidance, direction, and advice to project manager to ensure project delivers maximum value to the corporation.
- Approve project deliverables.
- Approve project major milestone deliverables and authorize the commencement of the next phase and release of funds.
- Approve use of project management and contingency funds.
- Approve changes to project schedule, costs, scope, and quality.
- Take accountability for delivery of the project on time within budget, quality, and scope.
- Provide direction and approve recommendations tabled by the project manager on major issues, changes, and risks affecting the project.
- Role: Approval committee.

Systems development manager–project manager

Coordination/leadership:

- Take responsibility for selecting the project team.
- Take direct accountability for the quality and effectiveness of the completed project.

- Plan, organize, control, and coordinate the project-related duties of the project team.
- Coordinate the activities of the project team with IT groups, other business groups, and external groups.
- Select and effectively apply project management tools and control techniques.
- Determine resources and cost estimates for the project.
- Recommend a staffing plan to users, BSG, and IT management.
- Take responsibility for all activities associated with project budget and work plan for the project.
- Maintain working knowledge of the system requirements, design concepts, and approaches and ensure that appropriate technologies are applied.
- Critically review and take responsibility for all project-related studies/recommendations completed by IT, BSG, and user management.
- Take responsibility for the timely and accurate completion of all systems development and operating documentation.
- Administer the change management process by ensuring that the suggested changes are initiated as a result of a business, technical, process, or people issues documented in the issues resolution log.
- Manage project risks by analyzing risks' impact on scope, cost, schedule, and quality.
- Produce regular progress reports for the steering committee.
- Obtain project funding authorization from the steering committee.
- Role: Project manager.

Systems architect

Coordination/leadership:

- Provide leadership to project team members on data, process/systems, technology, user interface, and documentation issues.
- Communicate with IT support functions on project progress requirements relative to planning and architectural activities.
- Establish network performance tuning criteria to ensure that the new system meets acceptable response times.
- Communicate with systems development manager on project progress and requirements.
- Take responsibility for ensuring that the project team members are aware of the significant activities of other groups on the project.
- Recommend resources and cost estimates for the particular project phase to the systems development manager.
- Keep the systems development manager, user group members, system project team members, and IT support staff informed of ongoing problems and their resolution.

- Ensure that IT standards and appropriate technologies and techniques are used during analysis, solution architecture, design, construction, and implementation phases.
- Administer the issues resolution process by ensuring that the issues are resolved efficiently and effectively.
- Lead the evaluation and selection of and recommend the acquisition of software tools and coordinate the implementation of these tools.
- Maintain a single point of contact with nonbusiness system resources (internal and external suppliers and vendors).
- Evaluate, recommend, and lead the implementation of software development methodologies and appropriate standards.
- Take responsibility for the overall design integrity, technical viability, documentation, and reusability of the development approach for the project.

Project execution:

- Plan, organize, control, coordinate, and maintain the project-related duties of the assigned project team members through the use of Microsoft Project management tools.
- Define project charter, scope, and conceptual design of the new system.
- Support the project team in defining an integrated representation of the business, data, application, and technology architectures.
- Support the IT technical staff in establishing the client/server technology solution in terms of hardware/software/network configuration.
- Develop and publish the existing and proposed technology architectures (TAs).
- Role: Systems architect.

Business architect

Coordination/leadership:

- Provide leadership to project team members on business process specifications and documentation issues.
- Communicate with business analysts on project progress requirements relative to business planning and architectural activities.
- Establish business deployment criteria (business process deployment, organizational and cultural change, training) to ensure that the new system meets acceptable business changes.
- Communicate with executive business manager on project progress and requirements.
- Take responsibility for ensuring that the business project team members are aware of the significant activities of other groups on the project.
- Recommend business resources and provide cost estimates to the systems development manager or project manager for each project phase.

- Keep the systems development manager, system architect, and executive business manager informed about ongoing problems and their resolution.
- Ensure that business process standards and procedures are used during analysis, solution architecture, and conversion/transition phases.
- Administer the issues resolution process by ensuring that business issues are resolved efficiently and effectively.
- Lead evaluation and selection of and recommend the acquisition of business process tools and techniques and coordinate the implementation of these tools.
- Maintain a single point of contact with business system resources (internal and external suppliers and vendors).
- Evaluate, recommend, and lead implementation of business process methodologies and appropriate standards.
- Take responsibility for the overall business design, acceptance testing, change management viability, documentation, and reusability of the business deliverables for the project.

Project execution:

- Plan, organize, control, coordinate, and maintain the business-related duties of the assigned business project team members through the use of business process modeling and project management tools.
- Define project charter, scope, and conceptual design of the new system from a business perspective.
- Support project team in defining an integrated representation of the business, data, and applications architectures (AA).
- Support business staff in establishing the acceptance testing, change management, and risk management criteria and solution.
- Develop and publish the existing and proposed business process architectures.
- Role: Business architect.

Billing system project team leader

The systems development manager or project manager is responsible for ensuring that the project team acquires the required business and technical skillsets and staffing requirements. This is done in concert with the business project team leader, IT project team leader, and billing systems project team leader. The billing system project team leader is assigned a lead role to ensure that the deliverables are produced according to the established project standards and that billing systems knowledge transfer and training is adequate. The project team, consisting of the BSG staff, IT support staff, and end users (on an as-required basis), is responsible for:

- Conducting analysis (business, data, process, technology, and user interface) through the use of models;

- Designing the solution architecture (data, application, technology);
- Designing, prototyping, and constructing and testing the new system;
- Converting and delivering the new system;
- Documenting the new system in the automated repository tool;
- Establishing standards, procedures, and plans for developing, documenting, and operating the new system;
- Performing the quality assurance (QA) role to ensure that the deliverables meet the standard criteria as defined in the standards and procedures/methodology documents;
- Identifying, defining, resolving, and documenting project-related issues according to the standard issue resolution process;
- Role: Project team leader–billing systems.

IT project team leader

The systems development manager or project manager is responsible for ensuring that the project team acquires the required business and technical skillsets and staffing requirements. This is done in concert with the business project team leader, IT project team leader, and billing systems project team leader. The IT project team leader is assigned a lead role to ensure that the deliverables are produced according to the established IT standards and that IT knowledge transfer and training are adequately attained. The project team consisting of IT support staff—data administrators, DBAs, software development staff, and technical support staff—is responsible for:

- Supporting the development and implementation of logical data models;
- Supporting the development and implementation of physical databases;
- Supporting the development and implementation of applications code;
- Supporting the installation and deployment of hardware, network, and systems software configuration;
- Supporting IT infrastructure standards, procedures, and plans for developing, documenting, and operating the new application;
- Performing the QA role to ensure that the deliverables meet IT standard criteria as defined in the project standards and procedures/methodology documents;
- Identifying, defining, resolving, and documenting IT infrastructure-support-related issues according to the standard issue resolution process;
- Role: Project team leader–IT.

Business project team leader

The systems development manager is responsible for ensuring that the project team acquires the required business and technical skillsets and staffing requirements. This is done in concert with business project team leader, IT project team leader, and billing systems project team leader. The business project team leader is assigned a lead role to ensure that the deliverables are produced according to the established standards and that business knowledge transfer and training are adequately attained. The project team, consisting of end users and business staff (on an as-required basis), is responsible for:

- Supporting the development and implementation of business processes;
- Supporting the development and implementation of business procedures;
- Supporting the development and implementation of training manuals;
- Supporting the deployment of business processes, organizational change, training, and business operating procedures;
- Supporting business process standards, procedures, and plans for deploying and operating the new application;
- Performing the QA role to ensure that the deliverables meet business process standards criteria, defined in the project standards and procedures/methodology documents;
- Identifying, defining, resolving, and documenting business-procedures-support-related issues according to the standard issue resolution process;
- Role: Project team leader–business.

Systems integrator

Coordination/leadership:

- Establish naming standards, policies, and procedures for use of the data repository during applications development.
- Lead the evaluation and selection of and the recommend acquisition of computer-assisted software engineering (CASE) tools and repository software and coordinate the implementation of these tools.
- Maintain a single point of contact with CASE/repository software vendor.
- Coordinate the deliverables and activities between analysis, solution architecture, design/prototype, construction, and implementation or conversion/transition phases through the use of the automated repository.
- Perform the QA role for the application system by ensuring that all databases, processes/objects, data windows (user interfaces, events), and technologies are designed, developed, distributed, implemented, and documented according to system specifications.
- Perform the QA role for the system documentation by ensuring that all project-infrastructure-type documentation (e.g., standards, procedures, guidelines, plans) is reasonably accurate, consistent, and reusable.

- Manage the applications change request process during development and production through the use of the data repository.

Project execution:

- Support the conversion coordinator in establishing migration/conversion plans and procedures to effectively and efficiently move from the existing to a new system.
- Support the project team to define an integrated representation of the business, event, data, application, and technology models.
- Support the IT technical staff in establishing, implementing, and installing the integrated development tools environments relative to ORACLE DBMS, applications development tools, CASE/repository tool, version control tool, testing tool, conversion tool, object libraries, and necessary interface requirements.
- Assemble and package the construction specifications for the construction team.
- Perform the role of repository administrator by ensuring that all data, process, user interface, and technology models are documented in the CASE repository, according to repository naming standards and procedures.
- Role: Systems integrator–repository administrator.

Data analyst/designer

Coordination/leadership:

- Establish data analysis and design standards and procedures based on IT existing data standards and procedures.
- Establish data retention standards for host and end-user computing.
- Establish data distribution, security, integrity, ownership, and access standards and strategies to support physical implementation of the databases.
- Consult with IT on data management tools and techniques and provide support to team members.
- Maintain a single point of contact with IT corporate data group.
- Perform leadership and coordination roles by leading joint applications design (JAD) sessions and coordinate the development of logical data models.

Project execution:

- Document logical data model in CASE repository according to established naming standards and procedures.
- Document physical database design model in CASE repository according to established standards and procedures, and use CASE tool to generate data definition language (DDL) syntax for the creation of ORACLE DBMS table spaces, tables, and indexes.

- Support project team in developing the logical data models of the business based on data analysis and design standards, policies, and procedures.
- Support IT DBA in converting logical data model to physical database design according to IT standards, policies, and procedures, including performing denormalization process, determining physical storage, database performance, data integrity, data control, and data distribution requirements.
- Support IT DBA in defining and implementing triggers and stored procedures to support data integrity and data/process distribution requirements.
- Support graphical user interface (GUI) coordinator in documenting and maintaining GUI component object libraries.
- Role: Data analyst.

Process analyst/designer

Coordination/leadership:

- Establish process and event analysis and design standards and procedures based on IT existing process standards and procedures, including establishing criteria for defining and documenting business events.
- Establish standards for defining and implementing triggers and stored procedures.
- Participate with the library management coordinator in establishing object version control and library management procedures.
- Develop process distribution standards and establish criteria for determining how to allocate processes and portions of processes between client and server.
- Establish system process distribution and implementation guidelines to support mapping of business processes and events to systems processes and events.
- Develop system performance tuning criteria to ensure adequate response times.
- Consult with IT on process development and implementation tools and techniques and provide support to team members.
- Maintain a single point of contact with IT corporate systems group.
- Perform leadership and coordination roles by leading JAD sessions and coordinating the definition and development of process models, business events, triggers and stored procedures, and process partitioning/distribution of the business or system based on existing standards, policies and procedures.

Project execution:

- Document process analysis and design deliverables in CASE repository according to repository naming standards and procedures.

- Document physical program structures (minispecs) in CASE repository according to established standards and procedures.
- Support the project team in developing the logical process models [data flow diagrams (DFDs), process decomposition, events] of the business based on process analysis and design standards, policies, and procedures.
- Support the project team in designing the program structures (minispecs, automated processes, events) to support the logical process and data models.
- Support the project team in converting logical process models to physical distributed program structures according to IT standards, policies, and procedures, including identifying and defining the processes that will be implemented on the client machines, application server machines, and database server machines.
- Support IT DBA in defining and implementing triggers and stored procedures to support data integrity and data/process distribution requirements.
- Role: Process analyst.

User interface (workstation/prototyping) analyst

Coordination/leadership:

- Establish prototyping walk-through standards and procedures to ensure that the GUI requirements are adequately supported, including establishing criteria for defining, developing, implementing, testing, and documenting the GUI requirements.
- Establish naming standards for defining, documenting, navigating, and maintaining GUI component objects and libraries.
- Develop GUI design standards and criteria for identifying reusable user interface and business objects and events.
- Establish criteria for implementing on-line help (context, ICON, general help) to ensure an effective user-friendly system.
- Develop GUI performance tuning criteria to ensure adequate response times.
- Support the developers in establishing procedures to map the business objects and events to the GUI objects and events (GUI development tools objects, events and script).
- Participate with the library management coordinator in establishing the object version control and library management procedures to support GUI objects.
- Consult with IT on GUI development and implementation tools and techniques and provide support to team members.
- Maintain a single point of contact with GUI software vendors.
- Perform leadership roles by leading GUI modeling sessions with users and BSGs.

- Coordinate the definition, development, and testing of user interface objects and events and reusable objects and events with project team in order to produce and test the user interface prototype, based on existing GUI standards, policies, and procedures, including ensuring that all development tools interfaces are working properly (e.g., applications development, ORACLE DBMS, CASE/repository).
- Coordinate purchase of reusable GUI objects (GUI object libraries).

Project execution:

- Train user group representatives in navigating the GUI design during the prototype.
- Document and maintain GUI component object library during the prototype phase.
- Document the mapping of the business objects and events to the GUI objects and events (GUI development tool objects, events, and script) during the prototype phase and communicate this mapping to the systems integrator.
- Support the user representatives/BSG operations staff in determining the GUI requirements based on prototyping and GUI standards and procedures.
- Support the developers in designing, building, and testing the prototypes based on user GUI requirements, database structures (from IT), and GUI standards and procedures.
- Develop users guide with users and conversion coordinator.
- Role: User interface analyst.

Construction team—contractors and BSG project team

Project execution:

- Review construction specifications with systems integrator, including the constructed database structure [e.g., table spaces, tables, indexes, referential integrity (RI) constraints], constructed GUI components, and identification of distributed processes (client, application, and database servers).
- Support IT in implementing database triggers and stored procedures on the ORACLE DBMS database server.
- Support project team in determining inheritance objects.
- Support project team in determining reusable components (objects, methods, scripts, object libraries).
- Build, test, and document applications against the server resident database according to construction module design specifications, including training the project team to support future development and maintenance tasks.
- Review conversion specifications with conversion coordinator.
- Build, test, and document data conversion and bridge routines.

- Support BSG systems administrator coordinator and IT network support group in installing, executing, and testing applications at user sites.
- Support BSG Systems administrator coordinator and IT network support group in removing the existing applications at user sites.
- Complete systems documentation and knowledge transfer and system developer training.
- Hand over completed systems documentation to BSG system developers.
- Role: BSG software developers.

BSG technology operations support

Coordination/leadership:

- Develop version control and library management standards and procedures during project development and production support to ensure that the correct versions of the application and the database are being developed among the developers and to ensure consistency and integrity during source code check-in/check-out and object version checking functions. [These standards and procedures will also be used to manage the libraries (application, databases, object components; class libraries) during the production environment.]
- Evaluate, select, recommend, and install version control software to support source code version tracking and control functions.
- Develop operations guide to assist the systems administrator and LAN support staff in executing and maintaining the application and using the appropriate hardware, software, and network technologies.
- Establish release management standards and procedures to ensure that the correct releases of the software are being installed and executed.
- Maintain contact with IT operations/network support for advice and technical support during installation and maintenance.
- Coordinate help desk activities between IT (technical operations support/DBA) and BSG to support end-user help.

Project execution:

- Perform the duties of a LAN administrator during development, installation, and production by installing system software and providing the necessary client/server support for client workstations, application server and database server machines.
- Perform the duties of a systems administrator during development and production by managing the application source code library through the use of a mechanized version control and tracking software, according to standards and procedures, including administration of security access to the application system.
- Support IT technical support staff in constructing, optimizing, tuning, and monitoring the network to ensure application systems availability.

- Support IT technical support staff in establishing disaster recovery plans (DRPs), network contingency plans, and database recovery procedures during development and production support.
- Support users, BSG, and IT staff in managing service agreements and application change requests.
- Support project team and IT staff in installing and distributing the production system (client, application server, database server) at appropriate user sites, according to the conversion or transition plan.
- Role: BSG systems analyst(s)–support.

End-user technical consultant

Coordination/leadership:

- Develop an end-user computing plan for next phase of development based on data, processing, and reporting requirements and the technology paradigm.
- Develop a migration plan to support migration to the end-user environment.

Project execution:

- Recommend solutions (PC/LAN level) to support end-user computing requirements based on user requirements or the technology paradigm.
- Provide overall consulting support to end users.
- Analyze end-user data, processing, and reporting needs and develop or recommend solutions to adequately meet the needs, including identifying and recommending processes that can be improved, standardized, or automated.
- Assist end users in migrating to the end-user environment.
- Provide end-user support in using end-users tools.
- Assist end user in executing the computing plan for the next phase of development.
- Assist the users in developing end-user computing systems standards to ensure consistency in performing end-user reporting and data manipulation functions.
- Role: BSG systems analyst–technical consultant.

System testing team

Coordination/leadership:

- Develop testing standards and procedures to provide consistency during system and acceptance testing, including working with the users in determining test acceptance criteria.
- Evaluate, select, and recommend testing software to assist in performing systems and user acceptance testing.

- Develop test plans (unit, system, integration, and user acceptance) based on testing standards and procedures.
- Coordinate the testing process and chair developers call on testing.
- Establish the testing environment by ensuring that all testing conditions are met.

Project execution:

- Establish test cases/data/results to support the test plans, including working with the end users in writing test scripts, preparing test data and test results, and performing pilot tests on testing tools.
- Perform unit, integration, and acceptance testing with the users, based on test cases/data/results, and acceptance criteria, including conducting parallel testing with the users.
- Perform data management and stress testing (network, hardware, software), based on test cases, test data, and acceptance criteria.
- Provide feedback to developers, project team, and users on the testing results, and obtain users approval for systems functionality.
- Role: BSG testers.

Data conversion/transition team

Coordination/leadership:

- Develop data conversion strategies or plan to convert data from an existing system to a new system, including working with the users in determining acceptable data conversion strategies/plans and developing strategies or plans for one-time conversion programs, bridge conversion programs, and batch load conversion processes.
- Evaluate, select, and recommend conversion utilities/tools to assist in performing data conversion.
- Coordinate the conversion process and chair the data conversion committee.
- Document the conversion/transition environment by determining the impact on the users' organization, documenting conversion contingency procedures, reviewing staging/transition procedures with users, and ensuring new systems functionality.
- Meet frequently with the users during the conversion or transition to ensure that all users' requirements are satisfactorily met.

Project execution:

- Document the mapping of the existing systems data (data elements, files, data definitions) to new systems data (columns and tables).
- Document detailed data conversion or transition plans and specifications based on conversion strategies and coded and tested by the developers.

- Support developers in testing and executing the conversion routines, including providing feedback to developers, project team, and users on the conversion results and obtaining users' approval for systems functionality.
- Develop users training requirements, schedule, and plan to provide for a smooth transition to new system, including designing computer based training (CBT) specifications and developing and coding training modules.
- Assist the training coordinator in training the trainers.
- Hand over documentation to users.
- Role: BSG systems analysts–deployment.

IT support team

Project execution:

- Install and maintain database software (ORACLE DBMS) and interfaces to applications development and CASE tools.
- Construct physical database structures (DDL statements) from the logical data model.
- Administer database security (grant, revoke), according to IT and business standards and procedures.
- Work with the data analyst/designer in verifying the correctness of the database structure, security, and integrity in accordance with user requirements.
- Work with data analyst/designer in promoting object reuse, based on user requirements.
- Maintain the accuracy of the data stored in the database through the use of DBMS integrity features.
- Monitor performance, tune database, and perform database backup and recovery, according to IT standards and procedures.
- Work with BSG operations support group in tuning applications for database access.
- Role: Project DA/DBA support.

Project execution:

- Provide consulting support to project team on client/server development and implementation standards, tools, and techniques, including providing technical support to BSG operations support staff.
- Role: Project software development support.

Project execution:

- Provide network-consulting support to project team on technology implementation techniques to support client/server technology (client, application server machines, database server machines) and

connectivity issues, including monitoring, tuning, and optimizing the network configuration.
- Role: Project technology–network support.

Business project team (end users)

Project execution:

- Work with the prototype coordinator in defining the GUI requirements to ensure that the new system will effectively support the functionality.
- Work with the project team (data and process analyst) in defining and verifying data and processing requirements.
- Work with the conversion/transition coordinator in defining and testing the conversion processes, user manual, and training requirements.
- Work with the testing coordinator in testing the functionality of the new system.
- Role: Business analyst support.

1.6.2 Project WBS

This section presents the WBS solution that was used as the key guiding principle in managing and delivering this major customer billing project. This WBS supports PMI fundamental principles and guiding definition of WBS, as defined in PMBOK: "A deliverables-oriented group of project elements that organizes and defines the total work-scope. Each descending level represents an increasing detailed definition of the project work."

The core deliverables of any IT project, whether or not the development approach is a custom-built or purchased-package solution, consist of data, process, user interface, and technology deliverables to support both business and IT processes. The key infrastructure deliverables that support the implementation of these core deliverables consist of project management, standards, training, and QA deliverables to ensure that these core deliverables fit within the context of the existing business and IT environments. The initiation, planning, execution, controlling, and closing of the project deliverables (core and infrastructure) are managed using the orderly project phased progression: planning/analysis, design/prototype, construction/development, and implementation as shown in Figure 1.3.

This WBS is a key foundation principle used in this book to demonstrate the model-centric or deliverables-oriented approach to managing one or more IT projects.

1.6.3 WBS hierarchy

The objective of this section is to show the WBS hierarchical principle presented in Figure 1.4, which formed the basis for the development of the

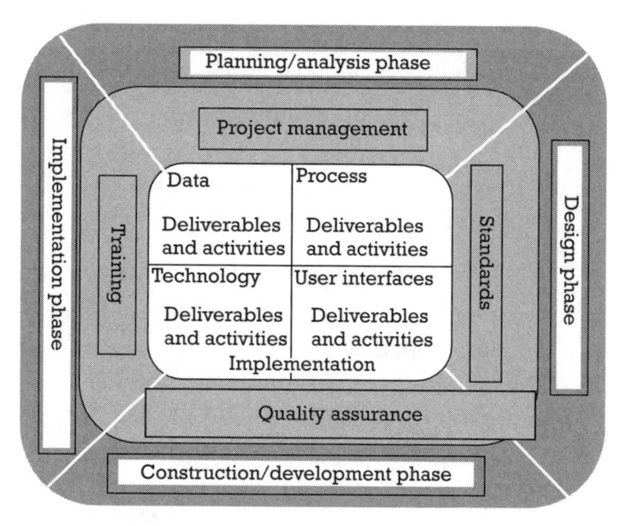

Figure 1.3 WBS solution.

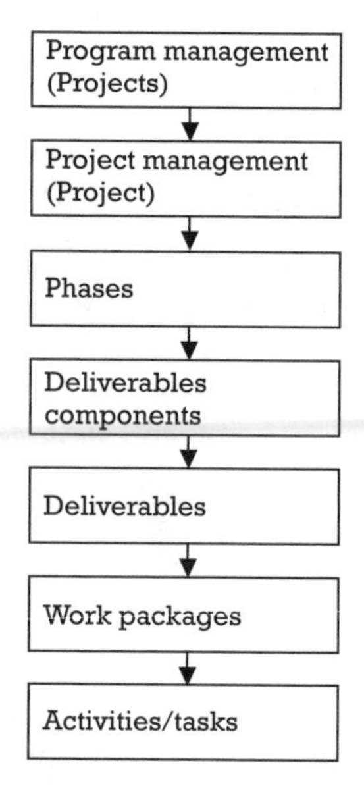

Figure 1.4 WBS hierarchy.

WBS structure. It is based on Harold Kerzner's[8] foundation principles for developing a work structure to ensure effective management of projects. These principles are sound and have been applied successfully on various IT projects. First, specify and break down the deliverables into valuable and measurable work packages. Second, assign major activities to each of the work packages. Finally, allocate effort, costs, and schedule to each of the work packages based on an allocation scheme to estimate the value of each deliverable.

8. Harold Kerzner: An author and world-recognized guru in project management.

1.6.4 Project schedule: Microsoft Project

It is simple to make something complex, and complex to make it simple.
—Fred Buterbaugh, Murphy's Technology Law

The real-world Microsoft Project schedule that was used as the tool to manage this major customer billing project is presented to demonstrate the project management tool implementation of this WBS major project management foundation principle. Figure 1.5 shows the tool-based solution to the WBS and the WBS hierarchy presented in Figures 1.3 and 1.4, respectively. The solution presented for the analysis phase shows the schedule at a given point in the project. Appendix A contains the detailed schedule for the phases specified in the WBS. The deliverables and activities were aligned with the work packages and roles and responsibilities presented in Section 1.5.2 using the project repository. This integrated alignment provided an effective means to optimize staff resource, cost, and schedule utilization. This model-centric, deliverables-based solution can be easily applied to the implementation of any major IT project, as a result of the flexibility, consistency, integrity, maintainability, and reusability of this unique and simple deliverables-based project schedule.

Figure 1.5(a–c) shows the deliverables, activities, schedule, and staff resources for the infrastructure deliverables components—project management, standards, and training—specified in the WBS. In this version of the project schedule, Training was integrated with standards to meet the integration requirements of these deliverables components.

Figure 1.5(c) is a continuum of the deliverables, activities, schedule, and staff resources for the infrastructure deliverables components—project management, standards, and training—specified in the WBS. In this version of the project schedule, training was integrated with standards to meet the integration requirements of these deliverables components

Figure 1.5(d–f) shows the deliverables, activities, schedule, and staff resources for the core deliverables components—data, process, user interface and technology—specified in the WBS. The deliverables, activities, schedule, and staff resources represent the schedule for the analysis phase at a given point in the project.

Figure 1.5(e) is a continuum of the deliverables, activities, schedule, and staff resources for the core deliverables components—data, user interface, process, and technology—specified in the WBS. The deliverables, activities, schedule, and staff resources represent the schedule for the analysis phase at a given point in the project.

Figure 1.5(f) is a continuum of the deliverables, activities, schedule, and staff resources for the core deliverables components—data, user interface, process, and technology—specified in the WBS. The deliverables, activities, schedule, and staff resources represent the schedule for the analysis phase at a given point in the project.

Figure 1.5(g) shows the deliverables, activities, schedule, and staff resources for the infrastructure deliverables components—QA—specified in

Microsoft Project - BSP.mpp

File Edit View Insert Format Tools Project Collaborate Window Help

	Task Name	Work	Duration	Start	Finish	% Complete	Resource Names
1	**BILLING SYSTEMS PROJECT**	**16,937.23 hrs**	**200 days**	**2/24**	**11/30**	**23%**	
2	**ANALYSIS**	**3,319.2 hrs**	**51 days**	**4/3**	**6/12**	**64%**	
3	**PROJECT MANAGEMENT**	**410.2 hrs**	**21 days**	**4/3**	**5/1**	**64%**	
4	**Managment Reports**	**32 hrs**	**2 days**	**4/3**	**4/4**	**100%**	
5	Prepare Management Reports	16 hrs	0.75 days	4/3	4/4	100%	SA[267%]
6	Conduct Client Meetings	16 hrs	2 days	4/3	4/4	100%	PM
7	**Project Plan**	**182 hrs**	**14 days**	**4/3**	**4/20**	**65%**	
8	Estimate Resource Requirements	14 hrs	1 day	4/20	4/20	33%	PM,PM1,SA
9	Prepare Skills Inventory	16 hrs	2 days	4/13	4/14	3%	PM1
10	Perform Risk Assessments	8 hrs	1 day	4/13	4/13	0%	PM1
11	Develop Work Strategy	32 hrs	2 days	4/3	4/4	100%	SI,BA
12	Update Plan & Schedule	112 hrs	10 days	4/3	4/14	80%	PM1,SA,PM
13	**Roles & Responsibilities**	**17 hrs**	**20 days**	**4/3**	**4/28**	**67%**	
14	Research Project Situation*	1 hr	1 day	4/3	4/3	100%	PM1[13%]
15	Update Roles & Responsibilities	16 hrs	2 days	4/27	4/28	50%	PM1
16	**Project Charter**	**131.2 hrs**	**21 days**	**4/3**	**5/1**	**42%**	
17	Obtain Existing Documents	8 hrs	0.38 days	4/3	4/3	100%	SA[267%]
18	Prepare/Update Project Charter	32 hrs	3 days	4/4	4/6	100%	SA[267%],PM[33%]
19	Define Glossary of Terms	91.2 hrs	6 days	4/24	5/1	10%	SI,SI
20	**Project Scope**	**48 hrs**	**1 day**	**4/16**	**4/17**	**100%**	
21	Update Business Processes	32 hrs	1 day	4/16	4/17	100%	SI,SA[267%],PM,SI
22	Update Context Level DFD	16 hrs	1 day	4/16	4/17	100%	SI,SI
23	**STANDARDS/PROCEDURES/TRAINING**	**1,427 hrs**	**51 days**	**4/3**	**6/12**	**62%**	

Ready

(a)

Microsoft Project - BSP.mpp

File Edit View Insert Format Tools Project Collaborate Window Help

	Task Name	Work	Duration	Start	Finish	% Complete	Resource Names
24	**Methodologies**	**216 hrs**	**18 days**	**4/3**	**4/26**	**53%**	
25	Develop/Purchase Analysis Techniques	48 hrs	6 days	4/3	4/10	100%	PM1
26	Develop/Purchase Design Techniques	40 hrs	5 days	4/10	4/14	0%	PM1
27	Develop/Purchase Construction Techniq	8 hrs	1 day	4/26	4/26	100%	PM1
28	Develop/Purchase Quality Techniques	8 hrs	1 day	4/26	4/26	100%	PM1
29	Develop/Purchase Transition Techniques	8 hrs	1 day	4/3	4/3	100%	PM1
30	Develop/Purchase Production Technique	8 hrs	1 day	4/26	4/26	100%	PM1
31	Prepare Deliverables List	16 hrs	2 days	4/3	4/4	100%	PM1
32	Describe Project Deliverables	80 hrs	10 days	4/13	4/26	24%	PM1
33	**Repository Standards & Procedures**	**64 hrs**	**6 days**	**4/3**	**4/10**	**100%**	
34	Document Repository Definition Types(M	16 hrs	1 day	4/3	4/3	100%	SI,SI
35	Document Standards(Naming) and Proc	16 hrs	1 day	4/3	4/3	100%	SI,SI
36	Update Data and Repository Stds & Proc	32 hrs	2 days	4/7	4/10	100%	SI,SI
37	**Data Standards**	**456 hrs**	**51 days**	**4/3**	**6/12**	**90%**	
38	Update Definition Type Standards	16 hrs	1 day	4/3	4/3	100%	SI,SI
39	Update Data Standards	440 hrs	50 days	4/4	6/12	90%	SI,SI
40	**GUI Standards**	**56 hrs**	**17 days**	**4/6**	**4/29**	**91%**	
41	Research Existing GUI Stds (IT/Industry)	8 hrs	1 day	4/6	4/6	100%	UI
42	Update GUI Stds	48 hrs	6 days	4/14	4/29	90%	UI
43	**Process Design Standards**	**80 hrs**	**9 days**	**4/3**	**4/13**	**100%**	
44	Research Existing Process Design Stds	8 hrs	1 day	4/3	4/5	100%	SA
45	Update Process Design Stds	72 hrs	9 days	4/3	4/13	100%	PA
46	**Testing Standards & Procedures**	**328 hrs**	**16 days**	**4/4**	**4/25**	**27%**	

Ready

(b)

Figure 1.5 Microsoft Project schedule: (a) project management, (b) standards.

the WBS. This figure also shows the deliverables, activities, schedule, and staff resources for the infrastructure deliverables components for the next phase—architecture or design/prototype—specified in the WBS to

Microsoft Project - BSP.mpp

	Task Name	Work	Duration	Start	Finish	% Complete	Resource Names
47	Research Existing Testing Stds	8 hrs	1 day	4/4	4/4	100%	TT
48	Identify how SA can help in testing(test	8 hrs	1 day	4/14	4/15	50%	TT
49	Update Acceptance Criteria	16 hrs	1 day	4/17	4/17	0%	TT,TT1
50	Update Testing Stds	200 hrs	15 days	4/5	4/25	31%	TT,TT1
51	Update Testing Approach/Procedures	32 hrs	2 days	4/14	4/17	0%	TT,TT1
52	Update Error Messages Stds	32 hrs	2 days	4/14	4/17	0%	TT,TT1
53	Conduct Structured Testing Walkthru	32 hrs	1 day	4/25	4/25	0%	TT,UI,CvT,SI
54	**Version Control Standards**	**48 hrs**	**19 days**	**4/3**	**4/27**	**0%**	
55	Research Version Control Stds *	16 hrs	2 days	4/3	4/11	0%	ST,CvT
56	Document Version Control Stds	32 hrs	4 days	4/24	4/27	0%	ST
57	**Change Control Stds & Procedures**	**41 hrs**	**16.25 days**	**4/11**	**5/3**	**0%**	
58	Research Existing Change Control Proce	1 hr	0.13 days	4/11	4/11	0%	ST
59	Document Change Control Procedures	40 hrs	4 days	4/28	5/3	0%	ST,PM
60	**Prototype/Walkthru Procedures**	**56 hrs**	**9.2 days**	**4/3**	**4/14**	**29%**	
61	Research Existing Walkthru Methodologi	8 hrs	1 day	4/3	4/3	100%	UI
62	Define Prototype & Walkthrough Methodc	8 hrs	1 day	4/12	4/12	100%	UI
63	Define Dynamic Mdl Prot Walkthru Metho	40 hrs	5 days	4/3	4/14	0%	PM
64	**Issue Resolution Procedure**	**64 hrs**	**14.13 days**	**4/3**	**4/21**	**8%**	
65	Administer Issues	16 hrs	2 days	4/3	4/4	0%	DT1
66	Resolve Issues	48 hrs	3 days	4/3	4/21	13%	SA,TT,SI,UI,PM1,CvT
67	**Training Plans**	**18 hrs**	**8 days**	**4/3**	**4/12**	**100%**	
68	Update Training Rqmts	9 hrs	1 day	4/3	4/3	100%	PM[13%],SA[267%]
69	Update Training Plan for Project Team	9 hrs	1 day	4/12	4/12	100%	PM[13%],SA[267%]

(c)

Microsoft Project - BSP.mpp

	Task Name	Work	Duration	Start	Finish	% Complete	Resource Names
70	**DATA COMPONENTS**	**392 hrs**	**7.88 days**	**4/3**	**4/12**	**84%**	
71	**Entity/Relationship Model**	**392 hrs**	**7.88 days**	**4/3**	**4/12**	**84%**	
72	Review Entity/Attributes	48 hrs	1 day	4/3	4/3	100%	SI,TT,UI,CvT,PA,DA
73	Update Primary Keys	40 hrs	1 day	4/3	4/3	100%	SI,UI,PA,TT,CvT
74	Normalize Data (Relationships, etc.)	40 hrs	1 day	4/3	4/3	100%	SI,UI,PA,TT,CvT
75	Update Entities & Attributes into SA Rep	8 hrs	1 day	4/4	4/4	100%	SI
76	Update Definitions in SA Repository	8 hrs	1 day	4/4	4/4	100%	SI
77	Update E/R Diagrams/Logical Model	88 hrs	5 days	4/3	4/7	100%	SI,UI,TT,CvT
78	Relate Data Model to Process Model	16 hrs	2 days	4/4	4/5	100%	SI
79	Resolve/Reverse data domains syntax	16 hrs	1 day	4/3	4/11	1%	SI,IT
80	Complete E/R Diagrams for Reference Ta	48 hrs	5.6 days	4/4	4/12	59%	SI,IT
81	DB/CDM Review	40 hrs	1 day	4/3	4/3	100%	SI,UI,PA,TT,CvT
82	Verify and Document LDM Model in Rep	40 hrs	1 day	4/10	4/10	100%	SI,UI,TT,PA,CvT
83	**PROCESS COMPONENTS**	**490 hrs**	**18 days**	**4/5**	**4/28**	**52%**	
84	**Event/Environmental Model**	**64 hrs**	**10 days**	**4/10**	**4/21**	**33%**	
85	Update System Events	16 hrs	2 days	4/10	4/11	0%	PA
86	Update Event/Analysis, Event Dependen	36 hrs	3 days	4/19	4/21	50%	SI,PA
87	Verify & Document Event Model in Repos	12 hrs	1 day	4/21	4/21	50%	SI,PA
88	**Process Model**	**249 hrs**	**18 days**	**4/5**	**4/28**	**43%**	
89	Update Functional Decomposition Diagra	64 hrs	5 days	4/17	4/21	23%	SI,UI,TT,CvT
90	Complete DFD & Process Summary	64 hrs	5 days	4/5	4/13	84%	SI,UI,TT,CvT
91	Update Error PNL needs	1 hr	1 day	4/5	4/5	100%	SI[13%]
92	Update EWP & include in LDM	8 hrs	1 day	4/5	4/14	11%	SI

(d)

Figure 1.5 (continued) (c) training and QA, (d) data deliverables and activities.

demonstrate the consistency, integration, and flexibility of the deliverables between the phases of the project. The QA deliverables and activities

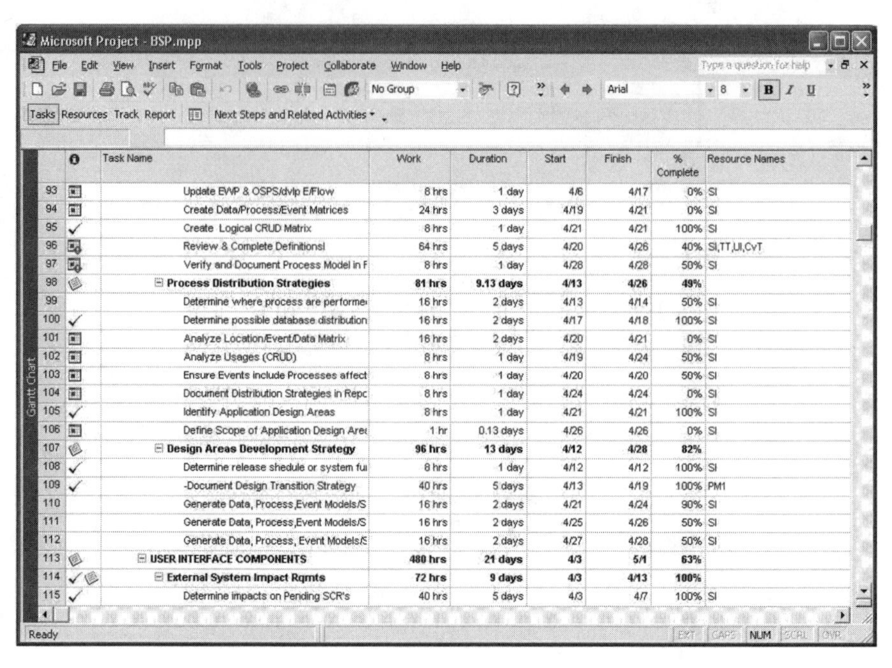

		Task Name	Work	Duration	Start	Finish	% Complete	Resource Names
93		Update EWP & OSPS/dvlp E/Flow	8 hrs	1 day	4/6	4/17	0%	SI
94		Create Data/Process/Event Matrices	24 hrs	3 days	4/19	4/21	0%	SI
95	✓	Create Logical CRUD Matrix	8 hrs	1 day	4/21	4/21	100%	SI
96		Review & Complete Definitionsl	64 hrs	5 days	4/20	4/26	40%	SI,TT,UI,CvT
97		Verify and Document Process Model in F	8 hrs	1 day	4/28	4/28	50%	SI
98		**Process Distribution Strategies**	**81 hrs**	**9.13 days**	**4/13**	**4/26**	**49%**	
99		Determine where process are performe	16 hrs	2 days	4/13	4/14	50%	SI
100	✓	Determine possible database distribution	16 hrs	2 days	4/17	4/18	100%	SI
101		Analyze Location/Event/Data Matrix	16 hrs	2 days	4/20	4/21	0%	SI
102		Analyze Usages (CRUD)	8 hrs	1 day	4/19	4/24	50%	SI
103		Ensure Events include Processes affect	8 hrs	1 day	4/20	4/20	50%	SI
104		Document Distribution Strategies in Repc	8 hrs	1 day	4/24	4/24	0%	SI
105	✓	Identify Application Design Areas	8 hrs	1 day	4/21	4/21	100%	SI
106		Define Scope of Application Design Are	1 hr	0.13 days	4/26	4/26	0%	SI
107		**Design Areas Development Strategy**	**96 hrs**	**13 days**	**4/12**	**4/28**	**82%**	
108	✓	Determine release shedule or system fu	8 hrs	1 day	4/12	4/12	100%	SI
109	✓	-Document Design Transition Strategy	40 hrs	5 days	4/13	4/19	100%	PM1
110		Generate Data, Process,Event Models/S	16 hrs	2 days	4/21	4/24	90%	SI
111		Generate Data, Process,Event Models/S	16 hrs	2 days	4/25	4/26	50%	SI
112		Generate Data, Process, Event Models/S	16 hrs	2 days	4/27	4/28	50%	SI
113		**USER INTERFACE COMPONENTS**	**480 hrs**	**21 days**	**4/3**	**5/1**	**63%**	
114	✓	**External System Impact Rqmts**	**72 hrs**	**9 days**	**4/3**	**4/13**	**100%**	
115	✓	Determine impacts on Pending SCR's	40 hrs	5 days	4/3	4/7	100%	SI

(e)

		Task Name	Work	Duration	Start	Finish	% Complete	Resource Names
116	✓	-Determine impacts on Employee DataBa	8 hrs	1 day	4/13	4/13	100%	SI
117	✓	-Determine Impacts on Mkt Mgt Data	8 hrs	1 day	4/13	4/13	100%	SI
118	✓	-Determin Impacts on Budget Data	8 hrs	1 day	4/13	4/13	100%	SI
119	✓	-Determine Communication Int TAP to CO	8 hrs	1 day	4/13	4/13	100%	UI
120		**Current Systems Interface Documetati**	**44 hrs**	**12 days**	**4/13**	**4/28**	**86%**	
121	✓	Update one application(MRB)	1 hr	1 day	4/13	4/13	100%	UI[13%]
122	✓	Update Issues/Problems with Existing Sy	1 hr	1 day	4/14	4/14	100%	UI[13%]
123	✓	Update Remaining Applications	1 hr	1 day	4/14	4/14	100%	UI[13%]
124		Update System Process Decomposition	40 hrs	5 days	4/20	4/28	75%	SI
125	✓	Update System Files	1 hr	1 day	4/14	4/14	100%	UI[13%]
126		**Protype (Watcom pilot user interfaces)**	**320 hrs**	**21 days**	**4/3**	**5/1**	**51%**	
127		Determine User Requirements	16 hrs	2 days	4/3	4/14	20%	UI
128		Map Entities to PowerBuilder Objects(Da	24 hrs	3 days	4/3	4/14	50%	UI
129		Map Business Events to PowerBuilder E	24 hrs	3 days	4/3	4/14	50%	UI
130		Determine GUI Requirements	16 hrs	2 days	4/3	4/16	50%	UI
131		Develop Screen Navigation & Flows	32 hrs	4 days	4/13	5/1	50%	UI
132		Determine Reusable Components	16 hrs	2 days	4/13	4/14	50%	UI
133		Design & Build Prototype System	160 hrs	10 days	4/3	4/14	80%	UI,CT1
134		Update Physical CRUD Matrices	8 hrs	1 day	4/24	4/24	0%	UI
135		Document Prototype Errors & Procedure	24 hrs	3 days	4/15	4/19	0%	UI
136		**User Reporting Requirements**	**44 hrs**	**19 days**	**4/4**	**4/28**	**20%**	
137		Update implementation of reporting	16 hrs	1 day	4/25	4/25	0%	TT,TT1
138		Update Dvlp report migration & report str	9 hrs	1 day	4/25	4/25	0%	TT,TT1

(f)

Figure 1.5 (continued) (e) process deliverables and activities, (f) user interface deliverables and activities.

ensure completeness of both the core and supporting infrastructure deliverables.

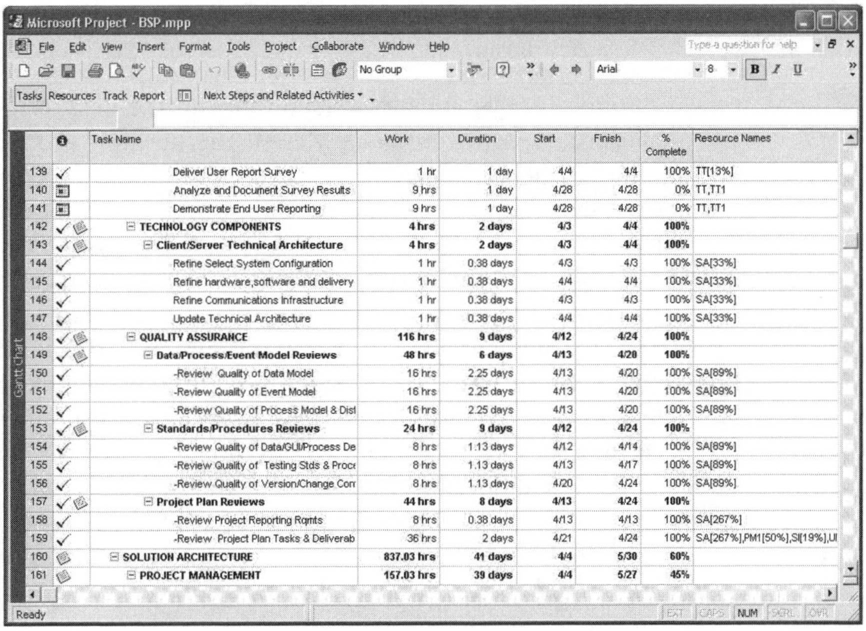

Microsoft Project - BSP.mpp

		Task Name	Work	Duration	Start	Finish	% Complete	Resource Names
139	✓	Deliver User Report Survey	1 hr	1 day	4/4	4/4	100%	TT[13%]
140		Analyze and Document Survey Results	9 hrs	1 day	4/28	4/28	0%	TT,TT1
141		Demonstrate End User Reporting	9 hrs	1 day	4/28	4/28	0%	TT,TT1
142	✓	**TECHNOLOGY COMPONENTS**	**4 hrs**	**2 days**	**4/3**	**4/4**	**100%**	
143	✓	**Client/Server Technical Architecture**	**4 hrs**	**2 days**	**4/3**	**4/4**	**100%**	
144	✓	Refine Select System Configuration	1 hr	0.38 days	4/3	4/3	100%	SA[33%]
145	✓	Refine hardware,software and delivery	1 hr	0.38 days	4/4	4/4	100%	SA[33%]
146	✓	Refine Communications Infrastructure	1 hr	0.38 days	4/3	4/3	100%	SA[33%]
147	✓	Update Technical Architecture	1 hr	0.38 days	4/4	4/4	100%	SA[33%]
148	✓	**QUALITY ASSURANCE**	**116 hrs**	**9 days**	**4/12**	**4/24**	**100%**	
149	✓	**Data/Process/Event Model Reviews**	**48 hrs**	**6 days**	**4/13**	**4/20**	**100%**	
150	✓	-Review Quality of Data Model	16 hrs	2.25 days	4/13	4/20	100%	SA[89%]
151	✓	-Review Quality of Event Model	16 hrs	2.25 days	4/13	4/20	100%	SA[89%]
152	✓	-Review Quality of Process Model & Dist	16 hrs	2.25 days	4/13	4/20	100%	SA[89%]
153	✓	**Standards/Procedures Reviews**	**24 hrs**	**9 days**	**4/12**	**4/24**	**100%**	
154	✓	-Review Quality of Data/GUI/Process De	8 hrs	1.13 days	4/12	4/14	100%	SA[89%]
155	✓	-Review Quality of Testing Stds & Proce	8 hrs	1.13 days	4/13	4/17	100%	SA[89%]
156	✓	-Review Quality of Version/Change Con	8 hrs	1.13 days	4/20	4/24	100%	SA[89%]
157	✓	**Project Plan Reviews**	**44 hrs**	**8 days**	**4/13**	**4/24**	**100%**	
158	✓	-Review Project Reporting Rqmts	8 hrs	0.38 days	4/13	4/13	100%	SA[267%]
159	✓	-Review Project Plan Tasks & Deliverab	36 hrs	2 days	4/21	4/24	100%	SA[267%],PM1[50%],SI[19%],UI
160		**SOLUTION ARCHITECTURE**	**837.03 hrs**	**41 days**	**4/4**	**5/30**	**60%**	
161		**PROJECT MANAGEMENT**	**157.03 hrs**	**39 days**	**4/4**	**5/27**	**45%**	

(g)

Figure 1.5 (continued) (g) technology deliverables and activities.

1.7 A real-world problem scenario: managing multiple IT projects

During my 28 years in IT in Canada and the United States, I have had the opportunity to experience similar types of project management issues and concerns at various companies. In this case, senior management is mostly concerned with the risks of managing multiple projects, rather than the single-project problems discussed previously. The complex project management problems discussed here require a much more comprehensive integrated solution than the single-project solution mentioned above.

Executive management (business and IT) from a large company mentioned:

> We had an urgent need to establish a program management plan to manage the delivery of multiple business and IT projects. Executive management made a decision to undergo a massive restructuring of the company's business. We contacted a consulting company with extensive industry experience to assist us in defining new business process reengineering models. This consulting company interviewed various executive managers and provided comprehensive documentation of the new business processes, which were approved in principle by executive management. However, there was

no program plan or recommended structure to direct the implementation of these new business processes.

Senior business managers responsible for the deployment of the newly established business processes viewed the business process reengineering solution as theoretical and impractical. They decided to perform this business reengineering effort using in-house staff and a local consulting firm. This local consulting firm, with business process reengineering expert knowledge, was contracted to deliver a program management solution with support from the client's business groups. The end result was another voluminous documentation of the future business processes in a new format and structure with contents similar to those of the previous consulting company's recommendations. The deliverables produced were presented at such a detailed procedural level that executive management was having great difficulty understanding the concepts and their applicability. A program plan was produced in the form of a project schedule with duplicated and redundant business and IT projects, inconsistent tasks and activities, and requirements for extensive business and IT resources without any quantifiable economic justification.

We, executive management, unsatisfied with the results produced so far, decided to hire an "expert" IT project management consulting firm to assist in managing the program. The objectives of the contract were to optimize the utilization of business and IT resources and to improve the program management communications reporting internal and external to the projects. We contracted a project management firm with "expert" knowledge of business processes, project management, and IT processes. A new jargon was now introduced including words like "architectures," "use-case analysis," "object-oriented project management," "reverse engineering," and the like, which compounded the existing communication problems. The desperate search for the silver bullet solution had started with the hope that this IT project management consulting company would magically deliver the solution to this program management problem. The expectation from executive management was that this consulting company would reuse the documentation produced so far and that this time around, progress would be communicated effectively.

This IT project management consulting firm, rightly so, suggested the use of a well-defined methodology. The end result was a set of extensive project management and architecture documentation based on the firm's self-proclaimed unique project management and architecture methodology, resulting in additional consulting resources, regular daily nonproductive status meetings, and more detailed, disintegrated documentation. The project team, again overly excited about the new project management and architecture methodology, adopted the consulting firm's fresh-start approach with great respect and enthusiasm. The result was another set of similar business, IT, and project management documentation, repackaged with the consulting company's logo and confidentiality and copyright terms appearing on the title page of each document.

The problem scenario discussed in Section 1.5, seems to be repeating itself, but to a much larger extent. This company has already expended extensive effort to document business processes, IT requirements, and project management standards and procedures, but it cannot justify the value of such extensive documentation. It had project plans and schedules with extensive activities/tasks and budget forecasting and tracking documents that were gradually becoming unmanageable. Enormous amounts of documentation were produced, which again looked impressive but lacked integration and consistency, resulting in the demand for additional IT, business, and project management resources. This company seemed to be going through another sort of information explosion crisis, but now to a larger extent, with unclear roles and responsibilities. Business and IT specifications were delivered and redelivered as different system specification with new labels, and project plans and schedules were managed by several enthusiastic but nonproductive project management teams. The familiar headless chicken environment seemed to be the project team's mode of operations, with reoccurring requests for approval appearing on the desks of executive management for additional systems architects, business analysts, systems and data analysts, developers, and project managers.

Now, a frustrating message, similar to the concerns expressed in Section 1.5, came from an IT executive with a program management perspective:

> The original need for a consistent and integrated program management plan to manage the identification and delivery of multiple IT projects is still nonexistent, with unclear linkages among project management, IT management and business management deliverables. An unexpected additional legal contracting issue as a result of the consulting firm confidentiality and copyrights terms appearing on the title page of each document now needs resolution. To add to the existing information jungle of unstructured documentation, the program schedule now contains additional projects and activities. The program is now way over budget and behind schedule, with scope and quality requirements being addressed with Victorian-novel-style documentation. A polished report from the consulting firm recommended a solution to the problems, which required organizational and cultural changes. This change recommended the establishment of two separate project management groups, one to support the business management functions and the other to support the IT management functions. This recommendation suggested the need for additional business and IT resources that escalated the program budget to a level that was difficult to justify to executive management committee.

The search for the silver bullet solution continues, and this IT executive seemed to be in more severe trouble, with valid excuses to executive management becoming nonexistent. The original questions of, What should I do to get an acceptable program management plan to manage these major multiple IT projects back to an acceptable schedule, budget, scope, and quality level? and What approach or technique should I apply? still remain unsolved with additional resource and legal complexities, as a result of another desperate search for the silver bullet solution.

This IT executive finally realized that there is no silver bullet solution and decided to approach the information jungle, inconsistency, and disintegration problems with the same IT project management consulting firm and search for a senior program manager. This individual needed to have experience with business management, IT management, and project management integration skills, as well as systems architecture and project management expertise. This IT executive continued his search for another solution, but in this case with a focus on risk management. The risks involved in searching for this senior program manager were minimal. His goal was to manage and resolve these project management problems by employing a senior project manager who had extensive experience in managing multiple projects and program management. This individual needed to have general business knowledge; specialized people, process, and technology integration skills; and excellent risk management skills.

This problem scenario shows some of the major project management errors that often occur, similar to those discussed in Section 1.5:

- Committing to a program schedule and budget, whether or not scope and quality can be effectively met.
- Establishing separate program management groups: one for business and one for IT, each group establishing different and unique project management policies, processes, and procedures.
- Holding business responsible for describing the business requirements with little or no IT involvement and IT responsible for development based on business requirements developed with little or no IT involvement. Business then gets involved during the integration and deployment/production phases. Integration only exists during the integration phases. This approach may work well using the traditional waterfall approach to software development, where business requirements are well known in advance. However, in most cases, business requirements are rarely well known and, as such, require a more iterative/prototyping method that demands the need for integration, conceptualization, and risk management.
- Assigning a program manager who lacks integration, conceptualization, and risk management skills.
- Addressing the communications issue with excessive status meetings.
- Ignoring the three major requirements of modern project management—conceptualization, integration, and risk management.
- Disregarding the root cause of communication problems—lack of consistency, completeness, and integration.
- Program manager's disregarding clear and articulated documentation of project team roles and responsibilities in a project structure environment to show how these roles and responsibilities are performed in the

project schedule, as well as the lack of an integrated program plan to guide the delivery of multiple projects.

- Program schedule's containing hundreds of activities without any WBS or an integrated master project schedule.
- Emphasizing human resources management with little consideration of business, IT, and project management components integration.

The solutions discussed in Section 1.8 demonstrate the value of business management, IT management, and project management integration to solve the root-cause communication problems resulting from the lack of consistency, completeness, and integration of these major components. It is worth mentioning that similar errors were discovered during the management of a major project, as were discovered during the management of multiple projects. However, the integrated solution to managing multiple IT projects is much more complex than managing a major IT project and, as such, will require program managers with excellent skills in conceptualization, integration, and risk management to ensure successful management and delivery of multiple IT projects.

1.8 A real-world solution scenario: managing multiple IT projects

Integration is not an option. If you don't integrate now, you are going to do so later, so you might as well think about it now.
—Marco Iansiti, Harvard business professor, at 2002 CIO Panel of Experts Award

This section introduces the recommended integrated program management solution presented in this book to solve the problems discussed in Section 1.7. This recommended solution is a compilation of my extensive project management experiences incorporated into a single volume. Various successful scenarios and lessons learned from less successful scenarios were modified accordingly to reflect modern project management principles and applicability.

The model-centric approach to managing multiple IT projects introduced in this section represents the foundation model detailed in this book as follows:

- Chapter 2—Integrated IT Project Management Model (Framework)—introduces the IPM-IT model, discusses the business management, project management, and IT management components of the model, and provides the roles and responsibilities of the program delivery manager during the management and delivery of a program or multiple IT projects.
- Chapter 3—Business Management Model—expands on the business management model introduced in Chapter 2, discusses the components of the model and provides the purpose, roles and responsibilities,

procedures—deliverable, process flow, and checklist (measurement criteria) templates for each of the components.

- Chapter 4—Project Management Model—expands on the project management model introduced in Chapter 2, discusses the components of the model, and provides the purpose, roles and responsibilities, procedure-deliverable, process flow, and checklist (measurement criteria) templates for each of the components.
- Chapter 5—IT Management Model—expands on the IT management model introduced in Chapter 2, discusses the components of the model, and provides the purpose, roles and responsibilities, procedures—deliverable, process flow, and checklist (measurement criteria) templates—for each of the components.
- Chapter 6—Integrated IT Project Delivery Life-Cycle Model—expands on the IT project delivery life-cycle (PDLC) model, discusses the components and phases of the model, and provides the linkage or horizontal integration to business management, project management, and IT management components.
- Chapter 7—Aligning PMI-PMBOK Processes with IBM Rational Corporation RUP—shows the integrative nature of these processes and their alignment with the IPM-IT framework presented in this book. This framework forms the basis for understanding the alignment or horizontal integration with RUP based on IBM Rational Corporation RUP life-cycle phases.

1.9 Gaining acceptance for integration

Misunderstandings sometimes occur because of differences in thinking preferences.
—James P. Lewis

One of the major obstacles that project managers often have to overcome is the lack of senior management support for modern integrated project management (IPM) practices and procedures. Application of IPM practices and procedures can be a frustrating experience, especially when senior management fails to recognize the need for such practices, or even worse, agrees in principle with integration, but questions the need for consistency and standardization. The result is the classic project manager's balancing act of standing up for what he believes is correct and risking loosing his position or hesitantly supporting senior management decisions, maintaining the status quo, and continuing with the nonintegrated, inconsistent, and resource-intensive approach to project management.

The IPM-IT framework discussed in this book can only be effective when properly applied with adequate support from senior management. The objective of this section is to present some guiding principles to assist project managers in obtaining management support for application of this IPM-IT framework.

Some senior business and IT management professionals resist integration practices and procedures because these procedures require consistency, uniformity, and standardization, which may block the progress, ambition, and ego of individual empire builders who usually strive to gain power through "single-view" management.

Often the arguments against project management integration by senior business and IT management professionals are formulated from a number of myths that are challenged throughout this book.

1. *Myth:* These methods are all theoretical; in this company things are done differently. Models and processes are theoretical and best suited for the academic environment.

   Reality: Any effective method requires some form of theory or principle as the foundation for execution; otherwise, chaos will result. Models are abstractions of reality.

2. *Myth:* Model-centric project management is no different from our current project management processes. The solution is executive management commitment, not enhanced project management processes.

   Reality: I agree that executive management commitment is necessary. However, no executive manager will commit to any process that does not demonstrate explicit linkage to business and IT management processes. Even if commitment is gained, it is the famous quote: "agreement in principle." Tools and techniques of the 1970s and 1980s when applied to modern software development processes will produce results suited for the 1970s and 1980s. We have just used technology to progress backwards.

3. *Myth:* Systems architects should not be project managers.

   Reality: Do not confuse construction industry projects with IT projects. According to Harold Kerzner, the three key skill requirements for modern project management are conceptualization, integration, and risk management, and it is the systems architect who has the conceptualization and integration skills necessary to manage IT-related projects. This is one of the heresies that have opened the gates for people with little or no IT training or knowledge to enter the IT field. In my view, the issue of the qualifications of the "skilled" project manager is one of the root causes of the communications problems that companies are ignoring or failing to address, probably because of political pressures from executive management with conflicting views on whom they perceived to be skilled project management resources.

   It is interesting to note that some senior project managers who have limited project management knowledge, skills, and qualifications based on PMBOK standards frequently manage the delivery of IT projects without any consideration of process or technology integration. They normally express the view that human resource management skills are the only qualifications needed of IT project managers. In my view, we have now promoted a human resources manager to the position of project manager and have ignored the widely accepted fundamental principles of modern project management as published by Harold Kerzner in [1].

4. *Myth:* Project managers are too formalized and focus on issue and change management as a means to gaining any recognized impact on the project.

   Reality: This is a true statement of project managers who normally lack the required training and knowledge of IPM and, as a result, focus on issue and change management formalized procedures. This is a typical example of the wrong project manager; these individuals are better suited as auditors, rather than project managers.

5. *Myth:* Scope creep is uncontrollable.

   Reality: Although this statement can be true in many cases, it is an example of negative project management. The essence of project management according to PMBOK is initiating, planning, executing, controlling, and closing. Integration of business management with project and IT management processes discussed in this book will provide some practical solutions for addressing this concern.

6. *Myth:* We have always developed software without these processes and procedures, and the products have worked, so why should we use these procedures now?

   Reality: The challenge here is to determine how successful these projects have been to the business and IT support staff, then to demonstrate that the integrated approach to project management is not a luxury, but a necessary and essential element for corporate success.

7. *Myth:* This is a business, not a university.

   Reality: Interesting enough, this statement usually comes from those who lack university training. My normal response is, "The cause of most IT failures is the inability of IT professional to apply fundamental principles and processes to solving business problems." The application of project management principles, as described in PMBOK and Kerzner's book, is a necessity to managing projects in order to provide value to both the business and IT environments.

8. *Myth:* These integrated methods are good, but our environment is not ready for such a change and will not be probably until some time in the future.

   Reality: This is a typical case of a "blocker" as described in Kerzner's book. The application of new and effective project management processes should be in the interest of the company.

9. *Myth:* None of our project managers is familiar with these integrated methods. It will take too much time, effort, and cost to retrain our project managers.

   Reality: This can be a valid statement if the IPM approach represents a major change in direction. The best response is to determine the maturity level of project management at the company and assess the ease of applying these new integrated methods. However, integration is one of the key requirements for modern project management and, as such, is necessary for linkage with the business and IT environments.

10. *Myth:* Integration is only necessary for very large projects.

    Reality: Nothing could be farther from the truth. The objective of integration is to ensure that the project, however success is defined, is a viable business investment and provides economic, political, technological, or environment value to the business (business management) that can be easily supported by the IT management infrastructure. All projects should fit these criteria; if they do not, then we have just spent time, effort, and costs on wasted and nonproductive initiatives. The application of new and effective project management processes should be in the interest of the company. The foundation principles of modern project management, based on essential skill requirements—business conceptualization; people, process, and technology integration; and risk management—are necessary to ensure that projects fully support the interests of the company.

It is important to note that these myths have persuasively swayed some professionals and senior managers whose genuine goals to strengthen the quality, consistency, integration, and reuse of project management integration principles, seem to have caused them to take an opposing viewpoint. These individuals, in particular, need examples and scenarios about the what, why, when, who, and how of IPM-IT, about alternatives, and about the consequences of disintegrated project management.

1.10 Summary

A major cause of most IT project management problems is the inability of certain IT executives/managers and senior project managers to apply basic IT concepts to solving specific business problems. Decisions are made based on emotions rather than objectivity because of the lack of adequate project management knowledge and training. Senior managers, rightly so, place emphasis on people skills—human resources management and communications management—but sometimes lack the fundamental project management knowledge and training to objectively assess the project risks and severity of quality and scope issues, resulting in added communications problems internal and external to the project.

Added to this set of issues, some senior project managers manage the delivery of IT projects by demanding excessive daily project meetings and status reporting with key project team members. Too much emphasis is placed on reporting of costs and schedule and time constraints, while the most important project management success factors are ignored—scope and quality constraints, the basic information needed to determine the costs, and schedule reporting information. In my view, we have just promoted a glorified secretary to the position of senior project manager to manipulate, produce, and report on project schedule and costs information using various disintegrated Microsoft Excel and Word documents—no wonder there are so many blunders and failures in delivering real successful projects.

The solution introduced in this chapter highlights some issues related to project management knowledge, skills, and experiences that surround the IT industry—project managers with specialized human-resource-management, auditor, or "glorified secretarial" expertise, without any formal industry-recognized project management qualifications. A recommended solution is presented in Figure 1.6 with further details included in the roles and responsibilities section of each chapter. I certainly hope that these suggested project management skill guidelines, based on the issues discussed, will be viewed by senior project managers at IT project management consulting firms as informative and necessary for project management excellence. In my view, the prerequisite to effective and efficient application of modern project management principles demands professional qualifications, in accordance with PMBOK or similar professional guidelines. The further we deviate from this goal, the more difficult it will be to close the gap between project management excellence and project management inconsistency, to increase progression to the SEI CMMI® project management maturity levels.

Please don't get me wrong. Soft skills such as communications management and human resource management are necessary and essential to implementing project management key objectives for scope, time, costs, and

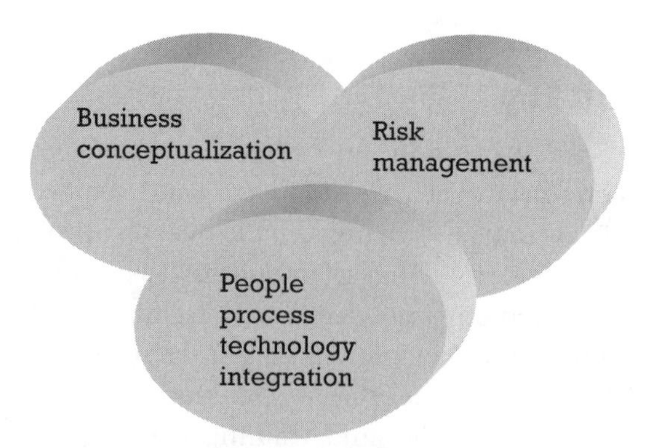

Figure 1.6 Modern IT project management skills framework: conceptual view.

quality. However, these skills cannot be effectively applied without clear, complete, and consistent understanding of the fundamental IT project management knowledge processes, which normally result from intensive IT project management training and experience. Executive management keeps searching for the silver bullet solution to project management, and the sooner they realize that there is no silver bullet solution, the less disappointment and frustration they will experience in their efforts towards establishing an efficient and effective project management environment in their organizations.

The solution is to establish project management policies, standards, and procedures that can be effectively applied by project managers who have the fundamental project management skills, training, qualifications, and experiences. Harold Kerzner mentions in his book that integration, conceptualization, and risk management are the three major skills that will be required of project managers to support modern project management requirements. If this observation, which I fully endorse, reflects a project management solution to our existing problems, I believe that industries, universities, and project management professional institutes may have some challenges ahead of them.

The contents of this book provide an IPM-IT framework solution in the form of information and process models that describe the purpose, policies and guiding principles, roles and responsibilities, and procedure-deliverable, process flow, and checklist (measure quality) templates—for business management, IT management, and project management integration. The overall goal is to present the components of this framework and to demonstrate how the models are implemented by resources having business conceptualization; people, process, and technology integration; and risk management skills to solve the most common and difficult root cause of project management problems—ineffective communications internal and external to the project. Figure 1.6 further elaborates Harold Kerzner's revealing insights into the skill requirements necessary and essential for modern project management implementation.

1.11 Questions to think about: management perspectives

1. Think about how your organization manages projects. How does your organization view integration? How would you align project management processes with business management and IT management processes?
2. Think about how your organization deals with IT project management problems. How do emotional, cultural, and political behaviors affect objectivity? Why have projects failed? How is success or failure aligned with business value? How is business value aligned with project deliverables?

3. Think about how your organization views standardization and consistency. How would you integrate projects with other projects? What are your major project communications problems and to what extent do the integration issues cause these problems?
4. Think about how your project is sponsored. What are the levels of involvement and commitment from business, IT, and project management executives? How is the formal or informal structure communicated? How do you know whether or not you still have executive management support?
5. Think about how you manage project expectations. What are your expectations of IT consulting firms? How do you manage success—with magic or measurement? How do you know that expectations are being met?
6. Think about how you manage project delivery. How do you structure project organization, roles and responsibilities, WBS, and schedule? How do you manage people, process, deliverables, and risk elements? How do you deal with the integration of these elements?
7. Think about how you manage multiple projects. How do you manage conflicting expectations and priorities? How do you deal with multiple consulting firms? How do you deal with communications issues and people resource scheduling?
8. Think about how you manage multiple-project delivery. How do you structure multiple-project organization, roles and responsibilities, WBS, and schedules? How do you manage people, processes, deliverables, and risk elements? How do you deal with the integration of these elements?
9. Think about how your organization addresses integration. What approach would you use to gain acceptance for integration? What are your skills requirements for an IT project manager managing a single project? What are your skills requirements for a project manager managing multiple projects? What are the major differences?

Reference

[1] Kerzner, H., *Project Management: A Systems Approach to Planning, Scheduling and Controlling*, 7th ed., New York: John Wiley & Sons, 2001.

Selected bibliography

Bennatan, E. M., *On Time, Within Budget, Software Project Management Practices and Techniques*, New York: McGraw-Hill, 1992.

Boehm, B., "Software Risk Management: Principles and Practices," *IEEE Software*, Vol. 8, No. 1, January 1991.

ESI International, *Project Framework—A Project Management Maturity Model,* Vol. 1, Arlington, VA: ESI International, 1999.

Fleming, Q. W., and J. M. Koppelman, *Earned Value Project Management,* 2nd ed. Newton Square, PA: PMI, 2000.

Lewis, J. P., *Project Planning, Scheduling, and Control,* 3rd ed., New York: McGraw-Hill, 2001.

Muller, R. J., *Productive Objects: An Applied Software Project Management Framework,* San Francisco, CA: Morgan Kaufmann Publishers, 1998.

Paulk, M. C., et al., *Key Practices of the Capability Maturity Model®,* Version 1.1, Pittsburg, PA: Software Enginering Institute, Carnegie Mellon University, 1993.

PMI, *Project Management Institute: A Guide to Project Management Body of Knowledge,* 2000 Edition, Newton Square, PA: PMI, 2000.

Thonsett, R., *Third Wave Project Management: A Handbook for Managing the Complex Information Systems for the 1990s,* Englewood Cliffs, NJ: Prentice Hall, 1989.

CHAPTER

2

Contents

Integrated IT Project Management Model (Framework)

The way the model is defined determines how we will attempt to implement it.
—Unknown

2.1 Introduction

Project management is a discipline that influences every aspect of a company's business. Projects, no matter how successfully viewed, that do not meet the needs of the business are not good investments of the company's resources. Business management, project management and IT management must all work together, with a common understanding of project management policies, roles and responsibilities, and procedures-deliverables, process flows, and checklists to successfully deliver projects. The components of business management, IT management and project management require clearly assigned responsibilities to effectively manage and deliver projects in a timely and cost-effective manner to successfully meet business needs. Any business management, IT management or project management group that fails to fulfil its responsibilities will negatively impact the outcome of the project.

The objective of this chapter is to introduce the IPM-IT models,[1] discuss the components of the conceptual, detailed

1. IPM-IT models: An integrated representation of business management, project management, and IT management component processes to manage the delivery of multiple IT projects. It is synonymous with IPM framework.

models shown in Figures 2.1 and 2.2, respectively, and provide the roles and responsibilities of the IT program delivery manager during the management and delivery of a program[2] or multiple IT projects. Further details on each of the three components are provided in succeeding chapters of this book.

Because both business and IT management are vital to the success of projects, the responsibilities are defined in the context of the company's IPM-IT framework. An overview of the conceptual framework is provided next.

Figure 2.1 is a representation of the IPM model that shows a conceptual view and the linkages with business management, project management, and IT management. Figure 2.2 is a more detailed representation of the IPM model, which shows the components of business management, project management, and IT management and the linkages among them. The succeeding chapters of this book provide further details on each of the three major components, with Chapter 6 focusing on the horizontal integration of these components, using the IT PDLC model shown in Figure 2.2. Chapter 7 discusses the implementation of the rational unified process (RUP).[3]

2.2 Program organizational structure: rationale

During my extensive IT career experiences, I have had the opportunity to evaluate, recommend, and implement project management organizational structures at many large corporations having varying project management maturity levels, from level 1—Inconsistency to level 5—Excellence, in accordance with ESI International[4] project management maturity levels.

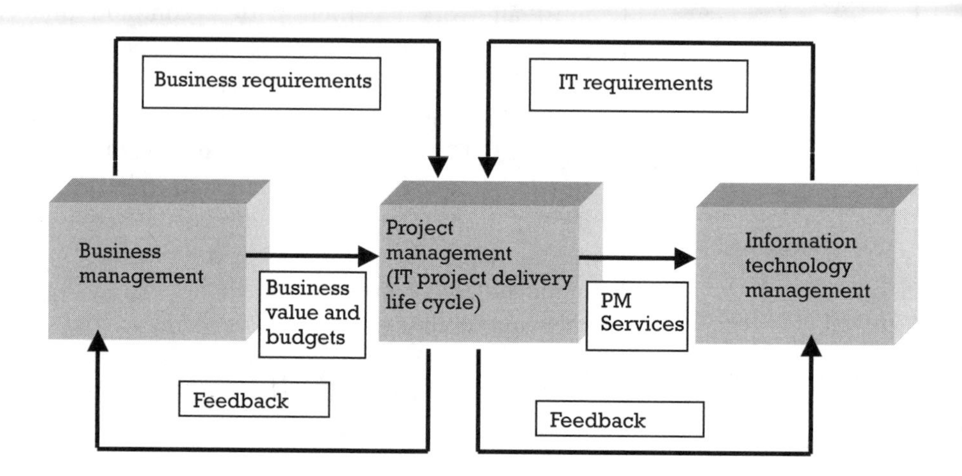

Figure 2.1 IPM-IT conceptual model.

2. Program: A portfolio of related projects with similar business functionality, managed using IPM models.
3. Rational unified process: A software engineering process framework from Rational Corporation.
4. ESI International: A company that provides project management and contract management training and consulting services.

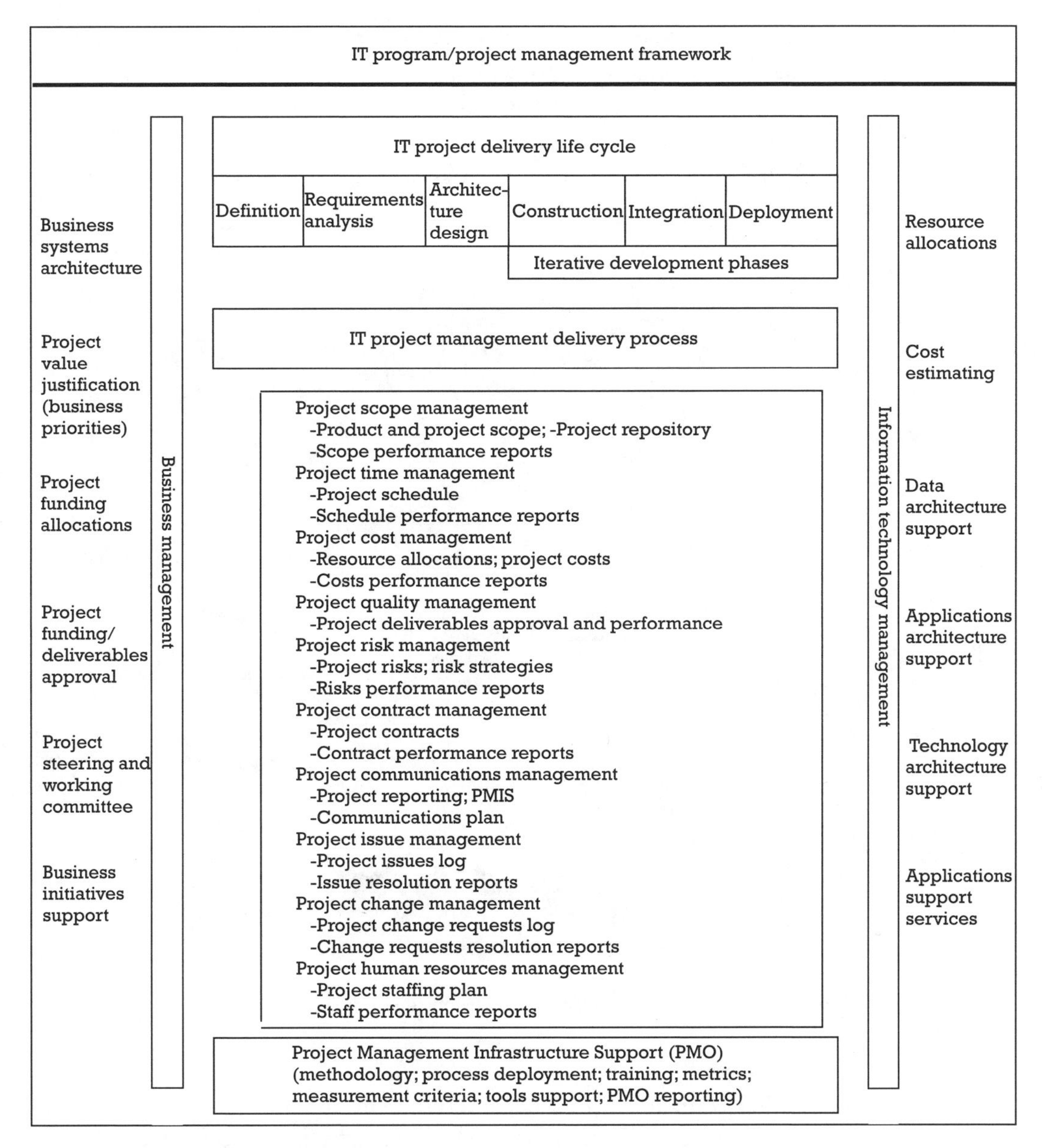

Figure 2.2 IPM-IT components model.

The recommendation in Figure 2.3 is based on the overwhelming need for an organizational structure to assist with the implementation of an integrated business management, project management, and IT management solution that can be objectively applied at most corporations with the goal of optimizing the utilization of IT resources during the management and delivery of multiple IT projects or during program management.

The program organizational structure in Table 2.1 expands on the roles identified in Figure 2.3. It is hoped that this recommendation will provide a more objective organizational structure than some of the traditional, emotional, politics-based structures, prejudiced by executive management or a

Table 2.1 Program Organizational Structure: Multiple IT Projects

BS: business support	PT: project team (applications delivery)	AS: applications support
		DS: development support
		TS: technical support

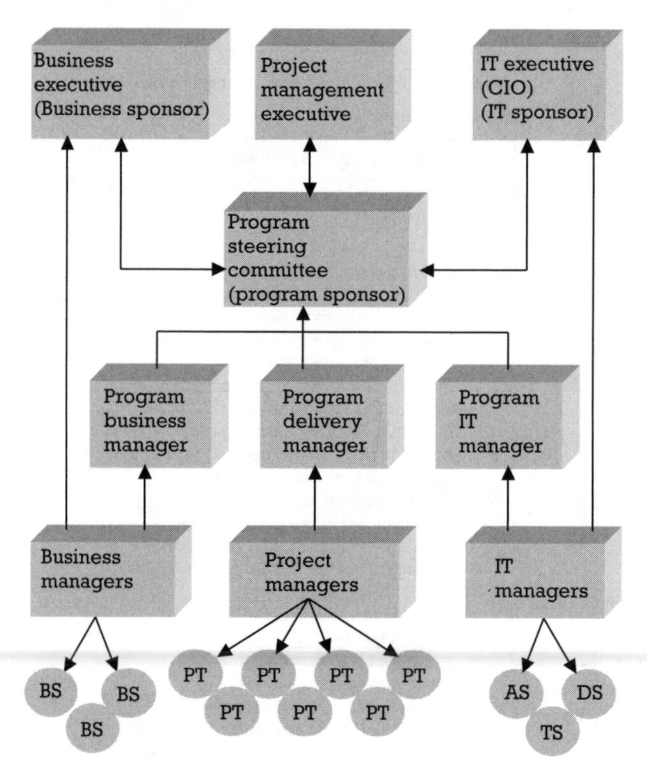

Figure 2.3 Program organizational structure solution.

consulting company's personal interests. One of the main root causes of project management failures is the partial and prejudiced organizational structure developed by certain executive management or consulting company's staff with self-interest that is politically motivated. Forcing alignment of an existing organizational structure to new project management processes or forcing alignment of existing project management processes to new organizational structures created for political reasons is a recipe for massive confusion and incompetence. In order to address these organizational structure issues at a large corporation, senior executives endorsed the recommendation for the creation of a project management executive position. This project management executive, who had the proper skills, knowledge, experiences, and professional qualifications, was highly effective in implementing an integrated program organization structure, similar to that in Figure 2.3, to support the management and delivery of the projects, based on an established program organizational framework. Project management executives with the right skills, knowledge, experiences, and professional qualifications can provide the necessary integrated solutions, especially to those corporations that have attained a reasonably mature project management cultural climate.

The integrated nature of this recommended solution is based on the effective execution of executive management, program management, and project management roles and responsibilities. A high-level discussion of the skills, knowledge, and qualifications of the team is presented. It highlights some real-world experiences of major issues, challenges, problems, and recommended solutions, which may interest those involved in project management practices.

The role of a project management executive–project position[5] was introduced, in addition to the existing functional business executive and IT executive positions to ensure that corporate project management processes are effectively communicated and consistently applied by both the business and IT staff. This role, if executed by the right person, will resolve most of the communication issues internal and external to the projects at the executive management level. This individual must demonstrate general business conceptualization knowledge; specialized people, process, and technology integration skills; and excellent risk management skills. In his or her specialized process integration skills, this individual must meet the criteria of having attained professional project management qualifications to ensure effective integration of project management processes with software development processes.

2.2.1 Real-world observations: project management executive

The productivity and competitive problems the IT industry faces result from ineffective top management, petrified in place, unwilling to accept change, and failing to provide wisdom and vision.
—Harvard Business Review

It is now time to share some real-world observations. A corporation attempted to fill this project management executive position with individuals having human resource–type project management skills, auditing-type project management skills, and glorified-secretary-type project management skills. It is interesting to note that neither of these individuals created any impact to justify the need for formal project management existence, and as a result the project management organizational structure was abandoned after some years of slow but moderately successful progress. The cause of this failure was attributed to the lack of an effective methodology.

5. Project management executive–project position: An executive project management position at the vice presidential level, held by project management professionals with generalized business knowledge; specialized people, process, and technology integration skills; and excellent risk management skills.

These project management executives, who seemed to have limited software applications development and project management training and experience, carefully examined this methodology deficiency problem and concluded that the existing project management methodology was ineffective. In this scenario, it seemed obvious that the qualifications and abilities of these project management executives to affect change and provide wisdom and vision on project management processes needed improvement.

This is a typical case of ineffective project management organizational structure at the executive level, which senior executives should attempt to understand and avoid in preventing the proliferation of communication problems and project management incompetence.

The role of the program steering committee–project position[6] consists of:

- Executive management (business executive–business sponsor, IT executive–IT sponsor, project management executive–PM processes champion);
- Program management (program business manager, program delivery manager, program IT manager);
- Project management (project managers for each of the projects).

2.2.2 Real-world observations: program steering committee

It is now time to share some real-world situations to emphasize the need for consistency and integration of project management processes. A large corporation, established a program steering committee, consisting of executive managers, program managers, and project managers whose major objective was to report on and communicate the status of the program to all members of this committee on a monthly basis. Each of the project managers produced individual status reports in different formats for the project that he or she was responsible for delivering and submitted them to the program manager on a weekly and monthly basis. The program manager, accountable to executive management, summarized and consolidated the inconsistent and sometimes incomplete projects status reports provided by each of the project managers. After various frustrating dialogues with the individual project managers, this program manager managed to produce the program status reports, which were delivered to the program steering committee meetings and business and IT executive stakeholders at regularly planned monthly status meetings.

During each of the program steering committee meetings, the participants spent most of their project time trying to explain inconsistencies, rather than reporting on the project statuses to improve communications internal and external to the projects. Unbelievably, at the end of each meeting, there were praises for everyone's efforts, with little consideration given to the secretary who spent her time nervously making conflicting and

6. Program steering committee–project position: A senior management committee position, held by executive and senior business, IT, and project management staff with project oversight and approval responsibilities.

sometimes inconsistent corrections to the program status reports, as suggested by the participants or the program steering committee chairman. This is the result of situation, which lacked consistent and integrated processes—a necessity for effective project management reporting.

This is again a typical case of project management reporting inconsistency at the executive and senior management level, which senior management should attempt to understand and avoid in order to prevent the proliferation of communication problems and project management incompetence during project management reporting. The need for a consistent, complete and integrated project management information system (PMIS), as suggested by PMBOK, that is accepted and endorsed by key stakeholders is a necessity for effective project reporting.

The role of program management and project positions[7] consists of:

- Program business manager (business manager/director to oversee management and delivery of the program for business requirements support);
- Program delivery manager (program manager/director to oversee management and delivery of the entire program for integrated business, project management, and IT delivery processes and must possess conceptualization, integration, and risk management skills);
- Program IT manager (IT manager/director to oversee management and delivery of the program for IT development support and project management processes, applications development processes, and systems architecture support, IT applications support, systems maintenance, and help desk, and IT technical support for hardware, system software, network, and facilities).

2.2.3 Real-world observations: program management

This same organization, rightly so, appointed a senior business manager to the position of program business manager, a senior project manager to the position of program delivery manager, and a senior IT manager to the position of program IT manager. There was no formal communications plan or any documented responsibility assignment matrix (RAM)[8]; as a result, each program manager assumed his or her responsibility based on the level of expertise that each of the program managers possessed—in this case, human resource management skills.

The program business manager, having an MBA, developed a budget and schedule for the overall program, using various disintegrated tools. The program delivery manager, having a masters in human resource management, developed an overall budget and schedule and a resource

7. Program management–project positions: Program management positions held by senior, business, IT, and project management staff with program management responsibilities and authorities.

8. RAM: A structure used to assign roles and responsibilities to particular people to produce specific deliverables. A RAM defines which individual is responsible for each WBS deliverable.

management plan for the entire program using various disintegrated tools, including project management tools, without any WBS—so you can imagine the chaos. The approach to project management seemed to focus on filling the working calendar with daily, weekly, and monthly meetings to justify the creation of a position for the "real" program delivery manager, who reported to him.

The program IT manager, a computer science graduate, developed technical architecture diagrams using various disintegrated tools, including Visio and CASE tools. At the end of each month, these three senior managers anxiously met to consolidate the information for the monthly steering committee meeting, celebrating with a sigh of relief and self-praise at completion. At this corporation, software applications development was viewed by senior management as the responsibility of IT and, as a result, should not be integrated with project management—an example of an ineffective project management organization, according to modern project management practices recommended by PMI, ESI International, and Kerzner.[9]

This is again, a typical case of program management inefficiency at the senior management level, which executive management should attempt to understand and avoid to prevent communication problems and project management incompetence during program management delivery. This style of project management of self-proclaimed program delivery mangers is typical of IT professionals who lack the proper IT project management skills, knowledge, and experiences. The need is great for qualified project management professionals with skills, knowledge, and experiences similar to those recommended by PMI; conceptualization, integration, and risk management skills are necessary and essential qualifications for practicing program delivery managers.

The role of project managers and supporting business and IT managers/directors consists of:

- Business support managers (functional business manager/director to oversee management and delivery of individual projects, for business requirement support);
- Project managers (project manager/director to oversee management and delivery of the individual projects for scope, time, cost, and quality to optimize resource utilization, while meeting stakeholders' expectation; must possess conceptualization, integration, and risk management skills);
- IT support managers (functional IT manager/director to oversee management and delivery of the individual projects for IT development support–project management and applications development processes, IT applications support–systems maintenance).

9. PMI, ESI International, and Kerzner: PMI, ESI project management training and consulting services, and Harold Kerzner's book on project management are excellent sources for project management reference materials.

2.2.4 Real-world observations: project managers

I never gave them hell. I just tell the truth, and they think it is hell.
—Harry S. Truman

In an organization every employee tends to rise to his level of incompetence.
—Laurence J. Peter, The Peter Principle

There are many obstacles that project managers are constantly faced with during their search for the right solution to delivering successful projects within the constraints of the existing project management cultural environment. In this scenario, the major reason executive management at a corporation endorsed project management principles was to justify the dollars given to a project management consulting firm to develop a project management methodology with the expectation that this methodology would be the substitute for their lack of involvement. Some functional business managers/directors and the functional IT managers/directors resisted any change to business or technical improvements, mainly because of their level of discomfort with operating in the existing environments, pessimistic or regressive perspectives on the job, and fear of the unknown.

Some project managers at this corporation had limited understanding of the fundamental principles of project management, such as WBS, and often lacked the required expertise to apply these basic concepts to any of their previous project management assignments. As a result of this knowledge and skills deficiency, these project mangers resorted to the negotiating aspects of project management, under endless frustration because of the lack of executive and senior management involvement and support. Remember! These executives firmly believed that a project management methodology was the substitute for their lack of involvement.

This project management methodology was delivered to executive and senior management by the consulting firm in three-ring glossy binders, in Victorian-novel-style format, suitable for the archives. The voluminous, unstructured, and boring contents created an instant distaste for project management by the business and IT professionals at this corporation. As a result, some senior and executive management were hesitant in enforcing the use of the methodology and adopted the nondecisive agreement-in-principle attitude. The bullish and emotional directive, Whatever it takes to deliver the project, whether or not a formal process was adopted, was the major theme from some senior and executive level managers.

The resistance-to-change attitudes by certain functional business and IT managers/directors were reinforced by these functional managers/directors as soon as they were informed of the lack of support from senior management for common, consistent, and IPM processes. The project managers and application developers were made responsible for delivering projects using any creative set of processes, and the difficulties in obtaining business and IT involvement and support have now reached that unresolved state. The project manager is now left to deliver a project solution with reluctant IT and business involvement, no accepted project management processes, and no support from senior management. Senior management's major directive to these project managers is to apply their negotiating skills. This directive is a result of their strong belief that project management is all about negotiating, another typical case of undesirable project management practices. I agree that negotiating and selling project management principles are necessary and essential to effectively apply these principles. However, it would be rather difficult to sell a concept, idea, principle, or theory without reliable involvement and support from senior management or sound knowledge of acceptable project management processes and practices.

In the absence of any accepted project management process, the individual project managers, especially the consultant project managers, adopted their so-called unique, proprietary methodology and totally ignored the client's project management processes—a mass confusion of political bickering and consensus building with consulting companies and project managers hoping for the magical silver bullet solution. The original objective of project management of managing scope, time, cost, and quality to optimize resource utilization, while meeting stakeholders' expectation, was transformed to people and contract negotiations, in line with some senior managers' views of project management.

I wish to emphasize that project management soft skills such as communications management, human resource management, and contract management negotiations are absolutely essential to the successful delivery of projects. However, application of these skills, without a consistent understanding of the processes, is analogous to shooting bullets in the dark, which frequently results in the project manager's spending unnecessary and nonproductive time and effort on people and contract negotiations. There exists a widespread need for a commonly accepted, consistent, and integrated understanding of applying IT project management processes by key decision makers and the project team to resolve the root cause of project management challenges—communication problems internal and external to the projects.

My view of this situation reveals that the existing project management methodology or processes were based on the Victorian-novel procedural style processes of the 1970s, which cannot be readily adapted to modern technology applications development and project management processes.

The scenario discussed requires the establishment of simple, flexible, and easy-to-use standards, policies, and processes based on a deliverables-based approach that demonstrates the value and benefits of project management principles and practices and shows the alignment with both business management and IT management processes.

The remaining sections of this chapter highlight the policies and responsibilities of the program delivery manager to support the functions of business management, project management, and IT management. Practicing program managers should find these practical guidelines very useful during the management and delivery of multiple IT projects, especially those who require an IPM framework solution.

2.3 Business management

The business management function, referenced in Figure 2.2, is introduced in this section to show the integrated nature of the program delivery manager's responsibilities. Chapter 3 discusses the integrated nature of business, IT, and project management executives, program managers, and project managers' roles and responsibilities to ensure optimum resource utilization. The program delivery manager manages and delivers multiple projects that support the following business management process components:

- Business systems architecture (BSA)/planning;
- Project value justification;
- Project funding allocation (PFA);
- Project funding/deliverables approval;
- Program steering and working committee;
- Business initiatives support (BIS).

These process components are integrated through the execution of business, project management, and IT management roles and responsibilities, and processes, tools, and techniques in accordance with ESI International's fundamental elements of integration—people, processes, and technology. Chapter 3 discusses these integrated components in greater detail, using a business management model as the communication medium.

I have introduced the process components (what) of business management. In the succeeding sections, I elaborate on the information contents for each of the process components, and present the responsibilities (how) of the program delivery manager who is accountable[10] for managing, communicating, and making decisions, based on the information contents. The responsibilities of the program delivery manager show the integrated nature of the business management component processes to support the management and delivery of the overall program.

2.3.1 Program BSA

All experience is an architecture to build upon.
—Henry Adams

The primary objective of this function is to establish the BSA by analyzing the strengths, weaknesses, opportunities, and threats (SWOT) of the business, identifying and prioritizing business strategies and objectives, and determining the priorities of business initiatives. The BSA is implemented by the program delivery plan,[11] which identifies the delivery projects and the

10. Accountability: accountability = responsibility + authority.

11. Program delivery plan: A plan that provides the purpose and scope (why and what), schedule (when), effort (who), cost (how much) for delivery of each project that constitutes the program. It is the key deliverable from the BSA.

interactions between them. The integrative features of BSAs include the development or reengineering of business processes and the integration of business initiatives with IT initiatives in order to optimize resource allocations and to minimize duplicate processes and efforts.

Executives shall use the BSA as the basis for communicating and making decisions. Business management decisions must be made based on sound objective judgements, rather than the emotions or office-based politics. Sound objective judgements require executive management understanding of the following elements of the company's products and services:

- Business strategic model: SWOT (competition), including business mission, goals, objectives, and strategies, and corresponding priorities;
- Business requirements model: Business problems, informational needs, and corresponding priorities that reflect business strategic directions;
- Business functional model: Business functions and processes that support the business strategic model to resolve the business functional model;
- Linkage to IT architectures: Data, applications, and technology to ensure that the technology solution is aligned with the goals, objectives, and strategies of the business, described in the BSA document;
- Program delivery plan: Delivery projects, corresponding priorities, and implementation sequences that support the business and IT architectures.

Program delivery manager responsibilities include the following:

- Document details of integrated business and IT business plans—SWOT.
- Document integrated prioritized business and IT direction, objectives, strategies, and supporting initiatives.
- Document architecture processes/methodology guidelines and standards, and provide guidance on applying architectural development methodology.
- Assemble and present integrated business and IT architecture (data, applications, technology) or program architecture documentation.
- Report to the program steering committee with direct accountability to project management executive.

2.3.2 Project value justification (business priorities)

The right project with the right scope, right time, right cost, and right quality delivers the right value.

The primary objective of the project justification function is to ensure that each of the projects within the program, identified in the BSA, can provide value and improve the performance of the business to support the objectives of the corporation. Decisions regarding IT investments must be driven by the BSA/planning initiatives. Business value statements shall be used to assess the value of projects based on alignment with prioritized business and IT benefits. Executive management shall determine the value of the projects to the business based upon six areas of value that align with the business mission statement:

1. *Operational efficiency:* Assess the value derived from areas such as reduction in staff, reduction in operating expenses, increased resource usage efficiency, increased staff efficiency, improvements in business process deployment, and improvements in customer services.
2. *Strategic business alignment:* Assess the extent to which the project aligns with the corporate strategic goals, and determine the value contributed to achieving the strategies.
3. *Competitive market advantage:* Assess the impact to which the project provides an advantage in the marketplace. A project that gains competitive advantage must:
   - Change the competitive marketplace structure, or influence the way buyers and suppliers compete in the marketplace.
   - Improve the company's products or services, or change the competitive nature of the business.
   - Use the information of the current business practices, competitive nature, and internal information processing capability to determine a new business opportunity.
4. *Organizational impact and risk assessments:* Assess the risk and organizational impact of undertaking or not undertaking the project. Assess the risk of organizational changes. Assess the risk of loosing market share, or determine the impact of a proactive strategy to prevent the competition from gaining a competitive edge.
5. *IT support:* Assess the value of management need for information on key performance indicators and efforts involved to realize this information's need to support the company's mission. Assess the degree to which the project assists in achieving the company's critical success factors for IT support.
6. *Strategic business and IT architectures support:* Evaluate the degree to which the project is aligned with the overall BSAs (business, data, applications, and technology). Projects that are an integral part of the architecture framework will be assigned a higher value.

   Program delivery manager responsibilities include the following:

   - Document details on integrated business processes and IT business plans—SWOT.
   - Document integrated prioritized business and IT direction, objectives, strategies, and supporting initiatives.

- Document architecture processes/methodology guidelines and standards and provide guidance on applying architectural development methodology.
- Assemble and present integrated business and IT architecture (data, applications, technology) or program architecture documentation.
- Integrate, assemble, and report to the program steering committee on project status, issues, change requests, and risks with direct accountability to project management executive manager.

2.3.3 Project funding allocations

The immediate judgement on a budget is almost invariably wrong.
—Iain Macleod

The primary objective of project funding is to control the allocation of funds within the program for projects that have been prioritized by the program steering committee meetings and to allocate funds in increments by reassessing the project cost and benefits. Projects with increased cost that exceed the expected benefits will be reassessed and reprioritized and their funding reevaluated. The program steering committee will allocate funds to a project as approved or reserved at specific phased milestones. The project managers can only spend approved funds with approval authority from the program delivery manager. Program budgets will be based on a fiscal budget year, and funds will be allocated to the portfolio of projects in two categories:

1. Approved, based upon definitive cost estimates: Project managers are authorized to spend approved funds. Definitive estimates will be within the range of –5% to +10% for the phase of the project being estimated. Definitive estimates must include the following supporting details:
   - Definitive estimates and detailed assumptions used to develop the estimates;
   - A project schedule integrated with a project planning and tracking tool and WBS;
   - Resources allocated in the schedule with efforts allocated for all project tasks;
   - Resources optimized or leveled for all project activities/tasks;
   - Technology and facilities resources allocated in the schedule;
   - A detailed budget report in a format similar to the monthly financial report, including allocation of reserved and approved funds.
2. Reserved, based upon program budget estimates and preliminary cost estimates: Funds are set aside for projects according to four categories until definitive cost estimates are available:

   a. *Contingency reserve:* The budget that the project managers have identified and set aside to accommodate known project risks.

b. *Management reserve:* The budget that the project executive sponsors have set aside to accommodate urgent scope changes for change requests.

c. *Program architecture reserve:* The budget that project executive sponsors have identified and set aside for program architecture development.

d. *Project definition reserve:* The budget that project executive sponsors have identified and set aside for project definition program and project managers.

Projects within the approved program will be prioritized and a project budget allocated, as approved or reserved, by the program steering committee. This committee will allocate the funds as reserved for each project, and the program delivery manager will distribute the allocated funds as reserved for each of the project phases. The program steering committee will allocate funds as approved for each project, to be spent by the project managers with approval from the program delivery manager at the following milestones in the PDLC:

- The conclusion of the project definition phase (PDP)–solution definition: The project charter will be approved and funds allocated as approved for the requirements analysis phase (RAP).
- The conclusion of the project RAP–solution assessment: The project will be revalidated with updated project costs and benefits and funds allocated as approved for the project design architecture–solution architecture/design phase.
- The conclusion of the project design–architecture phase: A project value analysis will be conducted for the iterative development phases (IDPs)—construction, integration, and deployment–solution implementation, and funds allocated as approved for the three IDPs. In certain situations, funds will be allocated as approved for each of the IDPs, based on the project value analysis.

Program delivery manager responsibilities include the following:

- Integrate and report prioritize opportunities for projects within the program and allocate budget based on the value of the project to the business and IT.
- Report on the portfolio of projects within the funds allocated by the program steering committee for the entire program using project management guidelines.
- Organize program steering committee meetings and track allocation of funds from program budget.
- Report on allocation of funds at various phases of the project life cycle.
- Report to the program steering committee with direct accountability to project management executive manager.

2.3.4 Project funding/deliverables approval

If you have no approvals, you have no support and commitment, by definition.
—James P. Lewis

Project deliverables approval consists of business management, project management, and IT management commitments to approve a project for the next phase of development. The key decision points will be during the following:

- Beginning of each project via program delivery plan: Deliverable from BSA or business definition statement in the absence of a BSA;
- Transition from project definition to project analysis phase: Prototype #1;
- Transition from project requirements analysis to project architecture: Prototype #2;
- Transition from project architecture to IDPs–construction, integration, and deployment;
- Transition from project construction to project integration phase, if required;
- Transition from project integration to project deployment phase, if required;
- Ending of each project.

Preliminary project funding and deliverables approval decisions, early in the project life cycle, before all requirements are known, are necessary for budgetary requirements. The purpose of preliminary estimates is to address these requirements to a –25% to +75% level of accuracy. I strongly believe that an urgent need exists for an incremental phased approach to project deliverables approvals to allow for an orderly development of project estimates that can be validated against business needs at critical points in the life cycle of a project. This phased approach to project deliverables approvals will also provide an understanding of the project scope, funding/cost, schedules, effort, and quality that are agreed upon and supported by both the business and IT leaders.

Program delivery manager responsibilities include the following:

- Submit phased deliverables to the program steering committee meeting for acceptance, approval, and funding from business and IT perspectives.
- Manage the portfolio of projects deliverables within the funds allocated by the program steering committee from business and IT perspective.
- Perform QA reviews and provide acceptance for project deliverables based on acceptance criteria/checklists within the funds allocated by the program steering committee from business and IT perspective.
- Assign business resources for PM process/methodology support.
- Report to the program steering committee with direct accountability to PM executive manager.

2.3.5 Program steering and working committee

Not even computers will replace committees because committees buy computers.
—Edward Shepherd Mead

The primary objective of these committees is to establish a business (major stakeholders), project management, and IT management committee with program approval responsibilities. The program steering committee will have overall responsibilities for the program and have joint accountability, with the program managers (business, project management, and IT), for the overall scope, effort, cost, schedule, and quality performance of the program. The program working committee, chaired by the program delivery manager, will have overall responsibilities for issue resolution, change requests approvals, risk strategies, communications reporting, human resources development, and contract administration. This working committee will have joint accountability with the project managers for scope, effort, costs, schedule, and quality performance of the projects.

As the project progresses through the iterative life-cycle phases, the program steering committee will approve the transition between phases and allocate approved funding to the project team. Approved funding will be based upon definitive estimates and will be allocated from funds that have been reserved for the project by the program steering committee. If the project requires additional funds that exceed the amount reserved by the program steering committee, a request for funding must be presented to and approved by the program steering committee.

The program steering committee must approve all changes to the approved program management plan[12] for each project plan. The program delivery manager will document and present all scope changes in the form a change request and the negative or positive impacts to cost, schedule, and quality for approval based on the fundamental project management formula $S = f(C, T, Q)$.[13] The program steering committee will meet, at a minimum, on a monthly basis to review the status of each project for the program. If the overall status of each project is Red or Yellow, more frequent meetings will be required. The program working committee will meet, at a minimum, on a scheduled biweekly basis to review the progress of each project.

These management committees will use consistent color-coded criteria to assess the progress of each project, provided by the program delivery manager, based on results of earned-value analysis.[14] The following is a high-level representation of the color-coded criteria; further details are provided later in this section.

12. Program management plan: A plan that provides the procedures (how) and organizational structure (who) to manage the delivery of the related series of projects identified in the program delivery plan.

13. $S = f(C, T, Q)$: Values are assigned to three constraints; the fourth represents the functional relationship or dependencies—$C = f(S, T, Q)$; $T = (C, S, Q)$; $Q = f(C, T, S)$—where S = scope, C = cost, T = time or schedule, and Q = quality constraints.

14. Earned-value analysis: Analysis of a project's schedule and financial progress compared to the original baseline plan.

- *Green:* Project adheres to scope, cost, schedule, and quality in accordance with approved project plan and any approved change request with cost performance index (CPI) and schedule performance index (SPI)[15] ≥ 1.
- *Yellow:* Project adheres to scope, cost, schedule, and quality with ≤10% variance, in accordance with approved project plan and any approved change request with CPI and SPI ≤ 0.8.
- *Red:* Project adheres to scope, cost, schedule, and quality with ≤20% variance, in accordance with approved project plan and any approved change request with CPI and SPI ≤ 0.7.

Program delivery manager responsibilities include the following:

- Attend program steering and working committee meetings, submit details on integrated business and IT direction, and provide advice to project managers to ensure the program delivers maximum overall business benefits to the company.
- Report on approved integrated business and IT funding to the program steering and working committee for each phase of the projects.
- Report on integrated business and IT changes to the approved integrated project plan.
- Provide integrated business and IT resource plans for business, IT, and PM support.
- Report to the program steering committee with direct accountability to PM executive, and keep program working committee informed of the current direction of the projects.

2.3.6 Business initiatives support

The first objective for a program delivery manager is to gain an integrated understanding of business and IT initiatives.

The primary objective of this function is to manage and track the business support initiatives, coordinate the development of business processes, and maintain integration of business initiatives with IT initiatives as determined in the BSA in order to optimize resource allocations and to minimize duplicate processes and efforts.

Real-world observations: program management plan

It is worth mentioning a real-world scenario at a company, which established a program management plan to manage the delivery of multiple IT-related projects. Business and IT resources were approved, budget allocated,

15. CPI and SPI: (1) The CPI ratio of budgeted costs to actual costs (BCWP/ACWP) is often used to predict the amount of a possible cost overrun or underrun using the formula BAC ÷ CPI = EAC. (2) The SPI ratio measures work performed against work scheduled BCWP/BCWS. The project is behind schedule if the SPI is less than 1.

and the program management plan was implemented according to the established schedule. During the program management plan implementation, there was a certain level of frustration, or political power struggles, which resulted in the business support areas identifying and executing various business-process-related projects, managed by an appointed business manager. At the same time, the IT support areas also identified and executed various IT-infrastructure-related projects, managed by an appointed IT manager.

However, neither of these IT or business process initiatives was integrated or aligned with the approved management plan, and as such there was an overwhelming demand for IT, project management, and business resources. This is a typical case of inefficient use of the company's most vital resources—people because of the lack of basic knowledge, skills, and application of project management integration. One of the recommended solutions presented in this book is based on the foundation principles of integrating business and IT support initiatives, defined in the program management plan, and establishing an integrated business, IT, and project management structure and process for implementing the program management plan. Since the focus of this section is on business support initiatives, an overview of the main contents is provided, with a highlight of the major responsibilities of the program delivery manager who is involved in executing this business support initiative process.

Business support initiatives include the following:

- Business process management: Development and transition;
- Business change management: Organizational and cultural changes;
- Business resource management: Staff assignment, training, and facilities management.

The program steering committee will have approval responsibilities for both the business and IT support initiatives and have joint accountability, with the program managers, for the overall scope, effort, cost, schedule, and quality performance of the business support initiatives.

Program delivery manager responsibilities include the following:

- Attend program working committee meetings and make project-related business and IT decisions.
- Manage the overall integrated program (business and IT support initiatives).
- Develop and maintain the program management plan for delivering the BSA:
  - Project charters;
  - Integrated program delivery plan—integrated WBS; integrated dependency diagram; integrated projects schedule; business organizational chart; business staffing profile; projects budget; contractors payment schedule;
  - Integrated transition plan (business and IT);

- Integrated communications plan;
- Integrated risk management plan;
- Integrated resource management plan.

- Develop integrated program schedule for IT and business support initiatives and maintain integrated representation of IT projects to support the BSA.
- Report to the program steering committee with direct accountability to the PM executive, and keep project managers informed of the current direction of the projects.

2.4 Project management

The project management function, referenced in Figure 2.2, is introduced in this section to show the integrated nature of the program delivery manager's responsibilities. Chapter 4 discusses the integrated nature of business, IT, and project management executives, program managers, and project managers' roles and responsibilities to ensure optimum resource utilization. The program delivery manager manages and delivers multiple projects to effectively support the following project management process components:

- IT PDLC;
- IT project management delivery processes:
  - Project scope management (PSM);
  - Project time management (PTM);
  - Project cost management;
  - Project quality management (PQM);
  - Project risk management (PRM);
  - Project procurement/contract management;
  - Project communications management (PCM);
  - Project issue management (PIM);
  - Project change management;
  - Project human resource management.
- PMO infrastructure support.

These project management process components are integrated through execution of business management, project management, and IT management roles and responsibilities, processes, tools, and techniques, in accordance with ESI's fundamental elements of integration—people, process, and technology. Chapter 4 discusses these integrated components in greater detail using a project management model as the communications medium.

I have introduced the process components (what) of project management. In the succeeding sections, I elaborate on the information contents for each of the process components and present the responsibilities (how) of the program delivery manager who is accountable for managing, communicating, and making decisions based on the information contents. The responsibilities of the program delivery manager show the integrated nature of the project management component processes to support the management and delivery of the overall program.

2.4.1 IT PDLC

The result of planning should be effective, efficient, and economical... that is, suitable for the intended purpose, capable of producing the desired results, and involving the least investment of resources.
—Clark Crouch

The primary objective is to ensure that projects are developed in an orderly manner of successive refinements from initiation, when a requirement is identified, to acceptance by the business representatives and sponsors, and delivery to business and IT production support services. Every project, whether business- or IT-oriented, has phases of development, and a clear understanding of these phases permits project managers to better control project schedules, budgets, scope, quality, resources, and efforts to achieve the desired goals of the company.

The IT PDLC discussed in this book focuses on an iterative or prototyping development approach,[16] based on successive refinements of deliverables from each phase. The phases of IT PDLC, also applicable to the business environment, include the following:

- Project definition: The primary deliverable from this phase is a program definition model (PDM), which consists of the following:
  - Project scope (conceptual-business, data, applications, technology);
  - Project management delivery and support processes (integrated project plan, project charter and project management plan).
- Project analysis/prototype #1: The primary deliverable from this phase is a requirements analysis model (RAM), which consists of the following:
  - Business assessments (operational, economic, and technical);
  - Business project design (detailed-business, data, applications, technology);
  - Demonstration prototype;
  - Updated project management delivery and support processes.
- Project architecture design/prototype #2: The primary deliverable from this phase is a Project Architecture Solution (PAS), which consists of the following:
  - Technical project design (refined—business, data, applications, technology);
  - Working or evolutionary prototype;

16. Prototyping development approach: An approach to software development that focuses on successive incremental/iterative refinements. It is based on the premise that it is impossible to completely determine a system's requirements before the start of development.

    - Updated project management delivery and support processes.
- IDPs—Iterations #1, #2, and #3: The primary deliverables from this phase are various iterative development solutions (IDSs), which includes iterative development prototypes during each of the iterative development subphases;
- Project construction: The primary deliverable from this phase is a constructed or executable set of the IT model, which consists of the following:
    - Iteration #1;
    - Executable models (executable-business, data, applications, technology);
    - Updated project management delivery and support processes.
- Project integration: The primary deliverable from this phase is an integrated and tested set of business and IT models, which consist of the following:
    - Iteration #2;
    - Integrated models (integrated/tested-business, data, applications, technology);
    - Test strategies, plans, cases, and results;
    - Updated project management delivery and support processes.
- Project deployment: The primary deliverable from this phase is a transition to production and warranty support or a deployed set of the business and IT model, which consists of the following:
    - Iteration #3;
    - Deployment executable models (deployed-business, data, applications, technology);
    - Business process deployment; organizations change; training; business and IT operations support; and facilities support services;
    - Updated project management delivery and support processes.

Program delivery manager responsibilities include the following:

- Submit details on PM project life cycle/phases enforcement processes to the program managers to control the project schedules and resources to achieve the desired goals.
- Report on integrated business and IT funding to the program steering committee for each of the project life cycle/phases within the program plan.
- Report on integrated business and IT changes to the approved project plan for each of the project life cycle/phases.
- Provide integrated business and IT resource plans for business, IT, and PM support.
- Report to the program steering committee with direct accountability to PM executive, and keep project managers informed of the current direction of the projects.

2.4.2 IT project management delivery processes

Leadership is getting the right people to do the right thing for the right reason in the right way at the right time at the right use of resources (costs).
—Clark Crouch

The primary objective is to ensure that projects are initiated, planned, executed, controlled, and closed using a set of consistent and integrated business and IT project management processes. Management of projects, whether business- or IT-oriented, requires consistent and effective project management processes. Every project, whether business- or IT-oriented, has phases of development, and a clear understanding of these phases permits project managers to better control project schedules, budgets, scope, quality, resources, and efforts to achieve the desired goals and to ensure integration with business and IT management processes.

The IT project management delivery processes, which can also apply to the business as discussed in this book, is based on PMI PMBOK standards. The nine knowledge areas[17] mentioned in PMI-PMBOK were extended to include issue management and change management. The knowledge area "integration management" defined in PMBOK was excluded from this IPM process framework because the major theme of this book is on integration, which is discussed throughout this text. The IT project management delivery processes, in accordance with extended PMBOK knowledge areas include the following:

1. *PSM:* Key deliverables are scope baseline, including product and project scope, integrated WBS, and supporting scope management plan.
2. *Project cost management:* Key deliverables are cost baseline, including budget and expense authorization and supporting cost management plan.
3. *PTM:* Key deliverables are schedule baseline, including integrated project schedule and costs and supporting schedule management plan.
4. *PQM:* Key deliverables are quality standards, including associated deliverables and supporting quality management plan.
5. *PRM:* Key deliverables are risks, risk strategies, and supporting risk management plan.
6. *Project contract/procurement management:* Key deliverables are contracts, and supporting contract management plan.
7. *PCM:* Key deliverables are communications reporting repository and PMIS[18] and supporting communications management plan.

17. Knowledge areas: A term used by PMI to represent project management processes.

18. PMIS: A project management information system and associated data and processes that provide automated support to gather, integrate, and distribute the output of project management processes.

8. *PIM:* Key deliverables are issues log and supporting issue management plan.
9. *Project change management:* Key deliverables are change requests log and supporting change management plan.
10. *Project human resource management:* Key deliverables are staffing plan and supporting human resource management plan.

2.4.2.1 Project scope management

If the scope is wrong, you will develop the right solution for the wrong problem.
—James P. Lewis

PSM as presented in this book focuses on four major subprocesses, namely requirements management, project repository management, project planning, and project progress tracking. The major responsibilities of the program delivery manager are highlighted to show why this individual requires the essential and necessary general business conceptualization knowledge; specialized people, process, and technology integration skills; and excellent risk management skills to ensure effective management of projects.

Requirements management

The purpose of requirements management is:

- To clearly and concisely define the scope of the project and associated objectives or constraints before budget plans are developed and approved funding committed;
- To document the project requirements and to ensure that each functional group affected by the project reviews the requirements.

The project scope/deliverables management process establishes a common understanding between the project sponsors and IT for the requirements that will be addressed by the project in the project charter. PSM establishes and maintains an agreement with the business, IT, and executive sponsors on the requirements of the project.

Requirements are initially identified at a high level in the project charter with additional details added as the project progresses through the life cycle. PSM focuses on defining specific project requirements during the project initiation and definition phase and managing the requirements during the remaining phases of the PDLC.

After the requirements have been approved, a change control process must manage all changes. The effect of the change on the project's scope, effort, cost, schedule, and quality must be documented and all plans and deliverables modified to meet the updated requirements. The components of project scope/deliverables management consist of the following:

- *Establishing a baseline:* Defining all business, technical, and nontechnical components of the project. This baseline forms the basis for planning and estimating, executing, controlling and tracking, verifying, and closing project activities during the PDLC.

- *Controlling change:* This entails making changes to the scope requirements in a controlled manner.

Program delivery manager responsibilities include the following:

- Provide sufficient resources to achieve the project deliverables.
- Provide sufficient tools and resources to support the activities of managing the deliverables.
- Verify that PMO guidelines are followed.
- Review project charters for integration, consistency, and completeness, and provide recommendations on the contents and structures.

Project master file/repository

The project master file provides a repository of project documentation. It shall be utilized to do the following:

- Provide transition information when there is a change in project managers.
- Inform project team members and other individuals on the current status of the projects.
- Provide a history of the project decisions made and the trade-offs that were evaluated.
- Provide information for QA reviews.
- Provide a history of project estimates that can be used in future projects.

Program delivery manager responsibilities include the following:

- Maintain a project file for all IT projects with the following program management structure:
- Program management;
- Project management deliverables:
  - Program management plan;
  - Project charter;
  - Integrated project plan[19] (integrated WBS; integrated dependency diagram; integrated schedule; integrated organizational chart; integrated staffing plan; integrated budget; integrated contractor's payment schedule);
  - QA/acceptance resolutions;
  - Contracts and contract agreement resolutions;
  - Risks and risk responses;
  - Issues and change requests logs and resolutions;
  - Project close-down recommendations;
  - Other correspondence.

19. Integrated project plan: A formal approved document consisting of business management, IT management, and project management plans, used as the baseline to guide project execution and control.

- PDLC deliverables:
  - Definition phase: project definition model (PDM);
  - RAP–project requirements analysis model (RAM);
  - Architecture design phase (ADP)–PAS;
  - IDPs-IDSs:
    Construction: project construction solution (PCS);
    Integration: project integration solution (PIS);
    Deployment: project deployment solution (PDS).

Project planning

Plans are nothing; planning is everything.
—Dwight D. Eisenhower

The purpose of project planning is to establish adequate project plans for delivering and managing the projects. It involves developing estimates, establishing commitments, and defining the plans to perform the work. Project planning is a disciplined process that establishes a plan to coordinate and direct resources such as time, people, and costs to achieve solutions to business priorities established by management. Emphasis is placed on the process of planning and the work required to produce the solution rather than focusing on the technical aspects of the project.

Project planning begins in the PDP with the development of the project charter and the integrated project plan, which provides a definition of the work to be performed, objectives, and constraints that define and scope the project. The PDP produces a project management plan that defines the steps to estimate the scope of the deliverables, resources needed, schedule, and costs, to identify and assess risks and to negotiate commitments. Executing these steps will be necessary to establishing the integrated project plan.

The planning effort results in the integration of the project scope, schedule, cost in human resources and capital assets, and quality. It is the primary tool for effective communication.

The deliverables from the PDP will provide the basis for delivering and managing the project activities and represents the commitments to the project stakeholders on the scope, schedule, costs, effort, and quality constraints of the project.

The benefits of an integrated project plan include the following:

- Providing the basis for effective communication with the project team and all stakeholders;
- Providing a check for ensuring that the project objectives are attainable with the time and resources available;
- Establishing the scope and the level of responsibilities and authority for all team members;
- Providing the basis for analyzing, negotiating, and recording scope changes and commitments of time, effort, and costs to the project,

which establishes a baseline for measuring progress, determining variances, and providing preventive and corrective actions;

- Minimizing the need for subjective narrative reporting; comparisons of the plan against actual performance in the form of graphics make reporting more efficient and effective and provide an audit trail and documentation of changes;
- Recording essential project data that can be used in planning future projects.

In summary, project planning provides the basis for delivering and managing the project activities and represents the commitments of project managers and stakeholders to the business, according to the resource, schedule, cost, and scope constraints of the project. The integrated project plan is based on the established project requirements baseline with support for the plans from the functional groups, business, IT, and consultants.

Program delivery manager responsibilities include the following:

- Provide tools and techniques to support the project planning activities.
- Verify that budgeting and all PMO administration are complete for project tracking.
- Ensure that the integrated project plan contains sufficient details and adheres to PMO guidelines.

Project progress tracking

Project planning is theory; project execution and control are practices.

The work is planned; now the plan must be worked. After the project has been properly planned in the PDP, the project manager enters the project management execution and control cycle, where the key questions to answer include the following:

- Where are we currently? An assessment of the current status of the project;
- Where do we want to be? A comparison of the actual progress made against the baseline project plan;
- How do we get there? A consideration of possible corrective actions to place the project back on track, if necessary, or keep it on track;
- Are we getting there? An analysis of the future and current impact that these corrective actions will have or are having on the project.

Here are six key steps to the executing and controlling process:

1. Update the plans including scope, effort, cost, schedule, and quality.
2. Update the status.
3. Analyze the risk impact.
4. Monitor and act on variances: scope, effort, cost, schedule, quality.

5. Publish the corrective actions, revisions, or changes: PM score card.[20]
6. Communicate with and inform stakeholders.

During the project management execution and control cycle, the approved baseline integrated project plan will be used as the basis for tracking the project, communicating status, and revising plans. Progress is primarily determined by comparing the actual scope, effort, cost, and schedule objectives to the baseline integrated project plan at selected milestones/deliverables or at month end. When it is determined that the integrated project plan is not being met, corrective actions are taken. These actions may include revising the plan to reflect the actual accomplishments and replanning the remaining work or taking actions to improve the performance objectives.

Program delivery manager responsibilities include the following:

- Verify that budgeting and all PMO administration are available for project tracking.
- Update integrated project plan and PM deliverables according to PMO guidelines.
- Review the performance of the project at appropriate milestones with respect to schedule, cost, effort, scope, and quality, and ensure adherence to PMO guidelines.
- Verify that the PM score card represents objective results.
- Verify that all IT PMO project tracking mechanisms have been completed.
- Report to the program steering committee with direct accountability to the PM executive, and keep project managers and stakeholders informed of the current scope of the projects.

2.4.2.2 Project cost management

Scope, schedule, and quality will always affect cost performance.
—James P. Lewis

The budget and expense authorization provides the framework for managing the program budgets and authorizing expenses. It is critical to managing the budget and is a major component of the integrated project plan that serves as a baseline for monitoring the budget and expenses and reporting on the costs status. If the budget and expense authorization changes, the integrated project plan and subordinate detail plans must change accordingly. The integrated project plan establishes the contract between the project teams and the major stakeholders.

Program delivery manager responsibilities include the following:

20. PM scorecard: A medium used to publish project management statistics and charts for project progress on scope, effort, schedule, cost, and quality objectives.

- Attend program steering committee meetings and present details on integrated business and IT budget and expense authorization to the program sponsors to control the overall program budget and expense authorization within budgetary constraints.
- Report on integrated business and IT budget and expense authorization to the program steering committee for each of the projects within the program plan.
- Report on integrated business and IT changes to the approved program budget.
- Provide integrated business and IT resource plan for business, IT, and PM budget and expense authorization support.
- Report to the program steering committee with direct accountability to the PM executive, and keep project managers informed of the current status of the program budget and expense authorization.

2.4.2.3 Project time management (integrated cost management)

Time is the scarcest resource, and unless it is managed, nothing else can be managed.
—Peter F. Drucker

To map out a course of action and follow it to an end requires some of the same courage that a soldier needs.
—Ralph Waldo Emerson

The major deliverable of PTM is the integrated schedule, and the major deliverable of project cost management is budget and expense authorization. These two processes require a high degree of integration; as a result, the discussion here focuses on PTM integrated with project cost management.

The integrated project plan–integrated schedule and budget and expense authorization are key components of the integrated plan that provide the framework for managing and integrating the projects. The plan includes the overall schedule, budget, and organizational resources that are critical to the success of the project. It serves as a baseline for measuring progress and reporting status and drives other project plans. If the integrated project plan changes, the subordinate detail plans must change accordingly.

Program delivery manager responsibilities include the following:

- Manage BSA program delivery plan. Forecast, using project management tools (Primavera-SureTrak, Microsoft Project) for project planning and tracking, and integrate forecasts with current integrated project plan and detailed project plans.
- Develop integrated WBS, dependency diagram, and schedule in accordance with PMO guidelines and ensure that project teams understand and follow guidelines. The integrated WBS provides a uniform structure for collecting resource expenditures in a consistent manner across projects. This allows for estimates of new projects to be compared with actual expenditures for previous projects. Figure 2.4 shows a graphical representation.

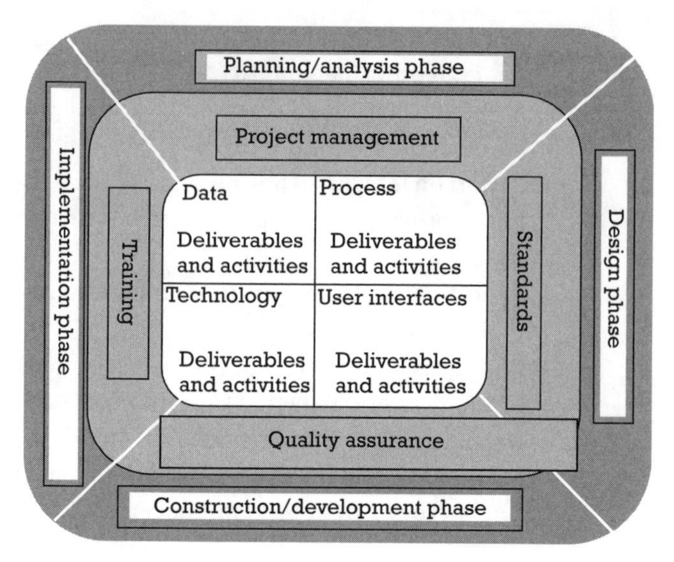

Figure 2.4 IPM-IT WBS.

- Report to the program steering committee for each of the projects within the program plan, on the forecasts, current status, and variances to schedule and costs updates.
- Report on integrated business and IT scope changes to integrated project plan.
- Provide forecasts, current status, and variances to integrated business and IT resource plan for business, IT, and PM support.
- Report to the program steering committee with direct accountability to the PM executive, and keep project managers informed of the current status of the integrated project plan.

Figure 2.4 is an integrated WBS to support the core data, process, user interfaces, and technology deliverables, and supporting project management, training, standards, and QA deliverables for each of the PDLC phases.

2.4.2.4 Project QA management

I don't worry whether something is cheap or expensive. I only worry if it is good. If it is good enough, the public will pay you back for it.
—Walt Disney

QA ensures that an appropriate level of quality has been applied to the project and that project quality has not been reduced to some inappropriate level because of political or other project pressures. QA verifies that the project adheres to the company's IT policies, standards, and procedures in areas such as:

- Adherence to organizational standards;
- Adherence to guidelines and standards to support the IT PDLC;

- Adherence to PMO guidelines and standards, appropriate for use by the project.

A minimum level of quality will be determined by the organization's IT policies, procedures, and standards. Specific project quality requirements may require higher quality levels, which will be described in the PQM management plan. The company's IT policies, procedures, and standards will be readily available to the project team. Project managers will determine additional project-specific quality requirements, which will be documented in the quality management plan. Templates and checklists will be developed and available as guidelines for major project deliverables for each phase of the IT PDLC. Templates and checklists will also be developed and available for project planning, executing, tracking, and management approval. QA reviews, for the appropriate IT, business management, and project management deliverables will be performed during each of the phases.

Program delivery manager responsibilities include the following:

- Conduct QA reviews of the overall quality of the project plans and deliverables.
- Verify that the project meets IT PMO policies, guidelines, and procedures.
- Verify that the project meets project-specific quality goals by utilizing the checklists.
- Report variances to the appropriate management team and recommend corrective actions.
- Change QA process as needed to keep it effective.
- Report to the program steering committee with direct accountability to PM executive, and keep project managers and stakeholders informed of the current quality of the projects.

2.4.2.5 Project risk management

What we anticipate seldom occurs.
What we least expect generally happens.
—Benjamin Disraeli

PRM identifies and quantifies risks during each phase of the project and develops risk response strategies and mitigation actions to maximize the results of positive events and minimize the consequences of adverse events. Each project will have a different tolerance to risk associated with it. The project manager should understand the risk tolerance level associated with the project. Project risk tolerance is influenced by project characteristics such as

- Is it a mission-critical project?
- Does it require operational capabilities (24 hours a day, 7 days a week)?
- Are optimum business process and technology performance necessary?
- What is the business impact of not being operational?

- What is the adverse effect on scope, schedule, cost, and quality?

The initial risk management plan will be developed during the PDP and updated and managed throughout each phase of the PDLC. The risk management plan will consist of identifying, analyzing, and mitigating the project risks. Risk checklists are available to help identify potential risk in three categories: known, predictable, and unpredictable. A risk identification template is also available for identifying potential risks. Risks will be analyzed to estimate the probability of the risk becoming a reality and the potential cost of the risk and to prioritize the risks so focus can be placed on the most critical items. Risk mitigation plans will be developed to identify the action to be taken to accept, eliminate, reduce, or control the project risks.

Program delivery manager responsibilities include the following:

- Define what the project managers can and cannot do when a risk occurs.
- Establish an agreed-upon process for submitting the risk and evaluating its impact on the current scope, schedule, cost, and quality baseline.
- Conduct risk reviews and facilitate risk response strategies based on issues and change requests in the issues and change request logs.
- Verify that the project team adheres to PMO risk management policies and procedures by utilizing the checklists included in the risk management plan.
- Manage the risk management plan by reporting on contents and corrective actions to the appropriate management team.
- Update risk management process as needed to keep it effective.
- Report to the program steering committee with direct accountability to the PM executive, and keep project managers and stakeholders informed of the current status of the risks.

2.4.2.6 Project contract management

No wind is favorable if we do not know into which port we are trying to sail.
—Rev. Dale Turner

Project contract management ensures that the company's guidelines and procedures are used to select and manage contractors for fixed priced and/or time-and-materials contracts. The contractor shall be selected based upon a balanced assessment of both technical and nontechnical criteria for the projects. The project contract requirements shall be determined and partitioned according to the IT PDLC. The contractor or subcontractor project charter or statement of work (SOW) shall be prepared, reviewed, agreed to, revised when necessary, managed, and controlled by both parties. The contractor or subcontractor's development plan for tracking progress and communicating status shall be reviewed and approved by the program manager. Any changes to the contractor SOW, contract terms and conditions, and other commitments shall follow the change management process.

The project manager and the contracting firm management staff shall conduct periodic status and technical reviews. Formal QA reviews shall be conducted to address the contractor's accomplishments and results at selected milestones. A review of the contractor's configuration management activities shall be conducted as appropriate. Acceptance testing shall be conducted on the delivery of the contracted deliverables, as defined in the acceptance test plan.

Program delivery manager responsibilities include the following:

- Conduct contract reviews based on the approved contract management plan.
- Verify that the project meets business and IT contract policies, guidelines, and procedures.
- Verify that the project meets the project-specific contract goals by utilizing the checklists.
- Report variances to the appropriate management team and recommend solutions.
- Change contract administration process as needed to keep it effective.
- Report to the program steering committee with direct accountability to the PM executive, and keep project managers and stakeholders informed of the current status of the contracts.

2.4.2.7 Project change management

Never let the future disturb you. You will meet it, if you have to, with the same weapons of reason, which today arm you against the present.
—Marcus Aurelius, Meditations 7:8

Project change management, as defined in this book, encompasses change request management and configuration management.

Project change request

The purpose of change requests is to ensure that changes made to project scope, consisting of plans, standards, schedules, budgets, effort, and deliverables are:

- Identified and documented;
- Evaluated and assessed for risk impact;
- Communicated to the affected groups and individuals;
- Tracked to completion in the change request log.

Changes to product scope refer to additions, modifications, or deletions made to the product or services deliverables. These changes to product scope affect the business processes, data, applications, and technology deliverables. The project manager or the person requesting the change of scope completes the change request template and defines the changes and its benefits. The change request is assigned a change number and the date it is received is recorded. The change is placed on the agenda of the next program steering committee meeting. The program steering committee reviews the change requests, and if they decide the request has merit, will agree to fund the changes, pending further investigations. The change request is assigned for investigation, and an analysis of the impact of the changes on the project will be conducted with resulting recommendations.

The program steering committee evaluates the potential impact of the change and decides whether or not to approve the change. If the change is approved, the committee sets priorities for the change. Funding and schedule changes are also approved, as appropriate. The project manager will communicate the changes to the affected groups and individuals.

Project configuration management

The purpose of project configuration management is to provide adequate controls, reviews, and approvals for tracking and controlling the changes to project and product deliverables. A change request management system (CRMS)–change request log is established to manage the versions of the changes to the deliverables. Changes to versions of the deliverables are managed using the change request system and are controlled by the change requests number and configuration auditing functions. Configuration management ensures that specific versions of the deliverables have a baseline, and the baseline is controlled and maintained for each change request with auditing functions. The project managers will inform all affected groups and individuals on the status, contents, and changes to the baseline, using the CRMS.

The components of configuration management include the following:

- Configuration identification;
- Configuration control and audit;
- Configuration status.

Program delivery manager responsibilities include the following:

- Define what the project managers can and cannot do when a change of scope occurs.
- Establish an agreed-upon process for submitting the change and evaluating its impact on the current baseline.
- Conduct change request reviews for the approved change requests in the CRMS log.
- Verify that the project team adheres to PMO change management policies, guidelines, and procedures by utilizing the checklists.
- Manage the change request log by reporting on the contents and corrective actions to the appropriate management team.
- Update the change request management process as needed to keep it effective.
- Report to the program steering committee with direct accountability to the PM executive, and keep project managers and stakeholders informed of current status of change requests.

2.4.2.8 Project issue management

When an affliction happens to you, you either let it defeat you, or you defeat it.
—Rosalind Russell

Issue management maintains a clear and objective focus on resolving issues by documenting, prioritizing, resolving, and reporting status on a periodic basis. The issues log shall be reviewed at team status meetings and used as a communication vehicle with team members. Issues will be prioritized, resolution plans developed, due dates established, and responsibilities assigned. Issues shall be managed and tracked using the same tracking system as the change requests log.

Program delivery manager responsibilities include the following:

- Define what the project managers can and cannot do when an issue occurs.
- Establish an agreed-upon process for submitting an issue and evaluating its impact on the current scope, time, cost, effort, and quality baseline.
- Conduct issue reviews and facilitate resolution to accepted issues in the issues log.
- Verify that the project team adheres to PMO issue management policies and procedures by utilizing the checklists included in the project management plan.
- Manage the issue management log by reporting on the contents and corrective actions to the appropriate management team.
- Update the issue request management process as needed to keep it effective.
- Report to the program steering committee with direct accountability to the PM executive, and keep project managers and stakeholders informed of the current status of issue requests.

2.4.2.9 Project communication/status reporting management

Misunderstandings sometimes occur because of differences in thinking preferences.
—James P. Lewis

A stupid man's report of what a clever man says can never be accurate because he unconsciously translates what he hears into something he can understand.
—Bertrand Russell

PCM provides formal written communications among the project team, senior management, and other stakeholders internal and external to the project. The project managers will compile the project biweekly and monthly status reports to formally communicate the project team progress. The reports for each project will be compiled by the PMO in a portfolio and distributed to the steering committees and to others as appropriate.

Program delivery manager responsibilities include the following:

- Define what the project managers can and cannot do when producing status reports.
- Establish an agreed-upon process for a submitting status report and evaluating its impact on the current communications management plan.
- Conduct project biweekly and monthly reviews through formal status reports according to the approved communications management plan.
- Verify that the project team adheres to PMO communications management policies and procedures.
- Verify that the project meets the project-specific communication requirements by utilizing the checklists included in the communications management plan.
- Manage the communications management plan by reporting on the contents and corrective actions to the appropriate management team.
- Update the communications management process as needed to keep it effective.
- Report to the program steering committee with direct accountability to the PM executive, and keep project managers and stakeholders informed of the current status of the project.

2.4.2.10 Project human resources management

You don't lead by hitting people over the head—that's assault, not leadership.
—Dwight D. Eisenhower

Project human resource management optimizes the utilization of people resources or stakeholders involved with the project. The major human resource management processes as described in PMBOK are:

- *Organizational planning:* Identifying, documenting, and assigning project roles, responsibilities, and reporting relationships;
- *Staff acquisition:* Getting the human resources needed assigned to and working on the project;
- *Team development:* Developing individual and group competencies to enhance project performance, including team building, dealing with conflicts, and other subjects related to dealing with the project team.

The program responsibility assignment matrix is compiled by the program delivery manager and represents the overall assignment of project team members for all projects within the program. It is part of the overall human resource management plan or staffing plan, which describes when and how human resources will be brought into and taken off the project team. Each project manager will be responsible for developing a responsibility assignment matrix. The human resource utilization reports for each project will be compiled by the PMO into a portfolio and distributed to the steering and working committee and to others as appropriate.

Program delivery manager responsibilities include the following:

- Define what the project managers can and cannot do when producing the RAM.
- Establish an agreed-upon process for submitting a staffing plan and evaluating its impact on the current baseline.
- Conduct project biweekly and monthly reviews for the RAM, according to the approved human resource management plan.
- Verify that the project team adheres to PMO human resource management policies, guidelines, and procedures.
- Verify that the project meets the project-specific human resources requirements by utilizing the checklists included in the human resource management plan.
- Manage the human resource management plan by reporting on the contents and corrective actions to the appropriate management team.
- Update the human resource management process as needed to keep it effective.
- Report to the program steering committee with direct accountability to the PM executive, and keep project managers and stakeholders informed of the current status of the project staffing plan.

Proper human resource management begins with understanding the organization's resource utilization and skills inventory. By consolidating multiple projects into a central repository, companies can gain the visibility needed to plan and balance their workload. PMO maintains a complete model of an organization's capacity, affected resources, skills inventory, total workload, and resource demand. It is the capacity and resource planning model that enables organizations to optimize skills usage and ensures that all mission-critical resources are productively aligned with high-priority projects.

Using the PMO, organizations can gain visibility on resource bottlenecks and realize unprecedented savings by answering the following questions:

- What are the company's total internal and external resource pool, capacity, and demand?
- Where can the necessary available skills and knowledge be found?
- What type of and how much additional resource capacity is needed for new initiatives?
- Are the most knowledgeable and skilled resources working on the most strategic initiatives?
- What is the availability of a given resource for a given period?
- How are resources performing in a particular project, program, or business area?
- What are the resource utilization, realization, and profitability?

This PMO introduction leads to the need for consistency and standardization of processes, deliverables, tools, and techniques to ensure that an IPM-IT environment is effectively implemented. The next major component, as presented in the Figure 2.2, is PMO support, which is the topic of discussion in the next section.

2.4.3 PMO processes

If what you are doing isn't working, you need to change the process by which it is done. A process is a way of doing things.
—James P. Lewis

Processes or methodologies can be SMART—specific, measurable, achievable, realistic, and timely—or DUMB—doubtful, unrealistic, massive, and boring.

Many IT consultants are providing leadership in helping organizations establish formal project management processes to support the delivery of project initiatives on time, within budget, and at an acceptable level of quality. These leadership roles are the result of urgent requests from clients to optimize the use of the company's resources, time, and budget. The ability to effectively deliver projects better, faster, and cheaper requires the implementation of common processes and practices across the entire organization. The result normally is a very short learning curve for the project manager and team members as they transition from one project to another.

The larger an organization gets and the more projects are executed at one time, the more difficult it becomes to enforce this organizational consistency. Without this consistency, the full value of implementing a common project management methodology cannot be realized. Some IT consulting firms recognize this and usually provide a common framework or a full methodology that most of their consultants execute in their consulting practices. However, this practice of a common, integrated, and consistent process is highly dependent on the maturity level of the client companies. This is where consulting firms—with the right skill levels and experience in applying project management processes—can offer tremendously useful support to their clients.

The right consultant can help to establish a centralized organization that is responsible for varying aspects of project management methodologies. This group can be called project office, enterprise project office, project management center of excellence, or the project management resource team. Here, I will use the term PMO. A PMO can offer many potential products and services, depending on the needs of the organization and the vision of the PMO sponsor (the person who is generally responsible for the PMO funding). Before the PMO can be successful, agreement must be obtained from the management team on the overall role and general expectations of the PMO. A typical PMO is responsible for deploying a consistent project management methodology within the organization, including processes, templates, and best practices. This is not a one-time event, but a broad initiative that could cover a number of years. While a PMO demands precious resources, the hope is that the investment in the PMO will be justified by implementing common practices that will allow every project within the organization to be completed better, more quickly, and more cheaply in meeting stakeholders expectations.

Certain IT consultants tend to be further along in their adoption of common project management processes that utilize model-driven or deliverables-based approaches. Many companies use these IT consulting services to assist them in their desperate search for a faster, cheaper, and better solution. Here is a list that shows the major responsibilities that the PMO will normally execute within an organization. It presents the responsibilities to show how the PMO can influence the client's business. The PMO value proposition can be established to provide a narrow or broad set of services. This list includes many of the common responsibilities that a PMO would normally perform:

- Establishing and deploying a common set of project management processes and templates, which saves each project manager or each organization from having to recreate these project management constructs. These reusable project management components help projects to start up more quickly and with less effort.
- Providing a methodology and the necessary updates to maintain improvements and best practices. For example, as new or revised processes and templates are made available, the PMO deploys them consistently to the organization.
- Facilitating improved project team communications by having common processes, deliverables, and terminology. Less misunderstanding and confusion occurs within the organization if everyone uses the same language and terminology for project-related work.
- Providing training (internal or outsourced) to build core project management competencies and a common set of experiences. If the training is delivered by the PMO, there is a further reduction in the overall training costs paid to outside vendors.
- Delivering project management coaching services to prevent project disasters. Projects at risk can also be coached to prevent further disasters.
- Tracking basic information on the current status of all projects in the organization and providing project visibility to management in a common and consistent manner.
- Tracking organizationwide metrics on the state of project management, project delivery, and the value being provided to the business. The PMO also assesses the general project delivery environment on an ongoing basis to monitor the improvements.
- Acting as the overall advocate for project management to the organization, including actively educating and selling managers and team members on the value gained through the use of consistent project management processes.

In summary, companies are finding that they need to standardize how projects are managed. They are recognizing that the process takes much more than just training the staff. It requires a holistic approach, covering many aspects of work and the company culture. However, in their search for this process solution, certain companies have experienced endless frustration and a lack of professional integrity, resulting in the expensive lessons-learned scenarios. Here is a real-world situation that executive management should avoid in the search for that ultimate process solution.

2.4.4 Real-world observations: PMO processes

A large corporation employed an IT consulting firm to develop and deploy a project management methodology with the expectation that this process would solve most of the communications problems previously experienced during the delivery of IT projects. The expectation was that this process solution would ensure a better, faster, and cheaper project solution. This consulting firm proudly produced a software development methodology that had the characteristics of a DUMB methodology, with volumes of disjointed texts.

Senior management at this corporation, overwhelmed and confused with the voluminous contents of the DUMB methodology, decided to establish a project management methodology similar to PMBOK processes. Another IT projects management consulting firm was contracted to deliver this solution. This consulting firm and the lead consultant believed that project management is a distinct and isolated discipline. The final deliverable from this consulting firm was a set of project management processes with no alignment with the previously developed software development methodology. In my assessment, this consulting firm produced a subset of PMI-PMBOK processes in the company's logo with no reference to how this process can be applied.

The IT development group, whose mandate was to deliver IT projects, adopted the software vendor's project management process solutions for all purchased-packaged solutions and totally ignored the methodology recommendations from the PMO. One of their main reasons for the nonacceptance of the PMO recommendation was the lack of business management integration with the DUMB methodology and the project management processes.

Senior management decided to addresses the IT development group's concerns by employing another consulting firm with business management expertise to deliver a business management methodology.

This consulting firm and the lead consultant believed that a BSA methodology was the process solution for this corporation. They convinced senior management to establish an enterprise architecture office (EAO). The resulting product was a business architecture methodology and recommended EAO structure with no clear alignment with project management processes and the software development methodology. The client company was now left with three useless methodologies that can be characterized as doubtful, unrealistic, massive, and boring—DUMB.

This is a real-world scenario, where these IT consulting firms have secretly turned client's problems to their personal goldmines—a common practice in the IT industry that senior executive management must be aware of and take action to avoid. PMO processes cannot be developed and deployed in isolation from business management and IT management processes. PMO process integration is a necessary and essential requirement for success, for delivering value to the corporation, and for ensuring that projects are delivered on time, within budget, with acceptable quality to effectively meet stakeholders' expectations.

I have introduced the process components (what) of the PMO. I will now present the responsibilities (how) of the program delivery manager who is accountable for managing, communicating, and making decisions based on the information contents. The responsibilities of the program delivery manager show the integrated nature of the PMO component processes to support the management and delivery of the overall program.

Program delivery manager responsibilities include the following:

- Provide policies, guidelines, best practices, and templates to enable cost effective and efficient project delivery.
- Maintain the project management information repository (PMIR).[21]
- Provide guidelines defining acceptable levels of analysis for estimating and cost/benefit analyses.
- Lead and coordinate the integration of all IT standards, templates, and checklists for projects into a single consolidated repository for access by project teams.
- Provide standard templates and tools for project management activities and project reporting.
- Provide guidelines for reporting frequency.
- Assist project teams in the use of project management templates, and provide the necessary training in the use of project management standards, practices, and templates.
- Provide information on past projects relevant to the current project.
- Provide guidance and counsel to the various project roles.
- Identify to the steering committee the appropriate trend analysis of the key indicators for performance monitoring in terms of effort, cost, schedule, scope, and quality.
- Conduct or arrange for externally conducted scheduled project reviews during project execution and postimplementation.
- Conduct or arrange for externally conducted steering committee directed reviews.

21. PMIR: The repository of all project documentation. It is the database that the PMIS uses to generate project management reports.

- Ensure that postproject reviews are completed and findings are executed.
- Communicate PMO policies and ensure project teams adhere to IT PMO methodology, processes, tools, and techniques.
- Manage the PMO processes by reporting on the contents and recommend solutions to the appropriate management team.
- Update PMO processes as needed to keep them effective.
- Ensure that project managers adhere to the guideline processes, deliverables, tools, and techniques established by the PMO to ensure project management consistency, completeness, and integration.
- Report to the program steering committee with direct accountability to the PM executive, and keep project managers and stakeholders informed of the effectiveness of the PMO processes.

2.5 IT management

The components of IT management, referenced in Figure 2.2, are introduced in this section to identify and define the major IT processes and to show how the program delivery manager executes these processes. Chapter 5 elaborates on these processes, including details on the integrated nature of the program delivery manager with supporting senior business, project management, and IT managers' responsibilities. The program delivery manager manages and delivers multiple projects to support the following IT management process components:

- Cost estimating;
- Resource allocations;
- DA;
- AA;
- Technology architecture;
- Application services support.

These major processing components are integrated through execution of business management, project management, and IT management roles and responsibilities, processes, tools, and techniques, in accordance with ESI's fundamental elements of integration: people, processes, and technology. Chapter 5 discusses these components in greater detail, using an IT management model as the communication medium.

I have introduced the process components (what) of IT management. In the succeeding sections, I will elaborate on the information contents for each of the process components and present the responsibilities (how) of the program delivery manager who is accountable for managing, communicating, and making decisions based on the information contents. The responsibilities of the program delivery manager show the integrated nature of the IT management component processes to support the management and delivery of the overall program.

2.5.1 Cost estimating

An exact estimate is an oxymoron. If you aren't careful, ballpark estimates can become targets.
—James P. Lewis

The primary objective is to estimate the costs of the project to the business, based on specified cost categories. The benefits to be realized must align with the corporation business and budgets. Decisions regarding IT investments must be driven from the business systems planning/architecture and budget estimates. The project IT investments shall include the full life cycle cost for the projects, to include the following:

- Labor costs:
  - IT internal staff;
  - IT external staff, consultants;
  - User staff.
- Technology costs:
  - Hardware and workstations;
  - Software;
  - Network/communications.
- Facilities/miscellaneous costs:
  - Telephone;
  - Meals;
  - Administration.
- Travel and training costs:
  - IT staff training;
  - IT staff travel;
  - User staff training;
  - User staff travel.

All costs for the projects are captured in the general ledger. These costs are used as the basis for providing a high, low, and most-likely estimate for each of the projects within the program. In the project analysis phase, cost estimates are further defined, and in the architecture phase, additional cost refinements are provided for construction, integration, and deployment. These cost estimates and assumptions are used to develop the definitive estimates. The costs estimates at the project definition or BSA phase are preliminary estimates and are in the range of a –25% to +75% level of accuracy.

Program delivery manager responsibilities include the following:

- Define what the project managers can and cannot do when producing cost estimates.
- Determine cost performance measurement to business and IT projects based on the value to the business within budgeting constraints.
- Establish an agreed-upon process for submitting a cost estimate and evaluating its impact on the current cost baseline.

- Verify that the project team adheres to PMO cost estimating policies, guidelines, and procedures by utilizing checklists included in the cost management plan.
- Manage the cost estimates by reporting on the contents and recommended solutions to the appropriate management team.
- Update cost estimating processes as needed to keep them effective.
- Report to the program steering committee with direct accountability to the PM executive, and keep project managers and stakeholders informed of the accuracy of cost estimates.

2.5.2 Resource allocations

Technology is dominated by two types of people: those who understand what they do not manage, and those who manage what they do not understand.
—Putt's Law

The primary objective is to allocate labor, technology, and facilities resources and to optimize the utilization of these resources. People skills/effort, technology, and facility resources are allocated at the beginning of each phase, based on the people skills/effort, technology, and facility requirements of the project. These resource allocations form the basis for cost estimating during each phase of the PDLC.

Proper resource allocations begin with an understanding of the organization's technology resource utilization and skills inventory, documented in the resource plan.[22] Technology resource capacity and staff planning enables organizations to optimize skills usage to ensure that all mission-critical resources are productively aligned with high-priority technology projects.

Program delivery manager responsibilities include the following:

- Define what the project managers can and cannot do when allocating resources.
- Allocate labor resources to business and IT projects based on the value to the business.
- Establish an agreed-upon process for allocating resources and evaluating the impact on the current and forecasted resource plan.
- Verify that the project team adheres to PMO resource allocations policies, guidelines, and procedures by utilizing checklists included in the resource plan.
- Manage the resource allocations by reporting on the contents and recommended solutions to the appropriate management team.
- Update resource allocations processes as needed to keep them effective.
- Report to the program steering committee with direct accountability to the PM executive, and keep project managers and stakeholders

22. Resource plan: A formal approved document consisting of labor, technology, and facilities resources, used as the baseline to allocate resources, estimate costs, and develop staffing plans.

informed of the availability of allocated resources and the resource plan.

2.5.3 Data architecture

Data is not information, information is not knowledge, knowledge is not understanding, understanding is not wisdom.
—Cliff Stoll and Gary Schubert

The primary development components of the IT management model divides the project into components parts—business, data, applications, and technology—and relationships to ensure that the components parts fit together. These components parts formed the basis for defining the deliverables-based WBS discussed in Section 2.4.2.3. The goals of data, application, and technology architectures are to:

- Understand and support the business strategy and align with business architecture.
- Focus on business processes required to support the business and IT strategies.
- Address change holistically by understanding the changes required in the other architecture components with a vision of the future.
- Plan implementation iterations as a series of small successes with each iteration delivering something of value to the business.
- Ensure adherence to the corporate architecture standards.
- Develop data, application, and technology architectures to meet the performance requirements of the business processes.

This section focuses on the data architecture (DA) components of the architecture processes discussed earlier. The primary objective of DA is to support the project teams in the management and delivery of databases, coordinate the development of business data, and ensure integration of business data with IT databases. The DA group will have approval responsibilities for both the business (logical) and IT (physical) data models and have joint accountability with the program delivery manager for the overall scope, effort, cost, schedule, and quality performance of these data models.

Program delivery manager responsibilities include the following:

- Review and communicate business and IT (physical) DAs/models and ensure linkage with business DAs.
- Communicate data models to project teams.
- Verify that the project team adheres to DA policies, guidelines, and procedures by utilizing checklists that are included in the DA deliverables.
- Manage the integrated DA by reporting on the contents and recommended solutions to the appropriate management team.
- Update DA processes as needed to keep them effective.

- Report to the program steering committee with direct accountability to the PM executive, and keep project managers and stakeholders informed of the effectiveness and usefulness of the DAs/models during projects implementation.

2.5.4 Applications architecture

The process of preparing programs for a digital computer is especially attractive, not only because it can be economically and scientifically rewarding, but also because it can be an aesthetic experience much like composing poetry or music.
—Donald E. Knuth

This section focuses on the AA components of the architecture processes discussed earlier. The primary objective of AA is to support the project teams in managing and delivering applications and to coordinate the development of applications to ensure the integration of business processes with IT applications. The AA group will have approval responsibilities for both the business (logical) and IT (physical) AAs and have joint accountability with the program delivery manager for the overall quality performance of these applications.

Program delivery manager responsibilities include the following:

- Review and communicate business and IT (physical) AAs/models and ensure linkage with business, data, and application architectures.
- Communicate AA policies and ensure project teams adheres to AA methodology, processes, tools, and techniques.
- Verify that the project team adheres to AA policies, guidelines, and procedures by utilizing checklists that are included in the AA deliverables.
- Manage the integrated AA by reporting on the contents and recommended solutions to the appropriate management team.
- Update AA processes as needed to keep them effective.
- Report to the program steering committee with direct accountability to the PM executive, and keep project managers and stakeholders informed of the effectiveness and usefulness of the AAs/models during projects implementation.

2.5.5 Technology architecture

Humanity is acquiring all the right technology for all the wrong reasons.
—R. Buckminster Fuller

The primary objective of this function is to support the project teams in managing and delivering technology (hardware, systems software, and network) solutions and to coordinate the installation of technology to ensure integration of technology with IT applications and data. The TA group will have approval responsibilities for both the business (logical) and

IT (physical) TAs and have joint accountability with the program delivery manager for the overall quality performance of these technology solutions.

Program delivery manager responsibilities include the following:

- Review and communicate business and IT (physical) TAs/models and ensure linkage with business, data, and application architectures.
- Communicate TA policies, and ensure project teams adheres to TA methodology, processes, tools, and techniques
- Verify that the project team adheres to TA policies, guidelines, and procedures by utilizing checklists that are included in the TA deliverables.
- Manage the integrated TA by reporting the contents and recommended solutions to the appropriate management team.
- Update TA processes as needed to keep them effective.
- Report to the program steering committee with direct accountability to the PM executive, and keep project managers and stakeholders informed of the effectiveness and usefulness of the TAs during projects implementation.

2.5.6 Applications support services

What happens is not as important as how you react to what happens.
—Thaddeus Golas

The primary objective of this function is to support the project teams in installing, implementing, and deploying the applications, coordinate the deployment applications, and ensure effective integration of existing and new IT applications. The applications support model[23] contains details on the service level agreement between IT, business, and PM during the warranty period and ongoing production support. The applications support services group will have approval responsibilities for the deployment and have joint accountability with the program delivery manager for the overall quality performance of these applications.

Program delivery manager responsibilities include the following:

- Manage the project deployment and closeout activities.
- Review and communicate the applications support model within the overall applications support services-level agreement, and ensure linkage with business, data, application, and technology architectures.
- Communicate application support services policies, and ensure project teams adheres to methodology, processes, tools, and techniques.
- Verify that the project team adheres to applications support services policies, guidelines, and procedures by utilizing checklists that are included in the applications support model deliverables.

23. Applications support model: A model used to develop service-level agreements among business, IT, and project management groups during warranty period and production support.

- Manage the applications support model by reporting on the contents and recommended solutions to appropriate management team.
- Update applications support model processes as needed to keep them effective.
- Report to the program steering committee with direct accountability to the PM executive, and keep project managers and stakeholders informed of the effectiveness and usefulness of the applications support model during a project's maintenance and warranty period.

2.6 Program delivery manager key responsibilities

Divide and rule, a sound motto. Unite and lead, a better one.
—Johann Wolfgang von Goethe

Generalized business conceptualization knowledge; specialized people, process, and technology integration skills; and excellent risk management skills are the major skills needed to effectively manage the delivery of multiple IT projects. The major responsibilities of the program delivery manager are highlighted in Tables 2.2 to 2.4 to show the knowledge and skills requirements of this individual.

Table 2.2 Program Delivery Manager Major Responsibilities: Business Management

Program Delivery Manager Responsibilities— Business Management What	*Approach to Delivery— How*	*Timing— When*
BSA	Assemble and present integrated business and IT architecture (data, applications, technology) or program architecture documentation.	Project definition
Project justification	Document integrated prioritized projects and justifications for allocated budget based on the integrated value of the project to the business and IT.	Project definition
Funding allocations	Report on allocation of funds at various phases of the project life cycle for the entire program.	Project execution
Deliverables approval	Manage portfolio of projects deliverables within the funds allocated by program steering committee.	Project execution
Steering committee	Report to the program steering committee with direct accountability to the PM executive, and keep program working committee informed of current direction of the projects.	Project execution
BIS	Develop integrated program schedule for IT and business support initiatives, and manage integrated representation of IT and business support initiatives.	Project execution

The major responsibilities of the program delivery manager in performing project management duties are highlighted in Table 2.3 to show PM knowledge and skills requirements.

The major responsibilities of the program delivery manager in performing IT management duties are highlighted in Table 2.4 to show the necessary knowledge and skills requirements.

Table 2.3 Program Delivery Manager Major Responsibilities: Project Management

Program Delivery Manager Responsibilities—Project Management What	*Approach to Delivery—How*	*Timing—When*
IT PDLC IT Project management delivery processes	Advise project managers on applying IT PDLC/phases and PM delivery processes	Project execution
Scope management Requirements management Project file Project planning Project progress tracking	Review project charters for integration, consistency, and completeness, and advise on contents and structures Maintain a project file/repository for all IT projects Manage projects using integrated project plan as baseline Monitor project performance at appropriate milestones for schedule, cost, effort, scope, and quality objectives	Project definition Project execution Project execution
Cost management	Manage integrated business and IT resource plan for business, IT, and PM budget and expense authorization	Project execution
Time management	Manage program delivery plan using project management software tools and integrate with integrated project plan	Project execution
Quality management	Conduct QA reviews for the approved quality management plans	Project execution
Contract management	Conduct contract reviews for the approved contract management plan	Project execution
Risk management	Mange risks and evaluate the impact on the current scope, schedule, cost, and quality baseline	Project execution
Change management	Conduct change request reviews for the approved change requests in the change management plan	Project execution
Issue management	Conduct issue request reviews for the approved issue requests in the issue management plan	Project execution
Communications management	Manage communications management plan and recommend corrective actions	Project execution
HR management	Manage PSM plan and recommend corrective actions	Project execution
PMO processes	Provide policies, guidelines, best practices, and templates to enable cost effective and efficient project delivery	Project execution

Table 2.4 Program Delivery Manager Major Responsibilities: IT Management

Program Delivery Manager Responsibilities—Technology Management What	*APPROACH to DELIVERY—How*	*TIMING—When*
Cost estimating	Manage the cost estimates by reporting on the contents and recommended solution to the appropriate management team.	Project execution
Resource allocations	Manage the resource allocations by reporting on the contents and recommended solutions to the appropriate management team.	Project execution
DA	Manage the integrated DA by reporting on the contents and recommended solutions to the appropriate management team.	Project execution
AA	Manage the integrated AA by reporting on the contents and recommended solutions to the appropriate management team.	Project execution
TA	Manage the integrated TA by reporting on the contents and recommended solutions to the appropriate management team.	Project execution
Applications support services	Manage the integrated applications support model by reporting on the contents and recommended solutions to the appropriate management team.	Project execution

2.6.1 Real-world observations: program delivery manager's responsibilities

Many IT project managers and senior IT managers, especially those who do not have the basic IT qualifications and skills, strongly express their opinion that IT professionals with technology qualifications, skills, and knowledge are not suited to perform the role of an IT program delivery manager. They openly make statements to the effect that these "techies" should be IT architects, not IT project managers. These senior IT managers firmly believe that the only skill required of IT program delivery managers is the ability to communicate. I wholeheartedly support this necessary communications skills requirement. However, one lesson learned as a result of my extensive experience in this technologically advancing industry is that project information reported by individuals without the necessary IT qualifications, skills, and knowledge is normally misunderstood, misrepresented, and poorly communicated.

I will end this real-world observation with three simple and powerful quotations, and leave the rest to the imaginations of readers.

A stupid man's report of what a clever man says can never be accurate because he unconsciously translates what he hears into something he can understand.
—Bertrand Russell

Managers who are skilled communicators may also be good at covering up real problems.
—Chris Argyris, Harvard Business Review, September 1986

Misunderstandings sometimes occur because of differences in thinking preferences.
—James P. Lewis

2.7 Applying integrated framework to other disciplines

It is common sense to take a method and try it. If it fails, admit it frankly and try another. But above all, try something.
—Franklin D. Roosevelt

Although this book is based on an IPM-IT model (framework) represented in Figure 2.2, this framework can be adapted to fit within the project management context of other business disciplines, such as engineering, manufacturing, and accounting. The basic functional components of business management and project management will not change drastically. However, the process components of the industry business discipline will change, based on the product/service life cycle of the industry business. Figure 2.5 shows a framework to manage projects in the petroleum engineering (PE) discipline.

2.8 Alternate project-based organizational structure

The more alternatives, the more difficult the choice.
—Abbé D'Allanival

There are certain companies where project management culture and maturity are not widely accepted by corporate executives to justify the need for a project management executive at the vice presidential level. As a result of this perspective, I decided to show a typical organization structure, represented in Figure 2.6, that these companies normally execute.

The alternate program organizational structure in Table 2.5 expands on the roles identified in Figure 2.6.

2.9 Summary

The IPM-IT model (framework) presented in this chapter shows the component processes of this framework. It also discusses the responsibilities of the program delivery manager to demonstrate the integrated nature of managing and delivering multiple IT projects. Companies that are in the process of managing and delivering multiple projects with the goal of integrated or enterprise project management should consider the following recommendations, as a framework, to guide them towards successful management and delivery of multiple IT projects:

- Develop a BSA that shows the alignment of business architecture with data, applications, and technology architectures. The end result is a

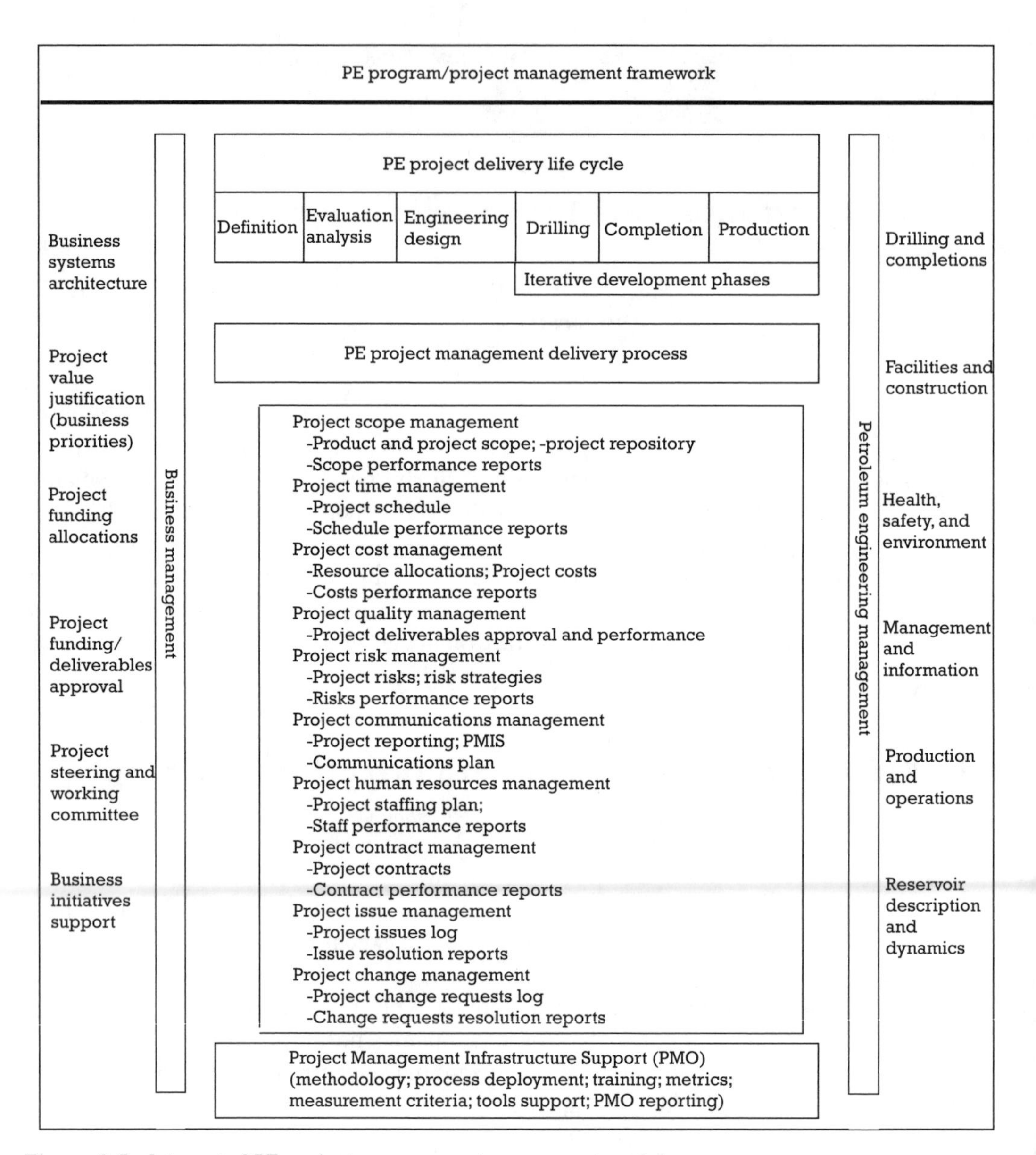

Figure 2.5 Integrated PE project management component model.

series of IT projects that are prioritized based on corporate business objectives and strategies and aligned with the business processes to achieve the goals, objectives, and strategies of the company's business.

- Define a portfolio of multiple related projects, and manage these multiple projects as a program.
- Establish a PMO methodology, processes, tools, and techniques and communicate them to business and IT staff.
- Define clear objectives, roles, and responsibilities for executive management, program management, and project management for business management, project management, and IT management.

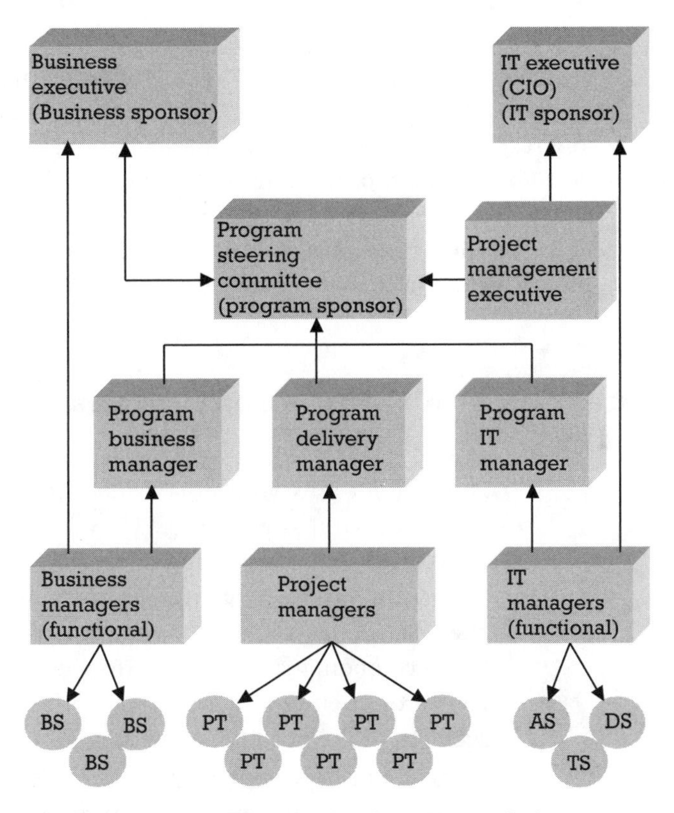

Figure 2.6 Alternate program organizational structure solution.

Table 2.5 Alternate Program Organizational Structure

BS: business support	PT: project team (applications delivery)	AS: applications support
		DS: development support
		TS: technical support

- Obtain approval and support from executive managers on the BSA deliverable and the overall IPM framework.
- Communicate and ensure project team members adhere to project management policies, processes, standards/guidelines, and procedures.
- Ensure that the program delivery manager is a project management professional with conceptual business knowledge; specialized people, process, and technology integration skills; and excellent risk management skills
- Ensure that the program delivery manager understands the linkage between IT management, business management, and project management, as presented in this chapter and detailed in the remaining chapters of book.

In this foundational chapter, I introduced an IPM framework model, discussed the applications of this framework in terms of the three functional components of business, IT, and project management, and highlighted the key responsibilities of the program delivery manager. Chapters 3, 4, and 5 elaborate on each of these three functional components; Chapter 6 shows the horizontal integration of these three functional components based on an IT PDLC model; and Chapter 7 discusses the implementation of this IPM model using the RUP from IBM Rational Corporation.

2.10 Questions to think about: management perspectives

1. Think about how your organization manages multiple IT projects. How does your organization view integration? How are projects structured? What is the key rationale for the organizational structure? How does personal greed for power, politics, and personal financial interests affect the organization structure? What are the major components of IPM-IT? How do these components relate to your project environment? What is the perception of senior management of the need for standard and consistent processes?
2. Think about how your organization involves business staff during project initiation and execution. What are the processes that align IT with the business during project initiation and execution? How are these processes communicated?
3. Think about how your organization involves project management staff during project initiation and execution. What are the project management processes that align IT with the business during project initiation and execution? How are these processes communicated? How does your organization integrate software development processes with project management processes? What is senior managements' perception of the responsibilities, needs, and benefits of the PMO? How does senior management in your organization promote PMBOK processes? What is the level of professional qualifications, skills, and experience of the program delivery manager? What is the level of professional qualifications, skills, and experience of the executive project management manager? What are the differences in thinking preferences?
4. Think about how your organization involves IT functional staff during project initiation and execution. What are the processes that align IT and PMO with the business during project initiation and execution? How are these processes communicated?
5. Think about how your organization involves PMO during project initiation and execution. What is the level of involvement and commitment from business and IT management executives? How is the

formal or informal PMO processes communicated? How do you know that you no longer have executive management support for the PMO?

Selected bibliography

Bennatan, E. M., *On Time, Within Budget, Software Project Management Practices and Techniques,* New York: McGraw-Hill, 1992.

Boehm, B., "Software Risk Management: Principles and Practices," *IEEE Software,* Vol. 8, No. 1, January 1991.

ESI International, *Project Framework—A Project Management Maturity Model,* Vol. 1, Arlington, VA: ESI International, 1999.

Fleming, Q. W., and J. M. Koppelman, *Earned Value Project Management,* 2nd ed., Newton Square, PA: PMI, 2000.

Kerzner, H., *Project Management: A Systems Approach to Planning, Scheduling, and Controlling,* 7th ed., New York: John Wiley & Sons, 2001.

Lewis, J. P., *Project Planning: Scheduling and Control,* 3rd ed., New York: McGraw-Hill, 2001.

Muller, R. J., *Productive Objects: An Applied Software Project Management Framework,* San Francisco, CA: Morgan Kaufmann, 1998.

Paulk, M. C., et al., *Key Practices of the Capability Maturity Model®,* Version 1.1, Pittsburgh, PA: Software Enginering Institute, Carnegie Mellon University, 1993.

PMI, *Project Management Institute: A Guide to Project Management Body of Knowledge,* 2000 Edition, Newton Square, PA: PMI, 2000.

Thonsett, R., *Third Wave Project Management: A Handbook for Managing the Complex Information Systems for the 1990s,* Englewood Cliffs, NJ: Prentice Hall, 1989.

CHAPTER

3

Contents

Business Management Model

3.1 Business management

The best business plans are straightforward documents that spell out the "who, what, where, why, and how much."
—Paula Nelson

The function of business management is to ensure that projects are identified, prioritized, communicated, and justified, that budget is allocated, and that deliverables and budget are approved based on the business value derived from the deployment of the business solution. It consists of the following business processes:

- Business systems planning/architecture;
- Project value justification;
- PFAs;
- Funding/deliverable approvals;
- Program steering and working committee;
- BIS.

Figure 3.1 is a graphical representation of the modeling concept that demonstrates the structure and relationships of the definitions or processing components of business management, from a project management perspective.

The intent of this chapter is not to provide detailed processes or procedures on the implementation of the business management process or similar business process improvement processes. There are many excellent books on business management processes, concepts, theory, and applicability, some of which are mentioned in the selected bibliography at the end of this chapter. The main objective of this chapter is to provide an overview of the contents (what), purpose (why), roles and responsibilities (who, what), and procedures (how) of the components of a typical business management process and to demonstrate how business management fits within the context

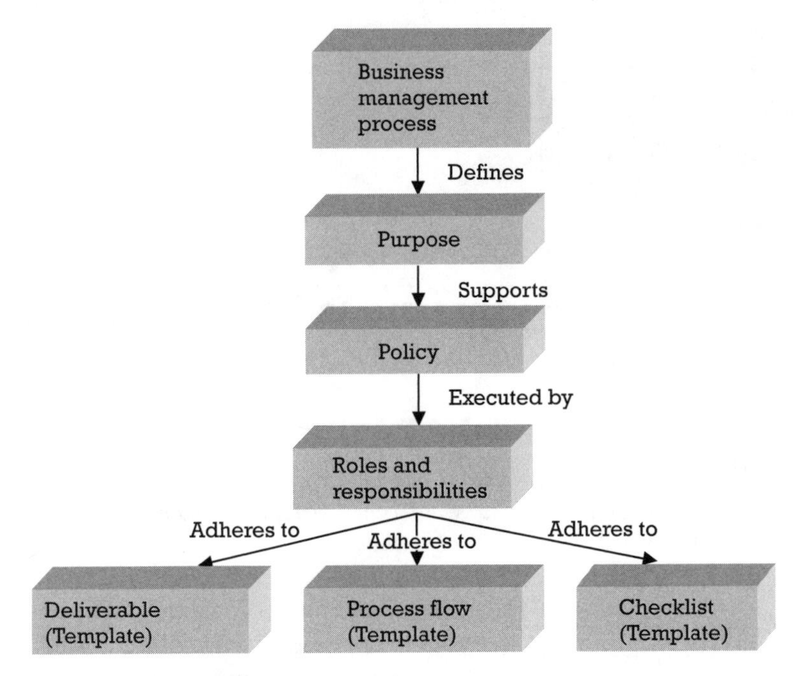

Figure 3.1 Business management model.

of the overall IPM-IT framework. The procedure sections for each of the business management component processes provide many baseline templates on the deliverables (what), process flow (how), and checklists (measurement criteria) for the practicing project manager during the management, delivery, execution, and integration of the business management component processes.

The materials presented in the following sections are organized to support the structure represented in Figure 3.1. This content structure will provide readers with a logical and consistent flow of information to enable them to gain a better understanding of and appreciation for the integrated nature of IT project management as presented in this book.

3.2 Business systems planning/architecture

Executives shall use the BSA[1] as the major source for communicating and making decisions. Business management decisions must be based on sound objective judgements, rather than emotions or office-based politics. Sound objective judgements require executive management understanding of the

1. Business systems architecture: An integrated representation of business and IT (data, applications, and technology) requirements at the conceptual level, which is used as the scope baseline to guide the prioritization, integration, and delivery of multiple IT projects to support the goals and objectives of the business. The end result is a program delivery plan, which is synonymous with the program implementation plan discussed in this book.

following elements of the company's products and services, as outlined in the BSA representation:

- *Business strategic model:* SWOT[2] (competition), including business mission, goals, objectives, strategies, critical success factors, and corresponding priorities;
- *Business requirements model:* Business problems, informational needs, and corresponding priorities that reflect business strategic directions;
- *Business functional model:* Business functions and processes that support the business strategic model to address/resolve the business requirements model;
- *Business architecture:* An integrated representation of the business functional model that supports the business strategic model to address/resolve the business requirements model;
- *Business architecture/IT architectures (data, applications, and technology):* Linkages to these that ensure the technology solution is aligned with the goals, objectives, and strategies of the business, described in the BSA document;
- *Program delivery plan[3]:* Delivery projects and corresponding priorities and implementation sequences that support the business and IT architectures;
- *Program deployment plan[4]:* Business and IT process deployments, change management, or organizational changes, training, and infrastructure transition strategies.

3.2.1 Purpose

The purpose of this process is to establish a BSA framework for the orderly prioritization and delivery of business and IT initiatives that will effectively support the goals and objectives of the company. This framework will form the baseline to manage and track the development projects, coordinate the development of business processes, and integrate business initiatives with IT initiatives, with the goal of optimizing resource allocations and minimizing duplicate processes and efforts. The project steering committee shall have approval responsibilities for both the business and IT projects and have joint accountability with the project managers for the overall scope, effort, cost, schedule, and quality performance of the projects defined in the program delivery plans.

2. SWOT: A method or process used to access the current and future state of a company's business by analyzing the strengths, weaknesses, opportunities, and threats of the business.
3. Program delivery plan: A document that describes by whom, when, and how the BSA will be delivered. It is synonymous with the program implementation plan.
4. Program deployment plan: A document that describes by whom, when, and how the program delivery plan will be deployed in the business and IT environments. It focuses on how process improvements, organizational change, training, and technology will be transitioned or deployed.

3.2.2 Policy

The following is the policy of the business planning systems/architecture:

- The development of the BSA shall adhere to the company's BSA methodology or processes.
- The program steering committee shall provide project approval responsibilities for the management and delivery of the BSA.
- All changes to the baseline BSA must be approved by the program steering committee. The program delivery manager shall document and present all BSA scope changes to the program steering committee for approval.
- The program steering committee shall meet, at a minimum, on a monthly basis to review the status of the BSA project. If the overall status of this project is Red or Yellow, more frequent meetings may be required.

3.2.3 Roles and responsibilities

Executive business manager responsibilities include the following:

- Provide details on business plan: SWOT.
- Prioritize business direction, objectives, and strategies.
- Identify program(s) and prioritize business initiatives for the program.
- Approve business architecture scope changes and overall integrated architecture.
- Act as the business sponsor for business systems planning/architecture engagement, and chair the program steering committee.

Executive IT manager responsibilities include the following:

- Provide details on IT business plan: SWOT.
- Prioritize IT direction, objectives, and strategies to support the program.
- Identify and prioritize IT initiatives for the program
- Approve IT architecture (data, applications, and technology) scope changes and overall integrated architecture.
- Act as the IT sponsor for BSA engagement.

Executive project management manager responsibilities include the following:

- Provide details on the integrated business and IT plans: SWOT.
- Identify and prioritize integrated business and IT direction, objectives, strategies, and supporting business and IT initiatives for the program.
- Provide processes/methodology guidelines and standards on architectural development.
- Approve integrated architecture in terms of adherence to project management standards and guidelines and alignment with business and IT architectures.

- Champion the project management business systems planning/architecture processes.

Program business manager responsibilities include the following:

- Document details on business plan: SWOT.
- Document prioritized business support direction, objectives, and strategies.
- Document business support initiatives and prioritized business initiatives.
- Assemble and present business architecture document.
- Report to the program steering committee on project status, issues, change requests, and risks, with direct accountability to the business executive manager.

Program IT manager responsibilities include the following:

- Document details on IT business plan: SWOT.
- Document prioritized IT support direction, objectives, and strategies.
- Document IT support initiatives and prioritized IT initiatives.
- Assemble and present IT architecture (data, applications, and technology) document.
- Report to the program steering committee on project status, issues, change requests, and risks, with direct accountability to IT executive manager.

Program delivery manager responsibilities include the following:

- Document details on integrated business processes and IT business plans: SWOT.
- Document prioritized integrated business and IT direction, objectives, strategies, and supporting initiatives.
- Document architecture processes/methodology guidelines and standards and provide guidance on applying architectural development methodology.
- Assemble and present integrated business and IT architecture (data, applications, technology) or program architecture documentation.
- Integrate, assemble, and report to the program steering committee on project status, issues, change requests, and risks, with direct accountability to project management executive manager.

Project team (project manager and team) responsibilities include the following:

- Project managers to attend program working committee meetings on a biweekly basis, and make recommendations to the program steering committee.

- Project managers to develop integrated project plan for IT and business initiatives and present integrated representation of business and IT projects.
- Project managers to manage the overall BSA development according to the established BSA methodology.
- Project managers to develop and maintain project plan for business system architecture development, consisting of:
  - Program charter;
  - Program delivery/implementation plan—WBS; projects dependency diagram; projects schedule; program organizational chart; program staffing profile; program budget;
  - Business and IT transition plan;
  - Program communications plan;
  - Program risk management plan;
  - Program issues and change management plan;
  - Program contract management plan;
  - Program quality management plan;
  - Program human resource management plan.
- Project manager to integrate, assemble, and report to the program working committee on project status, issues, change requests, and risks, with direct accountability to the program delivery manager;
- Project team to develop deliverables according to the BSA methodology.

3.2.4 Procedures: BSA

The roles and responsibilities discussed in Section 3.2.3 are brief because they are intended to serve as a reference as to who does what during the development of the BSA. Procedures are now introduced to demonstrate how this BSA process is executed using the major standard deliverables template (what), process flow (how), and checklists or measurement criteria (why). The process is intended to serve as a reference based on the English-language interrogatives: Who does what, how, and why. The deliverables, process flow, and checklist templates, based on real-world practical implementation, will serve as excellent references for practicing project managers during the execution of the BSA process.

The deliverables template in Table 3.1 highlights the key deliverable that the practicing project manager manages during the development of the BSA. The model or guideline in Figure 3.2, based on real-world practical implementations, will provide useful references for project managers. The BSA model diagram shows the foundation representation to effectively communicate the integrated components of this architecture deliverable to business, IT, and project management staff.

The process flow template in Figure 3.3 highlights the major processes to support delivery of the key BSA deliverables. These processes reference the

Table 3.1 Deliverables Template: Business Systems Architecture Report

Section 1—Introduction
This section includes an executive summary, describes the approach, and introduces the report.
Section 2—Models
This section highlights the components of the business strategic model, business requirements model, and business functional/use case model. These deliverables establish the initial baseline for further developments.
Section 3—Architectures
This section highlights the components of the business architecture, IT architectures—data, applications, and technology. These deliverables show the alignment with the models documented in Section 2 and alignment with business and IT architectures and establish the initial baseline for further developments of the delivery and deployment plans.
Section 4—Plans
This section highlights the components of the program and project management delivery plans, and program deployment plans. These plans show the alignment with the models documented in Section 2 and with architectures documented in Section 3 and establish the initial program baseline for managing the delivery of the projects.
Appendixes—Models, Architectures, and Plans
Appendixes provide further details on the models, architectures, and plans.

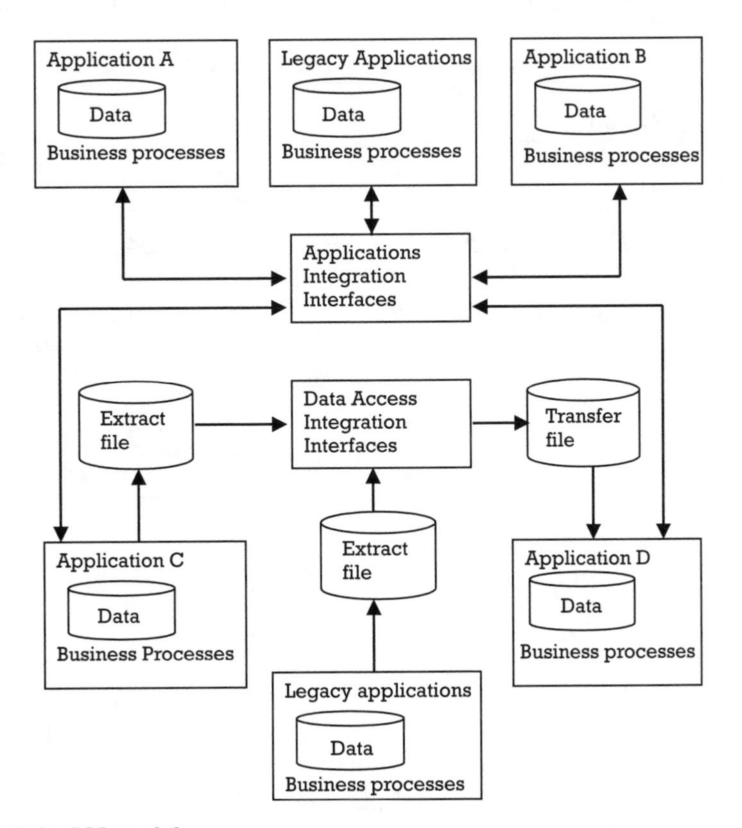

Figure 3.2 BSA model.

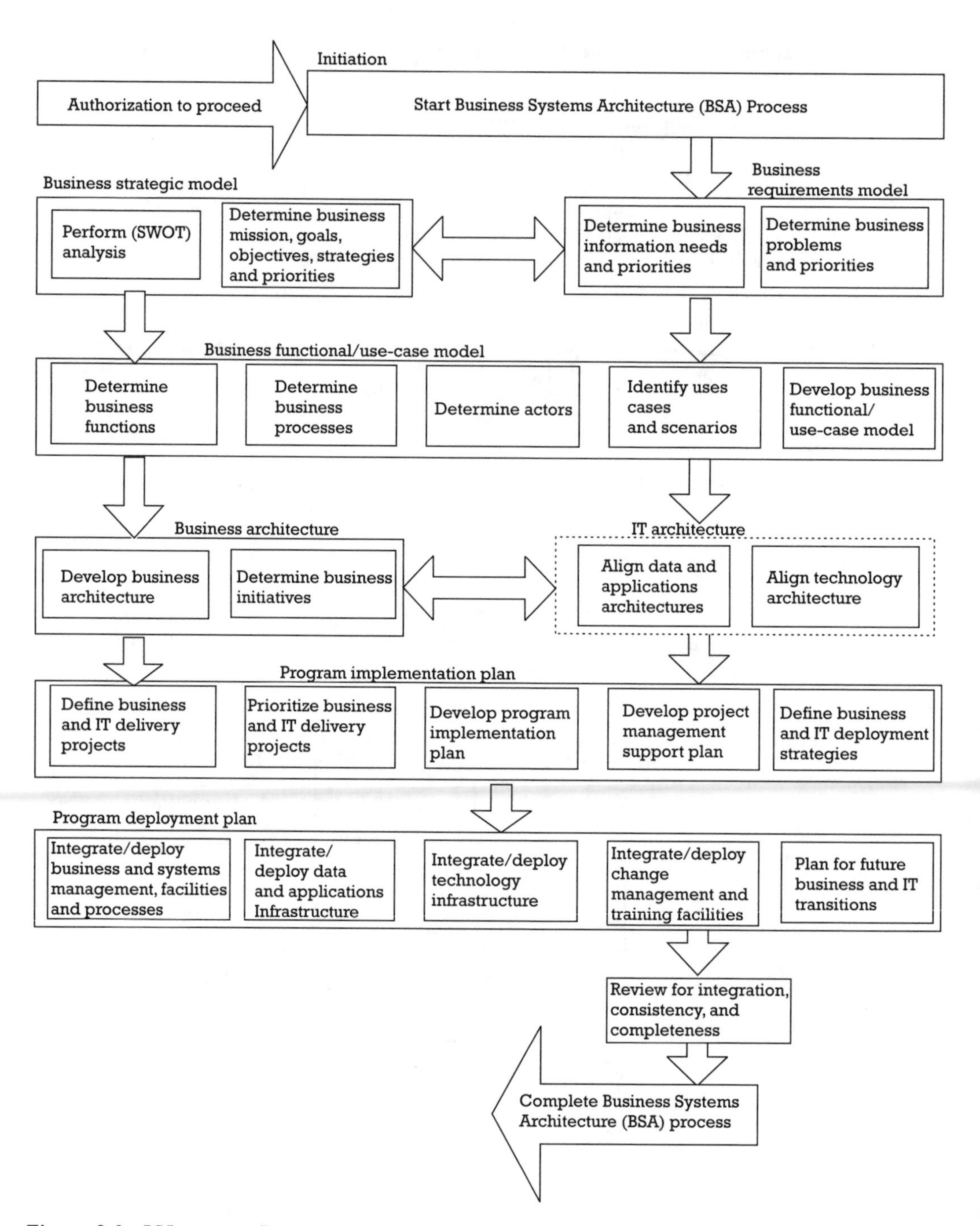

Figure 3.3 BSA process flow.

business plan and the IT architectures (data, applications, and technology) to show the integrated nature and dependencies of the business plan and IT architectures with the BSA process.

The checklist template in Table 3.2 highlights the major measurement criteria to ensure delivery of quality deliverables identified in the

Table 3.2 BSA Checklist

BSA Criteria	*Yes*	*No*	*BSA Criteria*	*Yes*	*No*
Initiation			*Business functional model*		
Is management committed?	❑	❑	Is there a list of existing business and IT processes?	❑	❑
Are a business sponsor and steering committee established?	❑	❑	Are deliverables identified?	❑	❑
Is there a program charter/plan or PMBOK guideline?	❑	❑	Are deliverables, activities, resources, schedule and costs included in project schedule?	❑	❑
Is there a project charter/plan for this BSA project?	❑	❑	Are deliverables approval criteria established?	❑	❑
Has a program delivery manager been assigned?	❑	❑	Is there a communications plan?	❑	❑
Does the assigned program delivery manager understand the BSA process?	❑	❑	Is alignment with the strategic and business model defined?	❑	❑
Business strategic model			*Business and IT architectures*		
Is there a business plan with prioritized business objectives?	❑	❑	Are there existing business and IT Architectures?	❑	❑
Are deliverables identified?	❑	❑	Are deliverables identified?	❑	❑
Are deliverables, activities, resources, schedule, and costs included in project schedule?	❑	❑	Are deliverables, activities, resources, schedule, and costs included in project schedule?	❑	❑
Are deliverables approval criteria established?	❑	❑	Are deliverables approval criteria established?	❑	❑
Is there a communications plan?	❑	❑	Is there a communications plan?	❑	❑
Is the alignment with the business requirement model understood?	❑	❑	Is the alignment with the business and functional models defined?	❑	❑
Business requirements model			*Program implementation and deployments plans*		
Is there a list of prioritized business and IT issues?	❑	❑	Is the list of business and IT issues prioritized by steering committee?	❑	❑
Are deliverables identified?	❑	❑	Are deliverables identified?	❑	❑
Are deliverables, activities, resources, schedule, and costs included in project schedule?	❑	❑	Are deliverables, activities, resources, schedule, and costs included in project schedule?	❑	❑
Are deliverables approval criteria established?	❑	❑	Are deliverables approval criteria established?	❑	❑
Is there a communications plan?	❑	❑	Is there a communications plan?	❑	❑
Is the alignment with the business strategic model understood?	❑	❑	Is the alignment with the models and architectures understood?	❑	❑

deliverables template and adherence to the process flows identified in the process flow template. The checklist is used to objectively determine success criteria during development of the BSA.

3.3 Project value justification (business priorities)

Executive management shall provide the expected value and priority of each project within the overall program to the business, based upon six areas of value that reflect the business mission statement:

1. *Operational efficiency:* Assess the value derived from areas such as reduction in staff, reduction in operating expenses, increased resource usage efficiency, increased staff efficiency, improvements in business process deployment, and improvements in customer services.
2. *Strategic business alignment:* Assess the extent to which the project aligns with corporate strategic goals, and determine the value contributed to achieving the strategies.
3. *Competitive market advantage:* Assess the extent to which the project provides an advantage in the marketplace. A project that gains competitive advantage must:
   - Change the competitive marketplace structure, or influence the way buyers and suppliers compete in the marketplace.
   - Improve the company's products or services, or change the competitive nature of the business.
   - Use information on current business practices, competitive nature, and internal information processing capability to determine new business opportunities.
4. *Organizational impact and risk assessments:* Assess the risk and organizational impact of undertaking the project. Assess the risk of organizational changes. Assess the risk of losing market share, or determine the impact of a proactive strategy to prevent the competition from gaining a competitive edge.
5. *IT support:* Assess management needs and value for information about key performance indicators and resources involved in realizing this information need to support the company's mission. Assess the degree to which the project will assist in achieving the company's critical success factors through IT support.
6. *Strategic business and IT architectures support:* Evaluate the degree to which the project is aligned with the overall BSAs (business, data, applications, and technology). Projects that are integral parts of the architecture framework will be assigned a higher value.

3.3.1 Purpose

The purpose of this process is to determine the value of the project to the business based on improved performance of the company's business. Decisions regarding IT investments shall be aligned with and driven by the six areas of project value justification,[5] defined earlier, to support the business mission statement.

3.3.2 Policy

The policy of project value justification is the following:

- Projects shall be prioritized and justified based on the six areas of business value and expected benefits to be derived from deployment of the business solution.
- The degree to which the project is aligned shall be evaluated within the context of the BSA framework. Projects that are integral parts of the architecture will be assigned a higher value. This is normally one of the major deliverables of the project RAM developed during the RAP of the PDLC model.
- Key business, IT, and project management stakeholders shall actively participate in and agree on the rankings and priorities of the projects. In most cases, a workshop shall be conducted with key stakeholders to formally agree on the rankings and priorities.

3.3.3 Roles and responsibilities

Executive business manager responsibilities include the following:

- Identify and prioritize projects within the program and justify budget based on the value of the project to the business.
- Review the business value of the projects at various phases of the project life cycle.
- Assign key business program management resources for the program.
- Approve the program management plan[6].
- Act as the overall program sponsor.

Executive IT manager responsibilities include the following:

- Identify and prioritize projects within the program and justify budget based on the value of IT support to the project.
- Review the business value of the projects at various phases of the project life cycle and the impact or risk of technology implementation.
- Assign key IT program management resources to the program.
- Approve the program management plan based on impact or risk of technology implementation.

5. Project value justification: A document that justifies the value of the project, based on the six areas of value: operational efficiency, strategic business alignment, competitive market advantage, organizational impact and risk assessment, IT support, and architecture support.
6. Project management plan: A document that describes by whom, when, and how the program/projects will be managed and how scope, effort, schedule, costs, and quality objectives will be controlled. It consists of the program delivery plan, program deployment plan, and various project management plans.

- Act as the program sponsor with IT accountability.

Executive project management manager responsibilities include the following:

- Integrate and prioritize projects within the program and justify budget based on the integrated value of the project to the business, with IT support.
- Review the business value of each project at various phases of the project life cycle, and the impact or risk of executing project management processes, tools, and techniques.
- Assign key PM program management resources to the program.
- Approve the program management plan based on impact or risk of PM implementation.
- Act as the program sponsor with PM accountability; champion PM processes, tools, and techniques.

Program business manager responsibilities include the following:

- Document the priority of projects within the program and justifications for budget based on the value of the project to the business.
- Document the business value of each project at various phases of the project life cycle and the impact or risk of executing business management processes, tools, and techniques.
- Assign business resources to projects within program.
- Assemble and present the program management plan from the business perspective.
- Report to the program steering committee with direct accountability to the business executive manager.

Program IT manager responsibilities include the following:

- Document the priority of projects within the program and justifications for budget, based on the value of IT support to the project in support of the business.
- Document the business value of each project at various phases of the project life cycle and the impact or risk of executing IT management processes, tools, and techniques.
- Assign IT resources to projects within the program.
- Assemble and present the program management plan from the IT perspective.
- Report to the program steering committee.

Program delivery manager responsibilities include the following:

- Document the priority of integrated projects and justification of allocated budget based on the integrated value of the project to the business, with IT support, for the program.

- Document the business value of each project at various phases of the project life cycle and the impact or risk of executing PM management processes, tools, and techniques.
- Assign PM resources to the portfolio of projects within the program.
- Allocate all resources to projects within the program.
- Assemble and present the program management plan from the PM perspective.
- Report to the program steering committee.

Project team (project manager and team) responsibilities include the following:

- Attend program working committee meetings on a biweekly basis, and make recommendations to working committee on project value justification.
- Develop project value justification for IT and business initiatives and present prioritized and integrated representation of business and IT projects.
- Manage the overall project value justification process.
- Develop and maintain a strategy for the development of the project value justification document based on the six areas of value:
  - Operational efficiency;
  - Strategic business alignment;
  - Competitive advantage;
  - Organizational impact and risk assessments;
  - IT support;
  - Strategic business and IT architectures support.
- Integrate, assemble, and report on the project RAM to the program working committee based on project value justification with direct accountability to the program delivery manager.
- Assign resources to business and IT projects based on the value to the business.

3.3.4 Procedures: project value justification

The roles and responsibilities discussed above are brief because they are intended to serve as a reference as to who does what during the development of the project value justification. Procedures are now introduced to demonstrate how this project value justification process is executed using the major standard deliverables (what), process flow (how), and checklist or measurement criteria (why) templates. This process is intended to serve as a reference based on the English-language interrogatives: who does what, how, and why. The deliverables, process flow, and checklist templates are based on real-world practical implementation and will serve as excellent references for practicing project managers during execution of the project value justification process.

The deliverables template in Table 3.3 highlights the key deliverables that the practicing project manager manages during the development of the project value justification. The model or guideline in Table 3.4 is based on real-world practical implementations and will provide useful references for project managers. The project value justification value/risk assessment model in Table 3.4 shows the degree of the value and risk impact based on the six areas of value analysis presented earlier.

The process flow template in Figure 3.4 highlights the major processes to support the delivery of the key project value justification deliverables. These processes reference six areas of value and risk impact to show the integrated nature and dependencies of business and technology value and risks derived during the project value justification process.

The checklist template in Table 3.5 highlights the major measurement criteria to ensure delivery of quality deliverables identified in the deliverables template and adherence to the process flows identified in the process flow template. The checklist is used to objectively determine success criteria during development of the project value justification process.

Table 3.3 Deliverables Template: Project Value Justification Report

Section 1—Introduction
This section includes an executive summary, describes the approach, and introduces the report.
Section 2—Business Value
This section highlights the value of operational efficiency, strategic business alignment, competitive advantage, and organizational and cultural impact. These value statements establish the initial baseline for project value justification from the business perspective.
Section 3—Technical Value
This section highlights the value of IT support and strategic IT architecture—data, applications, and technology. These value statements show the alignment with the business value, documented in Section 2, and establish the initial baseline for project value justification from the technology perspective.
Section 4—Risk Assessments Strategies
This section highlights the risk assessment in terms of risk avoidance, risk transference, risk mitigation, and risk acceptance strategies. These risk assessment strategies show the impact to the business value documented in Section 2 and to technical value documented in Section 3, and establish the initial risk management strategies for project value justification and for managing the delivery of the projects.
Section 5—Project Value Assessments
This section highlights the project value assessments in the form of business, technical, and risk assessment evaluation scores and forms the baseline for project value justification.
Appendixes—Business and Technical Value, Risk, and Project Value Assessments
Appendixes provide further details on the business value/evaluation scores, technical value/evaluation scores, risk assessments impact evaluations. and project value assessments.

Table 3.4 Deliverables template: Project Value Justification Model

Degree of Value/Risk	*Operational Efficiency*	*Strategic Business Alignment*	*Competitive Advantage*	*Organizational Process Efficiency*	*IT Efficiency*	*BSA Alignment*
Value-H	❑	❑	❑	❑	❑	❑
Value-L	❑	❑	❑	❑	❑	❑
Risk-H	❑	❑	❑	❑	❑	❑
Risk-L	❑	❑	❑	❑	❑	❑

Value		Risk: Low	Risk: High
	High	Quadrant 4	Quadrant 3
	Low	Quadrant 1	Quadrant 2

Quadrant 1: Low value and low risk—Minor impact
Quadrant 2: Low value and high risk—Definitely avoid
Quadrant 3: High value and high risk—Major impact
Quadrant 4: High value and low risk—Accept

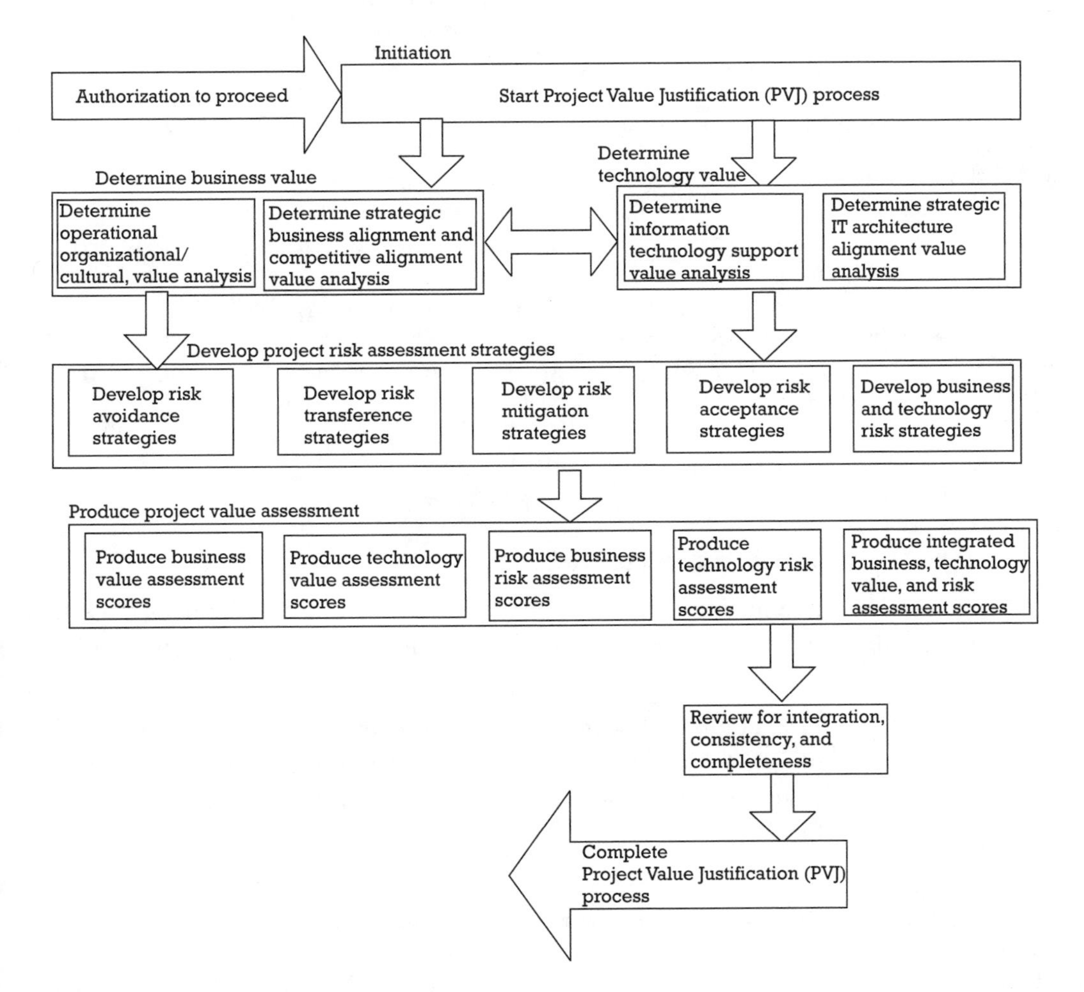

Figure 3.4 Project value justification process flow.

Table 3.5 Project Value Justification Checklist

Project Value Justification Criteria	*Yes*	*No*	*Project Value Justification Criteria*	*Yes*	*No*
Initiation			*Business and technology value*		
Is management committed?	❑	❑	Are there criteria to determine/score business operational efficiency value?	❑	❑
Are a business sponsor and steering committee established?	❑	❑	Are there criteria to determine/score business organizational/cultural value?	❑	❑
Are there project value justification guidelines available?	❑	❑	Are there criteria to determine/score strategic business alignment value?	❑	❑
Is there a project value statement for this project value justification?	❑	❑	Are there criteria to determine/score business competitive advantage value?	❑	❑
Has a program delivery manager been assigned?	❑	❑	Are there criteria to determine/score IT technology support value?	❑	❑
Does the assigned program delivery manager understand the project value justification process?	❑	❑	Are there criteria to determine/score IT architecture value?	❑	❑
Risk assessment strategies			*Project value assessments*		
Is there a list of potential business and technology risks in realizing values?	❑	❑	Are the business value assessment scores acceptable?	❑	❑
Are there risk avoidance strategies?	❑	❑	Are the technology value assessment scores acceptable?	❑	❑
Are there risk transference strategies?	❑	❑	Are the business risk assessment scores acceptable?	❑	❑
Are there risk mitigation strategies?	❑	❑	Are the technology risk assessment scores acceptable?	❑	❑
Are there risk acceptance strategies?	❑	❑	Are the integrated business and technology value and risk scores acceptable to sponsors?	❑	❑

3.4 Project funding allocations

Executive management shall allocate funds on a fiscal year basis to each of the projects within the overall program at the following milestones during the PDLC:

- *PDP:* Funds will be allocated as approved[7] for the PDP, after approval of a BSA, or approval of a business requirements statement (BRS) in cases where BSA is nonexistent. Funds will be allocated as reserved[8] for all phases after approval of the PDM that is completed at the

7. Approved funds: Funds approved by senior management for spending, based on definitive cost estimates. Project teams are only allowed to spend approved funds.

8. Reserved funds: Funds reserved by senior management for spending, based on preliminary and budget cost estimates.

conclusion of the PDP. Lack of executive management approval or commitment will result in the project being delayed or canceled.

- *RAP:* Reserved funds will be released for the RAP after approval of the PDM completed at the conclusion of the PDP. The project will be revalidated, based upon the updated project cost and benefits, and funds will be reallocated as approved for the project RAP. Lack of executive management approval or commitment will result in the project's being delayed or terminated.
- *ADP:* Reserved funds will be released for the ADP after approval of the RAM completed at the conclusion of the RAP. The project will be revalidated, based upon the updated project cost and benefits, and funds will be reallocated as approved for the ADP. Lack of executive management approval or commitment will result in the project's being delayed or terminated.
- *IDPs:* Reserved funds will be released for the IDPs after approval of the PAS completed at the conclusion of the project ADP. A project value analysis will be conducted for the IDPs—construction, integration, and deployment—and funds will be reallocated as approved for the three IDP phases. In certain situations, funds will be allocated, as approved for one of the IDP phases, based on the project value analysis. Lack of executive management approval or commitment will result in the project's being delayed or terminated.

3.4.1 Purpose

The primary objective of project funding is to control the allocation of funds to projects that have been prioritized by the program steering committee meetings and to allocate funds in increments based upon revalidating the project cost and benefits.

3.4.2 Policy

The policy of project funding allocations is as follows:

- The steering committee shall prioritize and allocate funds to projects at specific checkpoints/milestone as reserved and approved. Only approved funding shall be spent by the project team.
- Projects with costs that increase beyond the expected benefits will be reprioritized and funding reevaluated.
- Reserved funds shall be allocated based on preliminary and budget cost estimates and approved funds allocated for definitive cost estimates.

3.4.3 Roles and responsibilities

Executive business manager responsibilities include the following:

- Present, evaluate, and prioritize opportunities for projects within the program and allocate budget based on the value of the project to the business.

- Manage the portfolio of projects within the funds allocated by the program steering committee for program delivery.
- Chair the program steering committee meetings and track allocation of funds.
- Approve allocation of funds at various phases of the project life cycle.
- Act as the overall program sponsor with business accountability.

Executive IT manager responsibilities include the following:

- Present, evaluate, and prioritize opportunities for projects within the program and allocate budget based on the value of IT support for the project.
- Manage the portfolio of projects within the funds allocated by the program steering committee for IT support.
- Chair the program steering committee meetings and track the allocation of funds from the IT budget.
- Approve the allocation of IT funds at various phases of the project life cycle.
- Act as the program sponsor with IT accountability.

Executive project management manager responsibilities include the following:

- Integrate and prioritize opportunities for projects within the program and allocation of budget based on the integrated value of the project to the business, with IT support.
- Manage the portfolio of projects within the funds allocated by the program steering committee using project management guidelines.
- Organize the program steering committee meetings and track allocation of funds from the program budget.
- Approve allocation of program funds at various phases of the project life cycle.
- Act as the program sponsor with integrated business and IT responsibility and PM accountability; champion PM processes, tools, and techniques.

Program business manager responsibilities include the following:

- Report on opportunities for projects within the program and the allocation of budget based on the value of the project to the business.
- Report on the portfolio of projects within the funds allocated by the program steering committee.
- Report to the program steering committee meetings and track the allocation of funds.
- Report on the allocation of funds at various phases of the project life cycle.
- Report to the program steering committee with direct accountability to the business executive manager.

Program IT manager responsibilities include the following:

- Report on opportunities for projects within the program and the allocation of IT budget based on the extent of IT support.
- Report on the portfolio of projects within the funds allocated by the program steering committee for IT support.
- Report to the program steering committee meetings and track allocation of funds for IT support.
- Report on the allocation of funds at various phases of the project life cycle for IT support.
- Report to the program steering committee with direct accountability to the IT executive manager.

Program delivery manager responsibilities include the following:

- Integrate, prioritize, and report opportunities for projects within the program and allocation of budget, based on the value of the project to the business and IT for entire program.
- Monitor and report on the portfolio of projects within the funds allocated by the program steering committee, for the entire program, using project management guidelines.
- Organize the program steering committee meetings and track the allocation of funds from the program budget.
- Provide cost-estimating guidelines and report on the allocation of funds at various phases of the project life cycle for entire the program.
- Report to the program steering committee with direct accountability to the project management executive manager.

Project team (project manager and team) responsibilities include the following:

- Attend program working committee meetings on a biweekly basis, and make recommendations to the program steering committee on PFAs.
- Expend the project allocated funds and provide expense distribution and business value to the program working committee.
- Manage the portfolio of projects with the funds allocated by the program steering committee.
- Develop and maintain a strategy for managing the funds allocated by the program steering committee based on the milestones of the PDLC:
  - PDP;
  - RAP;
  - ADP;
  - IDPs.
- Integrate, assemble, and report to the program working committee on expending of the allocated funds with direct accountability to the program delivery manager.
- Allocate funds to project resources based on the value to the business.

3.4.4 Procedures: PFAs

The roles and responsibilities discussed in Section 3.4.3 are brief because they are intended to serve as a reference as to who does what during the development of the PFA. Procedures are now introduced to demonstrate how this PFA process is executed using the major standard deliverables (what), process flow (how) and checklist or measurement criteria (why) templates. The process is intended to serve as a reference based on the English-language interrogatives: who does what, how, and why. The deliverables, process flow and checklist templates are based on real-world practical implementation and will serve as excellent references for practicing project managers during execution of the PFA process.

The deliverables template in Table 3.6 highlights the key deliverables that the practicing project manager manages during the development of PFAs. The model or guideline in Table 3.7 is based on real-world practical implementations and will provide useful references for project managers. The PFA model in Table 3.7 shows how funds are reserved and approved during each PDLC phase.

The process flow template in Figure 3.5 highlights the major processes to support the delivery of the key PFA deliverables. These processes show how funds are reserved and approved for each of the PDLC phases and the

Table 3.6 Deliverables Template: PFA Report

Section 1—Introduction
This section includes an executive summary, estimating and funds allocation processes, and milestones and introduces the report.
Section 2—Project Definition Phase–Funds Allocation
This section provides a chronological history by date, phases, amount of approved and reserve funding, and approval signatures for the PDP for each of the projects based on the appropriate estimates. These funds establish the initial costs estimate baseline for project economic assessments and project costs management.
Section 3—Requirements Analysis Phase–Funds Allocation
This section provides a chronological history by date, phases, amount of approved and reserve funding, and approval signatures for the RAP for each of the projects, based on the appropriate estimates. These funds establish the initial costs estimate baseline for project economic assessments and project costs management.
Section 4—Architecture Design Phase–Funds Allocation
This section provides a chronological history by date, phases, amount of approved and reserve funding, and approval signatures for the ADP for each of the projects, based on the appropriate estimates. These funds establish the initial costs estimate baseline for project economic assessments and project costs management.
Section 5—IDPs–Funds Allocation
This section provides a chronological history by date, phases, amount of approved and reserve funding, and approval signatures for the IDPs for each of the projects, based on the appropriate estimates. These funds establish the initial costs estimate baseline for project economic assessments and project costs management.
Appendixes—PFAs by Phase Milestone
Appendixes provide further funding details on the funding history by date, amount of approved and reserve funding, and cost estimating guidelines for each of the PDLC phases for each of the projects within the program.

Table 3.7 Deliverables Template: PFA Model

Funds Allocation by Phases	*Definition Phase*	*Analysis Phase*	*Architecture Phase*	*IDPs*
Reserved (preliminary)	Funds reserved (preliminary cost estimates at completion of BSA/BRS)	Funds reserved (preliminary cost estimates at completion of BSA/BRS)	Funds reserved (preliminary cost estimates at completion of BSA/BRS)	Funds reserved (preliminary cost estimates at completion of BSA/BRS)
Reserved (budget)	Funds reserved (budget cost estimates at completion of project definition)	Funds reserved (budget cost estimates at completion of project definition)	Funds reserved (budget cost estimates at completion of project definition)	Funds reserved (budget cost estimates at completion of project definition)
Approved (definitive)	Funds approved (definitive cost estimates at completion of BSA/BRS)	Funds approved (definitive cost estimates at completion of project definition)	Funds approved (definitive cost estimates at completion of project analysis)	Funds approved (definitive cost estimates at completion of project architecture)

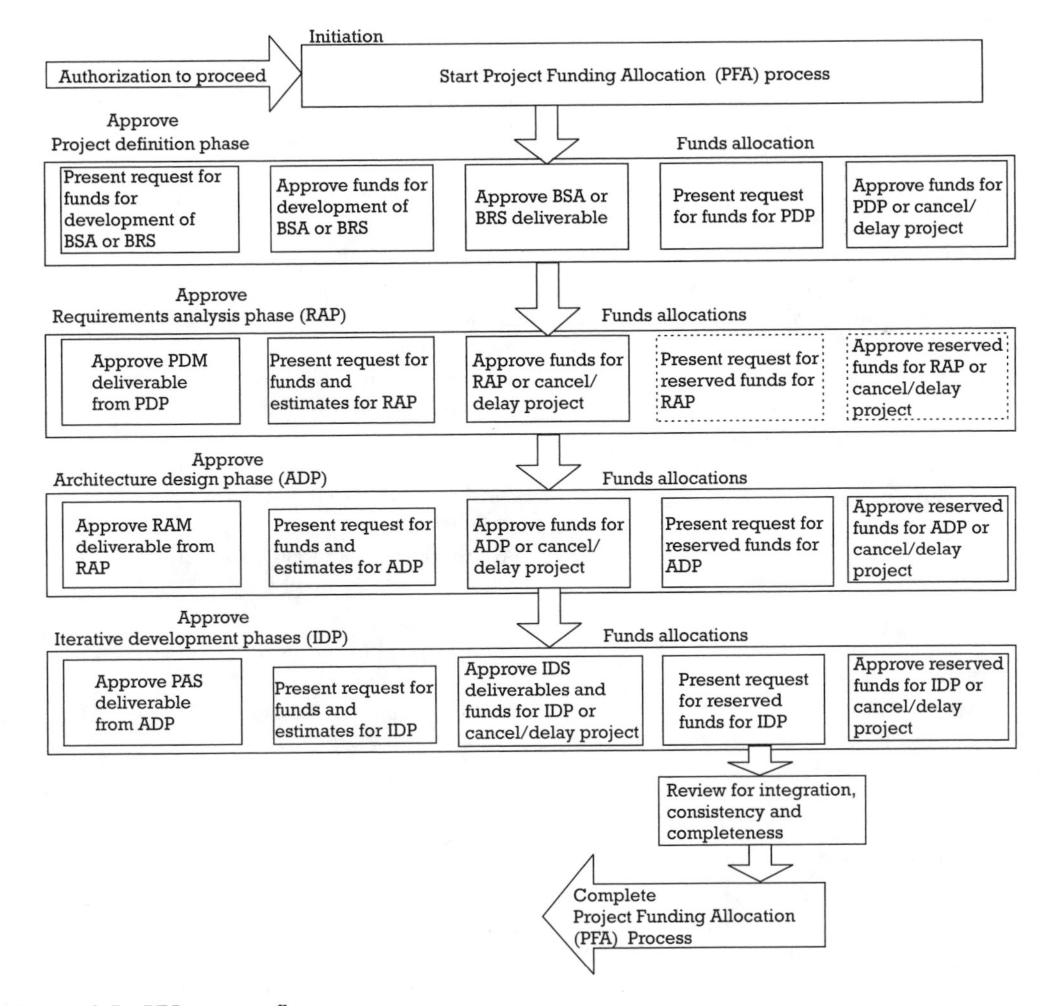

Figure 3.5 PFA process flow.

integrated nature and dependencies of the project deliverables and phased checkpoints during execution of the PFA process.

The checklist template in Table 3.8 highlights the major measurement criteria to ensure delivery of quality deliverables identified in the

Table 3.8 PFA Checklist

PFA Criteria	*Yes*	*No*	*PFA Criteria*	*Yes*	*No*
Initiation			*BSA/BRS*		
Is management committed?	❑	❑	Is the request for funds for BSA development approved?	❑	❑
Are a business sponsor and steering committee established?	❑	❑	Is the BSA process communicated properly?	❑	❑
Are PFA guidelines available?	❑	❑	Is the final BSA or BRS deliverable approved?	❑	❑
Are guidelines for approved and reserved funds understood?	❑	❑	Is there a list of criteria to measure success of BSA/BRS?	❑	❑
Has a program delivery manager been assigned?	❑	❑	Is senior management (business and IT) committed to BSA/BRS deliverables?	❑	❑
Does the assigned program delivery manager understand the PFA processes?	❑	❑	Is there a list of criteria to determine score/benefits/value of BSA?	❑	❑
PDP			*RAP*		
Is the request for funds for PDP development approved?	❑	❑	Is the request for funds for RAP development approved?	❑	❑
Is the PDP approval process communicated properly?	❑	❑	Is the RAP approval process communicated properly?	❑	❑
Is the final PDM deliverable approved?	❑	❑	Is the final RAM deliverable approved?	❑	❑
Is there a list of criteria to measure the success of PDP?	❑	❑	Is there a list of criteria to measure the success of the RAP?	❑	❑
Is senior management (business and IT) committed to deliverables?	❑	❑	Is senior management (business and IT) committed to deliverables?	❑	❑
Are there criteria to evaluate score/benefits/value of PDM?	❑	❑	Are there criteria to evaluate score/benefits/value of RAM?	❑	❑
ADP			IDPs		
Is the request for funds for ADP development approved?	❑	❑	Is the request for funds for IDP development approved?	❑	❑
Is the ADP approval process communicated properly?	❑	❑	Is the IDP approval process communicated properly?	❑	❑
Is the final PAS deliverable approved?	❑	❑	Is IDP deliverable approved?	❑	❑
Is there a list of criteria to measure the success of the ADP?	❑	❑	Is there a list of criteria to measure the success of the IDP?	❑	❑
Is senior management (business and IT) committed to deliverables?	❑	❑	Is senior management (business and IT) committed to deliverables?	❑	❑
Are there criteria to evaluate score/benefits/value of PAS?	❑	❑	Are there criteria to evaluate score/benefits/value of IDP?	❑	❑

deliverables template and adherence to the process flows identified in the process flow template. The checklist is used to objectively determine success criteria during the development of PFAs.

3.5 Project funding/deliverables approval

Executive management and major stakeholders shall provide both business and technical approvals required to commit a project to the next phase of development. The key decision points will be at:

- The beginning of the project via BSA or business definition statement;
- The completion of BSA or business definition statement;
- The completion of PDM during the PDP;
- The completion of project RAM during the RAP;
- The completion of PAS during the ADP;
- The beginning of each subsequent product release[9] of the IDPs—construction, integration, and deployment.

Preliminary project funding and deliverables approval decisions are necessary early in the life of a project before all requirements are known. An incremental approach to project deliverables approvals will allow for an orderly development of project estimates that can be validated against business needs at critical points in the life cycle of a project.

A phased approach to project deliverables approvals will also provide an understanding of the project scope, funding/cost, schedules, and effort that are agreed upon and supported by both business and IT managers.

3.5.1 Purpose

The purpose of this process is to obtain commitments or noncommitments from executive and senior management for project funding and deliverables approval to proceed or not proceed to the next phase of development.

3.5.2 Policy

The policy of the project funding/deliverables approval includes the following:

- The program steering committee shall approve the transition of a project from the PDP through to the project deployment phase at key decision points or at completion of key deliverables.
- Project deliverables shall be approved in accordance with the deliverables approval process.

9. Product release: The implementation of the product scope packaged as a release or iteration solution at the end of each IDP.

- Project deliverables shall be approved based on agreed-upon acceptance criteria from project managers, IT delivery managers, and contractors.

3.5.3 Roles and responsibilities

Executive business manager responsibilities include the following:

- Review phase deliverables prior to the program steering committee meeting for acceptance, approval, and funding from the business perspective.
- Manage the portfolio of projects deliverables within the funds allocated by the program steering committee from the business perspective.
- Perform high-level QA reviews and provide acceptance for project deliverables based on acceptance criteria/checklists within the funds allocated by the program steering committee from the business perspective.
- Recommend business resources for defining business scope and project benefit and cost estimates.
- Act as the overall program sponsor.

Executive IT manager responsibilities include the following:

- Review phase deliverables prior to the program steering committee meeting for acceptance, approval, and funding from the IT perspective.
- Manage the portfolio of projects deliverables within the funds allocated by the program steering committee from the IT perspective.
- Perform high-level QA reviews and provide acceptance for project deliverables based on acceptance criteria/checklists within the funds allocated by the steering committee from the IT perspective.
- Recommend IT support resource for defining IT scope and project cost estimates.
- Act as the program sponsor from the IT perspective.

Executive project management manager responsibilities include the following:

- Integrate and review phase deliverables prior to the program steering committee meeting for acceptance, approval, and funding from the IT and business perspective.
- Integrate the portfolio of projects deliverables within the funds allocated by the program steering committee from the IT and business perspective.
- Perform high-level QA reviews and provide acceptance for project deliverables based on acceptance criteria/checklists within the funds allocated by the program steering committee, in accordance with PM processes and guidelines.

- Recommend PM resources for defining business scope and project benefit estimates.
- Act as the program sponsor with integrated business and IT responsibility and PM accountability; champion PM processes, tools, and techniques.

Program business manager responsibilities include the following:

- Submit phase deliverables to the program steering committee meeting for acceptance, approval, and funding from the business perspective.
- Manage the portfolio of projects deliverables within the funds allocated by the program steering committee from the business perspective.
- Perform QA reviews and provide acceptance for project deliverables based on acceptance criteria/checklists within the funds allocated by the program steering committee from the business perspective.
- Assign business resources for defining business scope and project benefit estimates.
- Report to the program steering committee with direct accountability to the business executive.

Program IT manager responsibilities include the following:

- Submit phase deliverables to the program steering committee meeting for acceptance, approval, and funding from the IT perspective.
- Manage the portfolio of project deliverables within the funds allocated by the program steering committee from the IT perspective.
- Perform QA reviews and provide acceptance for project deliverables based on acceptance criteria/checklists within the funds allocated by the program steering committee from the IT perspective.
- Assign IT resources for defining IT scope and project cost estimates for IT support.
- Report to the program steering committee with direct accountability to the IT executive.

Program delivery manager responsibilities include the following:

- Submit phase deliverables to the program steering committee meeting for acceptance, approval, and funding from the business and IT perspectives.
- Manage the portfolio of projects deliverables within the funds allocated by the program steering committee from the business and IT perspectives.
- Perform QA reviews and provide acceptance for project deliverables based on acceptance criteria/checklists within the funds allocated by the program steering committee from the business and IT perspectives.
- Assign business resource for PM process/methodology support.

- Report to the program steering committee with direct accountability to the PM executive.

Project team (project manager and team) responsibilities include the following:

- Attend program working committee meetings on a biweekly basis, and make recommendations to the program steering committee on project funding and deliverables approvals.
- Submits phase deliverables to the program working committee for acceptance, approval, and funding.
- Submit phase deliverables and checklists to program delivery manager (PMO) for QA reviews and acceptance.
- Produce and manage the portfolio of project deliverables within the funds allocated by the program steering committee.
- Develop and maintain a strategy for managing the funds and deliverables approved by the program steering committee based on the milestones of the PDLC:
  - Start of program;
  - PDP;
  - RAP;
  - ADP;
  - IDPs.
- Integrate, assemble, and report to the program working committee on funding and deliverables approval with direct accountability to the program delivery manager.

3.5.4 Procedures: project funding/deliverables approval

The roles and responsibilities discussed in Section 3.5.3 are brief because they are intended to serve as a reference as to who does what during the development of project funding/deliverables approval. Procedures are now introduced to demonstrate how this project funding/deliverables approval process is executed using the major standard deliverables (what), process flow (how), and checklist or measurement criteria (why) templates. The process is intended to serve as a reference based on the English-language interrogatives: who does what, how, and why. The deliverables, process flow, and checklist templates are based on real-world practical implementation and will serve as excellent references for practicing project managers during execution of the project funding/deliverables approval process.

The deliverables template in Table 3.9 highlights the key deliverables that the practicing project manager manages during the development of project funding/deliverables approval. The model or guideline in Figure 3.6 is based on real-world practical implementations and will provide useful references for project managers. The project deliverables approval model in Figure 3.6 shows the process of how deliverables are submitted for approvals by stakeholders having approval responsibilities.

Table 3.9 Deliverables Template: Project Funding/Deliverables Approval Report

Section 1—Introduction
This section includes an executive summary, funding/deliverables approval processes, and milestones and introduces the report.
Section 2—Project Definition Phase–Funding/Deliverables Approval
This section provides a chronological history by date, phased deliverables, amount of funds approved for deliverables completion, and approval signatures for the PDP deliverable for each of the projects based on the appropriate deliverables quality measurement criteria. These approved funds and deliverables establish the initial costs estimate, scope, and quality baseline for the project.
Section 3—Requirements Analysis Phase–Funding/Deliverables Approval
(This section provides a chronological history by date, phased deliverables, amount of funds approved for deliverables completion, and approval signatures for the project RAP deliverable for each of the projects based on the appropriate deliverables quality measurement criteria. These approved funds and deliverables establish the initial costs estimate, scope, and quality baseline for project management.
Section 4—Architecture Design Phase–Funding/Deliverables Approval
This section provides a chronological history by date, phased deliverables, amount of funds approved for deliverables completion, and approval signatures for the project ADP deliverable for each of the projects based on the appropriate deliverables quality measurement criteria. These approved funds and deliverables establish the initial costs estimate, scope, and quality baseline for projects.
Section 5—IDPs–Funding/Deliverables Approval
This section provides a chronological history by date, phased deliverables, amount of funds approved for deliverables completion, and approval signatures for the project IDPs deliverables for each of the projects based on the appropriate deliverables quality measurement criteria. These approved funds and deliverables establish the initial costs estimate, scope, and quality baseline for project management.
Appendixes—Project Funding/Deliverables Approvals by Phase Milestone
Appendixes provide further funding details on the funding/deliverables approval history by date, amount of funds approved for deliverables completion, and deliverables measurement criteria for the deliverables produced during each of the PDLC phases for each of the projects within the program.

The process flow template in Figure 3.7 highlights the major processes to support the delivery of the key project funding/deliverables approval deliverables. These processes show how funds are approved for each of the PDLC phases and the integrated nature and dependencies of the project deliverables and phased checkpoints during execution of the project funding/deliverables approval process.

The checklist template in Table 3.10 highlights the major measurement criteria to ensure delivery of quality deliverables identified in the deliverables template and adherence to the process flows identified in the process flow template. The checklist is used to objectively determine success criteria during the development of project funding/deliverables approval.

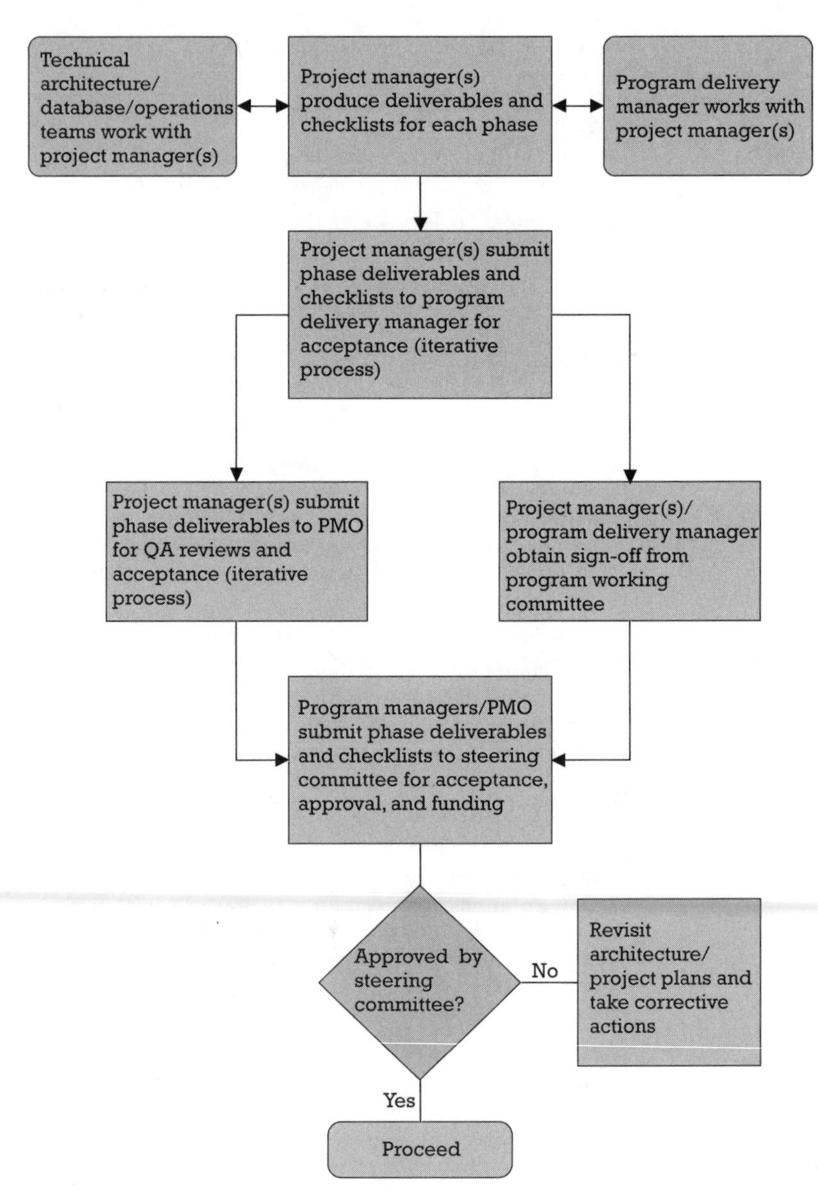

Figure 3.6 Project deliverables approval model.

3.6 Program steering and working committee

Executive management shall establish a program steering committee[10] with major business and IT senior management stakeholders having project approval responsibilities. This program steering committee will have overall responsibilities for the program and have joint accountability with the program manager for the overall scope, effort, cost, and schedule performance

10. Program steering committee: A committee consisting of executive management and senior management staff with program-/project-approval responsibilities.

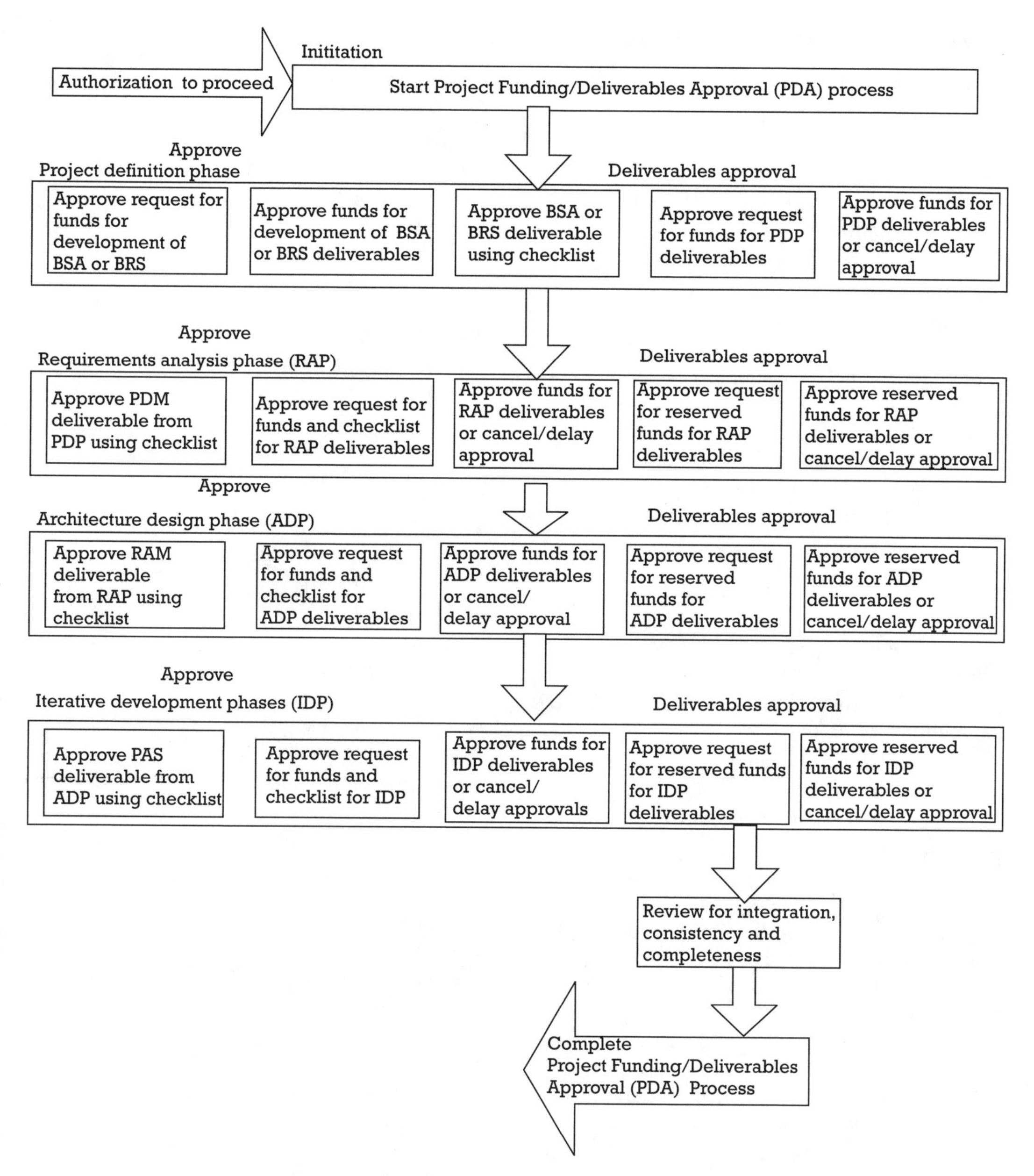

Figure 3.7 Project deliverables/funding approval process flow.

of the project. The representatives of this program steering committee shall include the following:

- *Business executive manager:* Program sponsor or chairman with vice president level of accountability, authority, and responsibility.
- *Executive IT manager:* IT sponsor with vice-president level of accountability, authority, and responsibility.
- *Executive project management manager:* Project management sponsor, preferably at the vice-president level.

Table 3.10 Project Funding/Deliverables Approval Checklist

Project Funding/Deliverables Approval Criteria	*Yes*	*No*	*Project Funding/Deliverables Approval Criteria*	*Yes*	*No*
Initiation			*BSA/BRS*		
Is management committed?	❑	❑	Is the request for funds for BSA development approved?	❑	❑
Are a business sponsor and steering committee established?	❑	❑	Is the BSA process communicated properly?	❑	❑
Are there funding/deliverables approval guidelines available?	❑	❑	Is the final BSA or BRS deliverable approved?	❑	❑
Are checklists for approving deliverables understood?	❑	❑	Is there a list of criteria to measure the success of the BSA/BRS?	❑	❑
Has a program delivery manager been assigned?	❑	❑	Is senior management (business and IT) committed to BSA/BRS deliverables?	❑	❑
Does the assigned program delivery manager understand the project funding/deliverables approval processes?	❑	❑	Are there criteria to determine score/benefits/value of BSA?	❑	❑
PDP			*RAP*		
Is the request for funds for PDP development approved?	❑	❑	Is the request for funds for RAP development approved?	❑	❑
Is the PDP approval process communicated properly?	❑	❑	Is the RAP approval process communicated properly?	❑	❑
Is the final PDM deliverable approved?	❑	❑	Is the RAM deliverable approved?	❑	❑
Is there a list of criteria to measure success of the PDP?	❑	❑	Is there a list of criteria to measure success of RAP?	❑	❑
Is senior management (business and IT) committed to deliverables?	❑	❑	Is senior management (business and IT) committed to deliverables?	❑	❑
Are there criteria to evaluate score/benefits/value of PDM?	❑	❑	Are there criteria to evaluate score/benefits/value of RAM?	❑	❑
ADP			*IDPs*		
Is the request for funds for ADP development approved?	❑	❑	Is the request for funds for IDP development approved?	❑	❑
Is the ADP approval process communicated properly?	❑	❑	Is the IDP approval process communicated properly?	❑	❑
Is the final PAS deliverable approved?	❑	❑	Are the IDP deliverables approved?	❑	❑
Is there a list of criteria to measure success of the ADP?	❑	❑	Is there a list of criteria to measure the success of the IDP?	❑	❑
Is senior management (business and IT) committed to the deliverables?	❑	❑	Is senior management committed to deliverables?	❑	❑
Are there criteria to evaluate score/benefits/value of PAS?	❑	❑	Are there criteria to evaluate score/benefits/value of IDS?	❑	❑

- *Program business manager:* Business leader with director level of accountability, authority, and responsibility.
- *Program IT manager:* IT leader with director level of accountability, authority, and responsibility.
- *Program delivery manager:* Program leader with director level of accountability, authority, and responsibility.

Program business management shall establish a program working committee[11] with major business, IT, and project management stakeholders, including project managers, having project-approval responsibilities. This program working committee will have overall delivery responsibilities for the program and have joint accountability with the project managers for the overall scope, effort, cost, and schedule performance of the projects. The representatives of this program working committee shall include the following:

- *Program delivery manager:* Program leader and chairman with director level of accountability, authority, and responsibility.
- *Program business manager:* Business leader with director level of accountability, authority, and responsibility.
- *Program IT manager:* IT leader with director level of accountability, authority, and responsibility.
- *Project managers:* Project leaders with manager level of accountability, authority, and responsibility.
- *Project team:* Key business, IT, and project management representatives.
- *External stakeholders:* Key external stakeholders with influence and interests.

3.6.1 Purpose

The purpose of these two committees is to provide project-approval responsibilities and accountability for the overall scope, effort, cost, schedule, and quality performance of the program.

3.6.2 Policy

The policy of the program steering and working committee is the following:

- The program steering committee shall provide project approval responsibilities.
- The program steering committee shall approve all changes to the approved project plan.
- The project manager will document and present all scope changes to the program steering committee for approval.

11. Program working committee: A committee consisting of senior managers and project managers with program/project accountability for the scope, effort, schedule, cost, and quality of the projects. Accountability = responsibility + authority.

- The steering committee shall meet, at a minimum, on a monthly basis to review the status of the project. If the overall status of the project is Red or Yellow, more frequent meetings may be required.

3.6.3 Roles and responsibilities

Executive business manager responsibilities include the following:

- Attend program steering committee meetings and provide business guidance, direction, and advice to the program managers to ensure the program delivers maximum overall business benefits to the company.
- Allocate approved business funding for each project phase within the program.
- Approve all business changes to the approved program management plan.
- Provide business resources as identified in program management plan.
- Act as the overall program sponsor and chairman of the program steering committee.

Executive IT manager responsibilities include the following:

- Attend program steering committee meetings and provide IT support guidance, direction, and advice to the program manager to ensure the program delivers maximum overall business benefits to the company.
- Allocate approved IT funding for each project phase within the program.
- Approve all IT support changes to the approved program management plan.
- Provide IT resources as identified in program management plan.
- Act as the program sponsor from the IT perspective.

Executive project management manager responsibilities include the following:

- Attend program steering committee meetings and provide project management guidance, direction, and advice to the program managers to ensure the program delivers overall business benefits to the company.
- Integrate approved business and IT funding for each project phase within the program.
- Integrate all business and IT support changes to the approved program management plan.
- Recommend PM resources as identified in program management plan.
- Act as the program sponsor with integrated business and IT responsibility and PM accountability; champion PM processes, tools, and techniques.

Program business manager responsibilities include the following:

- Attend program steering committee meetings and submit details on business guidance, direction, and advice to the program delivery manager to ensure the program delivers maximum overall business benefits to the company.
- Report approved business funding to the program steering committee for each phase of the project within the program management plan
- Report on all business changes to the approved project plan.
- Assign business resources for business support and project benefit estimates.
- Report to the program steering committee with direct accountability to the business executive, and keep the program working committee informed of the current direction of the projects.

Program IT manager responsibilities include the following:

- Attend program steering committee meetings and submit details on IT guidance, direction, and advice to the program delivery manager to ensure the program delivers maximum overall business benefits to the company.
- Report approved IT support funding to the program steering committee for each phase of the project within the program management plan.
- Report on all IT changes to the approved project plans.
- Assign IT resources for IT support.
- Report to the program steering committee with direct accountability to IT executive, and keep the program working committee informed of the current direction of the projects.

Program delivery manager responsibilities include the following:

- Attend program steering committee meetings and submit details on integrated business and IT guidance, direction, and advice to the project managers to ensure the program delivers maximum overall business benefits to the company.
- Report on integrated approved business and IT funding to the program steering committee for each phase of the project within the program management plan.
- Report on integrated business and IT changes to the approved project plan.
- Provide integrated business and IT resource plan for business, IT, and PM support.
- Report to the program steering committee with direct accountability to the PM executive, and keep the program working committee informed of the current direction of the projects.

Project team (project manager and team) responsibilities include the following:

- Attend program steering committee meetings on a biweekly basis, and keep the committee informed of the progress of the projects.
- Submit status reports for IT and business initiatives, document critical issues from issues log, and present change requests for decisions by the committee.
- Develop deliverables and submit them to the program steering and working committees for approval.
- Manage the change requests and issue resolution processes.
- Develop and maintain a strategy for management and delivery of the approved project plan, based on the PMBOK extended knowledge areas:
  - Scope management;
  - Time management;
  - Cost management;
  - Quality management;
  - Risk management;
  - Change management;
  - Issue management;
  - Contract management;
  - Communications management;
  - Human resource management.
- Integrate, assemble, and report to the program working committee on project progress with direct accountability to the program delivery manager.
- Allocate resources to business and IT projects based on the projects' progress.

3.6.4 Procedures: program steering and working committee

The roles and responsibilities discussed above are brief because they are intended to serve as a reference as to who does what during the establishment of the program steering and working committee. Procedures are now introduced to demonstrate how this program steering and working committee is established using the major standard deliverables (what), process flow (how), and checklist or measurement criteria (why) template. The process is intended to serve as a reference based on the English-language interrogatives: who does what, how, and why. The deliverables, process flow, and checklist templates are based on real-world practical implementation and will serve as excellent references for practicing project managers during establishment of the program steering and working committee.

The deliverables template in Table 3.11 highlights the key structure and responsibilities of the program steering and working committee. The model or guideline in Table 3.12 is based on real-world practical implementations and will provide useful references for project managers. The program steering and working committee model in Table 3.12 shows the structure of the stakeholders having approval responsibilities.

The process flow template in Figure 3.8 highlights the major processes to support the establishment of the program steering and working committee. These processes show how responsibilities are approved for the two committees and the integrated nature and dependencies of these committees.

The checklist template in Table 3.13 highlights the major measurement criteria to ensure that the program steering and working committee consists of skilled, knowledgeable, and experienced management staff, as described in the deliverables template, and adheres to the process flow identified in the process flow template. The checklist is used to ensure that the right staff resources are assigned to the program steering and working committees.

Table 3.11 Deliverables Template: Program Steering and Working Committee Report

Section 1—Introduction
This section includes an executive summary and program steering and working committee structures and introduces the report.
Section 2—Program Steering Committee Structure
This section provides the steering committee roles and responsibilities for the overall program. These approved roles and responsibilities establish the initial communications baseline for effective execution of the communications management plan.
Section 3—Program Working Committee Structure
This section provides the working committee roles and responsibilities for the overall program. These approved roles and responsibilities establish the initial communications baseline for effective execution of the communications management plan.
Appendixes—Project Steering and Working Committee Skills
Appendixes provide further funding details on the skills, knowledge, qualifications, and experiences of the representatives on these committees. The skills, knowledge, qualifications, and experiences of the project manager and the project team for each of the projects within the program, are detailed in this report.

Table 3.12 Program Steering and Working Committee Model

Program Steering Committee
• Executive business manager: chairman at VP level
• Executive IT manager: VP level
• Executive project management manager: VP level
• Program business manager: director level
• Program IT manager: director level
• Program delivery manager: director level
↑ ↓
Program Working Committee
• Program delivery manager: Chairman at director level
• Program business manager: Business manager at director level
• Program IT manager: IT manager at director level
• Project managers: Managers with project management expertise
• Project team: IT and project management staff
• External stakeholders: Other influential and interested personnel

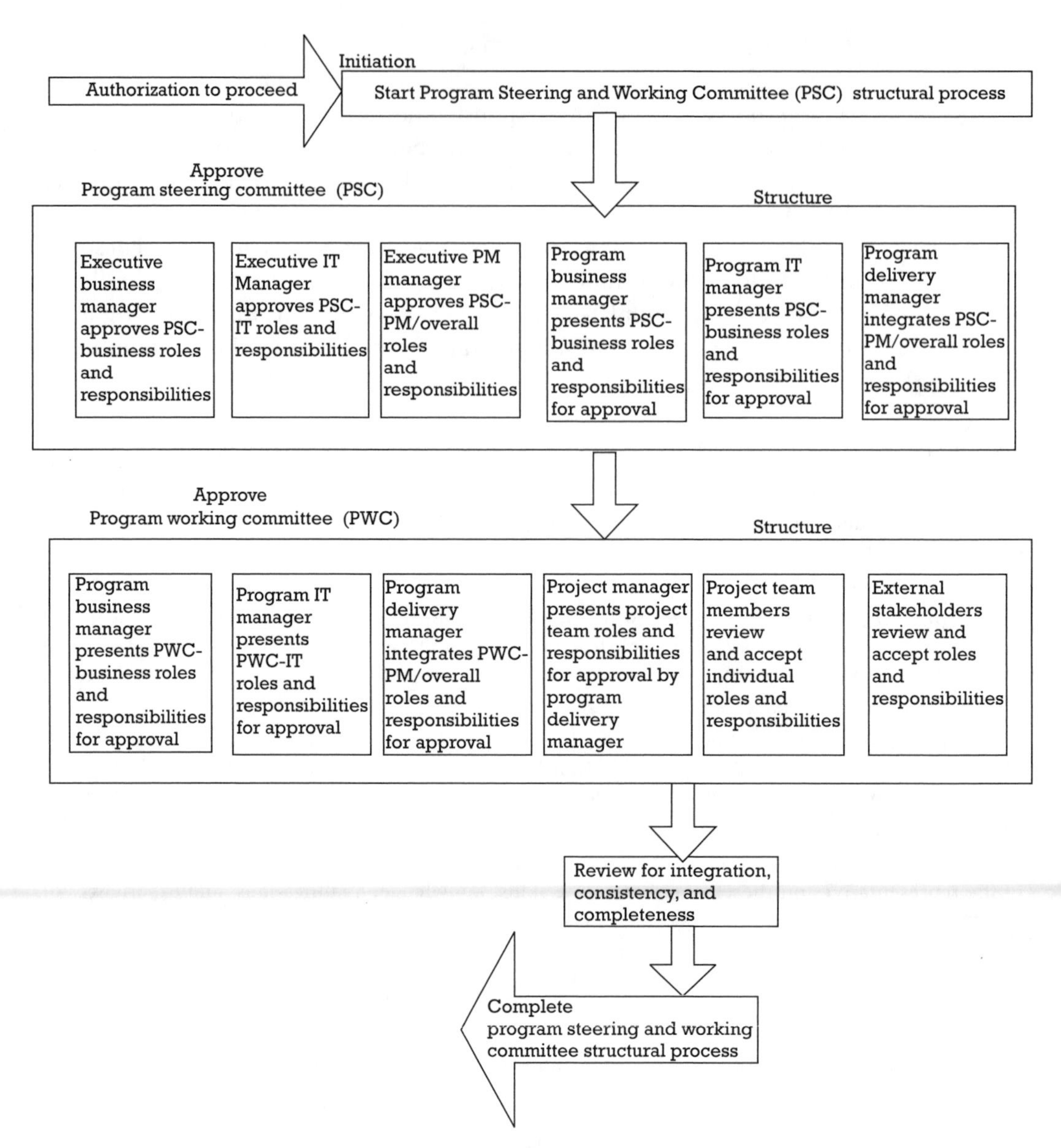

Figure 3.8 Program steering committee process—how.

3.7 Business initiatives support

Executive business management shall establish business initiatives[12] support strategies with the primary objective of this function being to manage and track the business projects, coordinate the development of business processes, integrate business initiatives with IT initiatives in order to

12. Business initiative: A business requirement or issue that needs resolution to improve the performance of the business.

Table 3.13 Program Steering and Working Committee Checklist

Program Steering and Working Committee Criteria	*Yes*	*No*	*Program Steering and Working Committee Criteria*	*Yes*	*No*
Initiation			*Program steering committee*		
Is management committed?	❑	❑	Did the executive business manager approve business roles and responsibilities	❑	❑
Are a business sponsor and steering committee established?	❑	❑	Did the executive IT manager approve IT roles and responsibilities?	❑	❑
Are there roles and responsibilities guidelines available?	❑	❑	Did the executive PM manager approve overall/PM roles and responsibilities?	❑	❑
Are checklists for approving roles and responsibilities understood?	❑	❑	Did program business manager present business roles and responsibilities for approval?	❑	❑
Has a program delivery manager been assigned?	❑	❑	Did the program IT manager present IT roles and responsibilities for approval?	❑	❑
Does the assigned program delivery manager understand the PSC structural processes?	❑	❑	Did the program delivery manager integrate overall/PM roles and responsibilities for approval?	❑	❑
Program working committee			*Appendixes: Skills Matrix*		
Did the program business manager present business roles and responsibilities for approval?	❑	❑	Is there a skills matrix for executive management?	❑	❑
Did the program IT manager present IT roles and responsibilities for approval?	❑	❑	Is there a skills matrix for program management?	❑	❑
Did the program delivery manager integrate overall/PM roles and responsibilities for approval?	❑	❑	Is there a skills matrix for each of the project managers?	❑	❑
Did the program delivery manager approve the project manager roles and responsibilities?	❑	❑	Is there a skills matrix for each individual on the project team?	❑	❑
Did the program delivery manager approve the project team roles and responsibilities?	❑	❑			
Did the program delivery manager approve external stakeholders' roles and responsibilities?	❑	❑			

optimize resource allocations and to minimize duplicate processes and efforts. The program steering committee will have approval responsibilities for both the business and IT initiatives and have joint accountability with the project managers for the overall integration of these initiatives to optimize the scope, effort, costs, and schedule performance of these projects.

The BIS strategies can be implemented through the establishment and delivery of a BIS plan,[13] consisting of the following components:

- Business support initiatives;
- Business support plan—WBS; dependency diagram; business support initiatives schedule; business organizational chart; business staffing profile; business support budget; consultant payment schedule;
- Business support transition plan;
- Business support communications plan;
- Business support risk management plan;
- Business support resource management plan.

3.7.1 Purpose

The purpose of this process is to manage and track the business support initiatives, coordinate the development of business processes, maintain integration of business initiatives with IT initiatives as determined in the BSA, and optimize resource allocations and minimize duplicate processes and efforts.

3.7.2 Policy

The policy of the business initiatives report is the following:

- The program steering committee shall provide project approval responsibilities.
- The BIS plan shall be integrated with the overall integrated program management plan.
- The program steering committee shall approve all changes to the BIS plan. The project manager shall document and present all scope changes to the program steering committee for approval.
- The program steering committee will meet, at a minimum, on a monthly basis to review the status of the business initiatives projects. If the overall status of the project is Red or Yellow, more frequent meetings may be required.
- BIS shall be managed using the business enhancement processes[14] and the procedures for tier-1 (production support–business focus), tier-2 (new development support–technical focus) and tier-3 (strategic support–strategic focus).

13. BIS plan: A plan that provides the procedures (how), cost (how much), organizational structure (who), and schedule (when) to manage the delivery of the business support initiatives. It is integrated with the program delivery plan.

14. Business enhancement processes: Processes and procedures to support tier-1 (production support–business focus), tier-2 (new development support–technical focus) and tier-3 (strategic support–strategic focus) enhancement requests.

- Business managers shall manage the priority of the tier-1, tier-2, and tier-3 requests based on business needs and integration with the BSA (business, data, applications, and technology).

3.7.3 Roles and responsibilities

Executive business manager responsibilities include the following:

- Provide business management guidance, direction, and advice to the program managers to ensure the business initiatives integrate within the overall program management plan to ensure effective utilization of business resources.
- Approve the business support initiatives within the overall program management plan.
- Approve the business support initiatives plan, if required, by the business program manager in managing the business support initiatives.
- Approve the business support initiatives schedule, if required, in managing the business support initiatives.
- Act as the overall program sponsor and chairman of the program steering committee.

Executive IT manager responsibilities include the following:

- Provide IT management guidance, direction, and advice to the program managers to ensure the IT initiatives integrate within the overall program management plan to ensure effective utilization of IT resources.
- Approve the IT support initiatives within the overall program management plan.
- Approve the IT support initiatives plan, if required, by the IT program manager in managing the IT support initiatives.
- Approve the IT support initiatives schedule, if required, in managing the IT support initiatives.
- Act as the program sponsor from the IT perspective.

Executive project management manager responsibilities include the following:

- Provide project management guidance, direction, and advice to the program delivery manager to ensure the business and IT initiatives integrates within the overall program management plan to ensure effective utilization of business and IT resources.
- Integrate the business and IT support initiatives within the program.
- Approve the structure of the integrated program plan by the program delivery manager in managing the approved program management plan.

- Approve the structure of the integrated program schedule as identified in the program management plan.
- Act as the program sponsor with integrated business and IT responsibility and PM accountability; champion PM processes, tools, and techniques.

Program business manager responsibilities include the following:

- Attend program working committee meetings and make business-related decisions.
- Manage the business support initiatives.
- Develop and maintain the following components of the project plan for business support initiatives, if required:
  - Business support initiatives;
  - Business support delivery plan—WBS; dependency diagram; business support initiatives schedule; business organizational chart; business staffing profile; business support budget; contractor payment schedule;
  - Business support transition plan;
  - Business support communications plan;
  - Business support risk management plan;
  - Business support resource management plan.
- Develop a schedule for business support initiatives and maintain linkage with integrated project schedule to support the BSA.
- Report to the program steering committee with direct accountability to the business executive, and keep project managers and business functional managers informed of the current business direction of the projects.

Program IT manager responsibilities include the following:

- Attend program working committee meetings and make project-related IT decisions.
- Manage the IT support initiatives.
- Develop and maintain the following components of the project plan for IT support initiatives, if required:
  - IT support initiatives;
  - IT support plan—WBS; dependency diagram; IT support initiatives schedule; IT organizational chart; IT staffing profile; IT support budget; consultant payment schedule;
  - IT support transition plan;
  - IT support communications plan;
  - IT support risk management plan;
  - IT support resource management plan.
- Develop schedule for IT support initiatives and maintain linkage with integrated project schedule to support the BSA.

- Report to the program steering committee with direct accountability to the IT executive, and keep project managers and IT functional managers informed of the current IT direction of the projects.

Program delivery manager responsibilities include the following:

- Chair program working committee meetings and make project-related business and IT decisions.
- Manage the overall integrated program (business and IT support initiatives).
- Develop and maintain the following components of the integrated program plan for BSA:
  - Project charters;
  - Integrated business project plan—integrated WBS; integrated dependency diagram; integrated projects schedule; business organizational chart; business staffing profile; projects budget; contractor payment schedule;
  - Integrated transition plan (business and IT);
  - Integrated communications plan;
  - Integrated risk management plan;
  - Integrated resource management plan.
- Develop an integrated program schedule for IT and business support initiatives, and maintain integrated representation of IT projects to support the BSA.
- Allocate resources to business support initiatives.
- Report to the program steering committee with direct accountability to the PM executive and keep project managers informed of the current direction of the projects.

Project team (project manager and team) responsibilities include the following:

- Attend program working committee meetings on a biweekly basis, and keep the committee informed of the progress of the business support initiatives.
- Submit status reports for business support initiatives, document critical issues from issues log, and present change requests for decision by the committee.
- Develop BIS deliverables and submit them to the program steering and working committee for approval.
- Manage the BIS change requests and issues resolution processes.
- Develop and maintain a strategy for management, integration, and delivery of the approved BIS plan.
- Integrate, assemble, and report to the program working committee on business support initiatives with direct accountability to the program delivery manager.

- Assign resources to the development of business support initiatives.

3.7.4 Procedures: BIS

The roles and responsibilities discussed above are brief because they are intended to serve as a reference as to who does what during the development and support of business initiatives. Procedures are now introduced to demonstrate how this BIS process is executed using the major standard deliverables (what), process flow (how), and checklist or measurement criteria (why) templates. The process is intended to serve as a reference, based on the English-language interrogatives: who does what, how, and why. The deliverables, process flow, and checklist templates are based on real-world practical implementation and will serve as excellent references for practicing project managers during the execution of the BIS process.

The deliverables template in Table 3.14 highlights the key deliverables that the practicing project manager manages during the development and support of business initiatives. The model or guideline in Table 3.15 is based on real-world practical implementations and will provide useful references for project managers. The BIS model in Table 3.15 shows how business initiatives requests are managed.

The process flow template in Figure 3.9 highlights the major processes to support delivery of the key business initiatives. These processes show how business initiatives are supported, managed, and integrated with the BSA and integrated program plan.

The checklist template in Table 3.16 highlights the major measurement criteria to ensure delivery of quality deliverables identified in the deliverables template and adherence to the process flows identified in the process

Table 3.14 Deliverables Template: Business Initiatives Support Report

Section 1—Introduction
This section includes an executive summary, describes the approach, and introduces the report.
Section 2—Initiatives
This section highlights the components of the business support initiatives and alignment with the associated IT support strategies. These deliverables establishes the initial baseline initiatives for further developments.
Section 3—Architectures
This section highlights the updates to the BSA and IT architectures—data, applications, and technology. These updates show the alignment with the initiatives documented in Section 2 and establish the initial architecture baseline for further developments to the implementation and deployment plans.
Section 4—Plans
This section highlights the components of the BIS implementation plan and shows the alignment or integration with the existing program implementation and deployment plans. The BIS implementation plan shows the alignment with the initiatives documented in Section 2 and the architectures documented in Section 3 and establishes the initial business support program baseline for managing the business support initiatives.
Appendixes—Initiatives, Architectures, and Plans
Appendixes provide further details on the initiatives, architecture updates, and plans.

Table 3.15 Deliverables Template: Business Initiatives Support Model

Business Initiatives Support Requests
Section 1—Enhancement Requests
Request ID: Request issue date:
Requestor name: Request implementation date:
Requestor area: Request approval date:
Request reason: ❑ IT problem ❑ Business problem ❑ PM problem
❑ Performance ❑ Maintenance ❑ New ❑ Procedure
Request severity: ❑ High ❑ Medium ❑ Low
Request change type: ❑ Business ❑ Applications ❑ Data
❑ Policy ❑ Process ❑ Policy ❑ Process ❑ Policy ❑ Process
Request change description:
Business opportunity statement:
Section 2—Enhancement Assessments
Estimated business time/effort to implement change:
Estimated IT time/effort to implement change:
Business priority ranking: ❑ 1-Urgent ❑ 2-High ❑ 3-Medium ❑ 4-Low
Business testing level: ❑ Business testers ❑ IT testers
IT phases affected: ❑ Definition ❑ Analysis ❑ Design ❑ Iterative development
Support level: ❑ Tier-1 ❑ Tier-2 ❑ Tier-3
Section 3—Risk Assessments
Impact on business: ❑ 1-High ❑ 2-Medium ❑ 3-Low
Impact on technology: ❑ 1-High ❑ 2-Medium ❑ 3-Low
Impact on people: ❑? 1-High ❑? 2-Medium ❑? 3-Low
Impact on processes: ❑? 1-High ❑? 2-Medium ❑? 3-Low
Impact on organization: ❑ 1-High ❑ 2-Medium ❑ 3-Low

flow template. The checklist is used to objectively determine success criteria during the development and support of business initiatives.

3.8 Summary

The business management model presented in this chapter describes the component processes of business management. The purpose provides a statement of the derived result, and the policy provides various rules stating the minimum requirements for achieving the purpose statement for each of the business management component processes. The roles and responsibilities of executive management, program management, and project delivery (the project team) are presented to demonstrate the integrated nature of these processes in managing and delivering multiple IT projects. The key business management deliverables, process flow, and checklist templates provide some excellent real-world implementation guidelines, which can be referenced and applied to business management processes.

Business organizations that are in the process of managing and delivering multiple IT projects with the goal of integrated or enterprise project

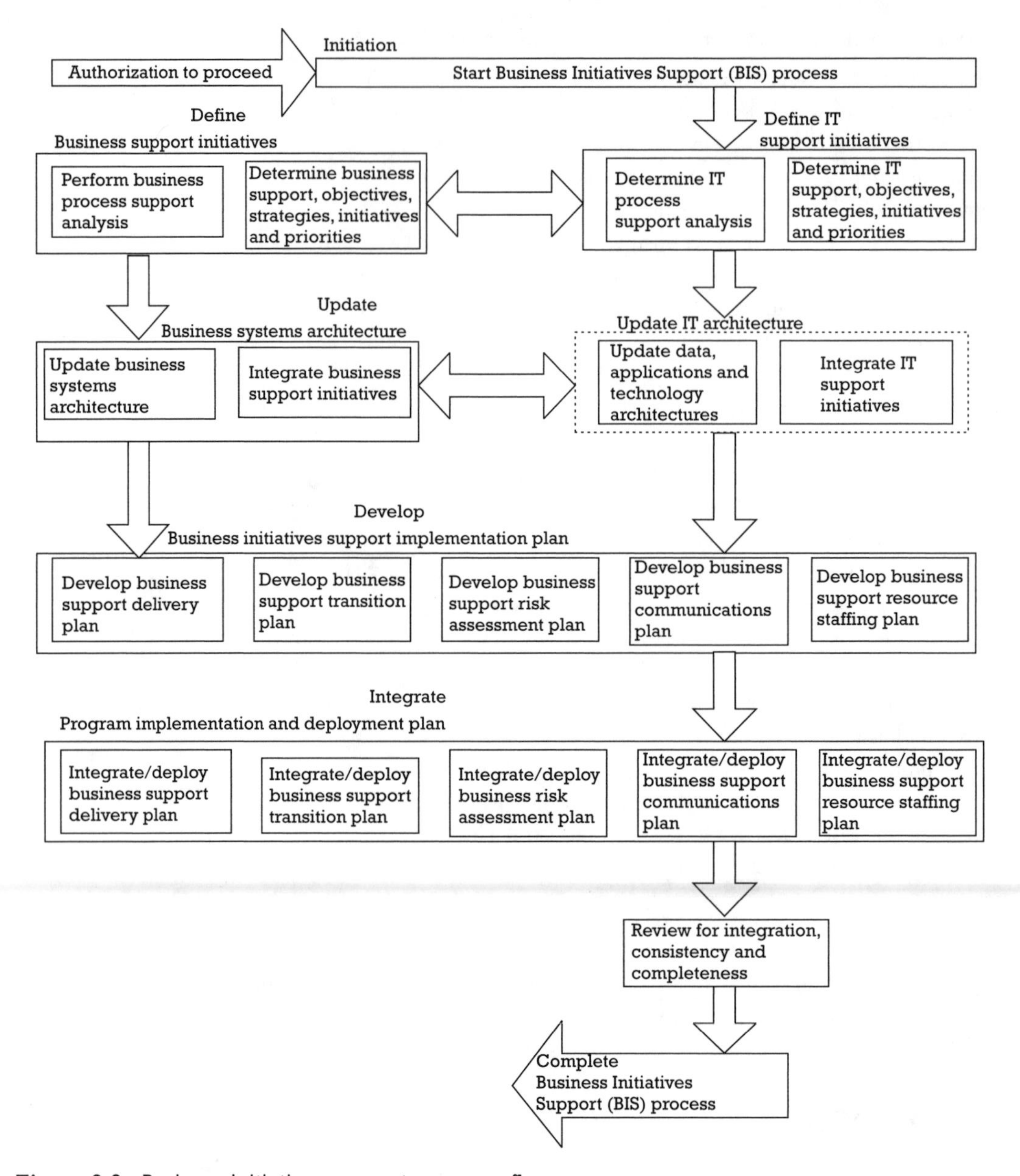

Figure 3.9 Business initiatives support process—flow.

management should consider the following business management recommendations as a framework to guide them towards successful management and delivery of multiple IT projects:

- Establish a BSA framework that will guide the prioritization and delivery of business and IT initiatives to effectively support the goals and objectives of the company's business.
- Determine the value of the project to the business based on improved performance of the company's business for the six areas of value, the project value justification.

Table 3.16 Business Initiatives Support Checklist

BIS Criteria	*Yes*	*No*	*BIS Criteria*	*Yes*	*No*
Initiation			*Initiatives*		
Is management committed?	❑	❑	Is there a list of prioritized business and IT initiatives?	❑	❑
Are a business sponsor and steering committee established?	❑	❑	Are deliverables identified?	❑	❑
Is there a project charter/plan or PMBOK guideline?	❑	❑	Are deliverables, activities, resources, schedule, and costs included in project schedule?	❑	❑
Is there a project charter/plan for business support initiatives?	❑	❑	Are deliverables approval criteria established?	❑	❑
Has a program delivery manager been assigned?	❑	❑	Is there a communications plan?	❑	❑
Does the program delivery manager understand the business support initiative process?	❑	❑	Is the alignment with business and IT architectures established?	❑	❑
BSA and IT architecture updates			*BIS implementation plan*		
Is there an existing business and IT architecture?	❑	❑	Is there a business support delivery plan?	❑	❑
Is the existing business architecture updated?	❑	❑	Is there a business support transition plan?	❑	❑
Is the existing IT architecture updated?	❑	❑	Is there a business support communications plan?	❑	❑
Are business support initiatives aligned with business architecture?	❑	❑	Is there a business support risk assessment plan?	❑	❑
Are the IT support initiatives integrated with IT architecture?	❑	❑	Is there a business support resource-staffing plan?	❑	❑
Is the integration with the BIS implementation plan understood?	❑	❑	Is the alignment with the existing program implementation plan understood?	❑	❑
Program implementation and deployments plans			*Appendixes: Initiatives matrix*		
Is the business support delivery plan integrated?	❑	❑	Is there an integrated matrix that shows the business support initiatives and alignment with the BIS implementation plan, existing architectures, and the existing implementation and deployment plans?	❑	❑
Is the business support transition plan integrated?	❑	❑			
Is the business support communications plan integrated?	❑	❑			
Is the business support risk assessment plan integrated?	❑	❑			
Is the resource-staffing plan integrated?	❑	❑			
Is the integration with existing delivery projects understood?	❑	❑			

- Control the allocation of funds to projects that have been prioritized by the program steering committee meetings and allocate funds in increments by revalidating the project cost and benefits for phased deliverables.

- Obtain commitments or noncommitments from executive and senior management for project funding and deliverables approval at key decision points prior to proceeding to the next phase of development.
- Establish two separate committees—the program steering committee meetings and the program working committee—to provide project approval responsibilities and accountability for the overall scope, effort, cost, schedule, and quality performance of the program.
- Manage and track the business support initiatives, coordinate the development of business processes, maintain integration of business initiatives with IT initiatives as determined in the BSA, optimize business and IT resource allocations, and minimize duplicate processes and efforts.

The main focus of this chapter is to provide readers with further details on the business management component processes of the IPM framework model. By now, readers should have gained a fairly good understanding of the integrated nature of these processes. Chapter 6 includes further discussions to show the horizontal integration of these business management processes based on an iterative and incremental IT PDLC model. Business managers, especially those with keen interests in business process improvements, will appreciate the value of these horizontal and vertical integration processes in better preparing them to understanding project developments in this dynamically changing IT industry.

3.9 Questions to think about: management perspectives

1. Think about how your organization identifies, prioritizes, and manages multiple IT projects. How does your organization manage conflicting business and IT priorities? What are the key rationales for determining the value of projects? What are the major components of business management? How do these components relate to your project environment? What is the perception of senior management of the need for business and IT architectures to guide the delivery of projects?
2. Think about how your organization justifies projects. What are the six value areas for project justification? How do these six value areas relate to your project justification approach?
3. Think about how your organization allocates funds to projects. How are reserved and approved funds allocated? How does this funds allocation process relate to your funds allocation approach?
4. Think about how your organization approves funds to projects. How does your organization approve deliverables? What is the funding/deliverables approval process discussed in this book? How does this process relate to your funds approval approach?

5. Think about how your organization establishes project organization structures. How do personal greed for power, politics, and personal financial interests affect the organization structure? What is the project organizational structure recommended in this book? How does this recommendation relate to your approach?
6. Think about how your organization manages business support initiatives. How do these business initiatives impact projects' development? How are these business support initiatives integrated into the project environment as discussed in this book? How does this process relate to your approach to managing business support initiatives?

Selected bibliography

Business Process Modeling Language (BPMI Initiatives), 2001, http://www.bpmi.org.

Ericksson, H.-E., and M. Parker, *Business Modeling with UML*, New York: John Wiley & Sons, 1999.

Jacobson, I., M. Ericsson, and A. Jacobson, *The Object Advantage: Business Process Engineering with Object Technology*, Reading, MA: Addison-Wesley, 1994.

Royce, W., *Software Project Management—A Unified Framework*, Reading, MA: Addison-Wesley, 1998.

CHAPTER

4

Contents

Project Management Model

4.1 Project management

The function of project management is to ensure that projects are delivered in a continuous stepwise progression, from concept when an initial need is expressed, to sign off by the user representatives/sponsor and hand over to production support. Every project has phases of development, and a clear understanding of these phases will permit managers to better control the resources to achieve the expected result. Project management consists of the following processes:

- IT PDLC;
- IT project management delivery processes;
  - PSM;
  - PTM;
  - Project cost management;
  - PQM;
  - PRM;
  - Project contract management;
  - PCM;
  - PIM;
  - Project change management;
  - Project human resources management.
- PMO infrastructure support.

Figure 4.1 is a graphical representation of the modeling concept that is used to demonstrate the structure and relationships of the definitions or processing components of project management from a project management perspective.

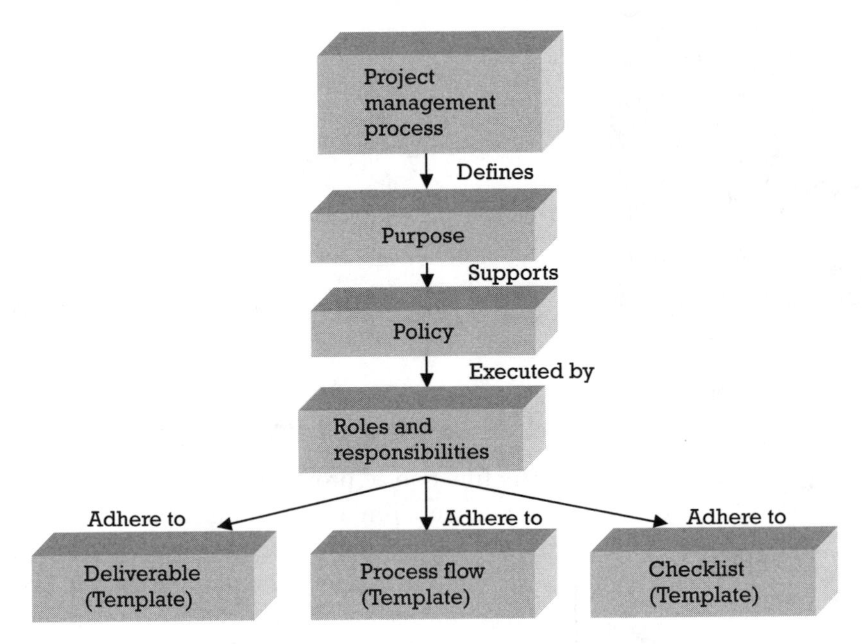

Figure 4.1 Project management model.

4.2 IT PDLC

Even the longest journey begins with a single step.
—Chinese proverb

Executive management shall use the phases of the IT PDLC to control the funding of the projects. Funds are approved or disapproved at the completion of each phased deliverable, having met or not met the established success criteria, using approved checklists as the measurement criteria to make go or no-go decisions. The IT PDLC presented in this book focuses on an iterative and incremental approach, consisting of the following phases:

- *Enterprise or PDP:* The major deliverable is a PDM[1] which includes an integrated project plan.
- *RAP:* The major deliverable is a RAM,[2] which includes a demonstration prototype.
- *ADP:* The major deliverable is a PAS,[3] which includes a working or evolutionary prototype.

1. Project definition model (PDM): The major deliverable produced from the PDP. It is an integrated representation of business management, IT management, and project management deliverables.
2. Requirements analysis model (RAM): The major deliverable produced from the project RAP. It is an integrated representation of business management, IT management, and project management deliverables and a demonstration prototype
3. Project architecture solution (PAS): The major deliverable produced from the project design/architecture phase. It is an integrated representation of business management, IT management, and project management deliverables and a working or evolutionary prototype.

- *IDPs:* The major deliverables are various IDSs,[4] which include iterative development prototypes during each of the iterative development subphases:
  - *Construction:* The major deliverable is Iteration #1 and supporting documentation—models, reports, and plans.
  - *Integration:* The major deliverable is Iteration #2 and supporting documentation—models, reports, and plans.
  - *Deployment:* The major deliverable is Iteration #3 and supporting documentation—models, reports, and plans.

The intent of this section is not to provide detailed processes or procedures on the implementation of any specific software development process or similar applications development process. There are many excellent books on software development processes, concepts, theory, and applicability, some of which are mentioned in the selected bibliography at the end of this chapter. The main objective of this section is to provide an overview of the contents (what), purpose (why), roles and responsibilities (who, what), and procedures (how) of the components of an iterative and incremental IT PDLC process and to demonstrate how this process fits within the context of the overall IPM-IT framework. The procedure sections of the IT PDLC process provide real-world baseline templates of the deliverables (what), process flows (how), and checklists (measurement criteria) that practicing project managers can readily retrofit and apply during the management, delivery, execution, and integration of the IT PDLC phases.

The materials presented in the following sections are organized to support the structure represented in Figure 4.1. This model structure provides a logical and consistent flow of information to enable the reader to gain a better understanding and appreciation for the integrated nature of IT project management as presented in this book.

4.2.1 Purpose

The purpose of this process is to establish an IT PDLC model consisting of well-defined phases for the orderly management, approval, control, and delivery of IT projects. This model will form the baseline to manage the scope, cost, time, effort, and quality objectives of the development projects. The program steering committee will use these phases as critical milestones during the approval or disapproval of funds.

4.2.2 Policy

The policy of the IT project delivery cycle is the following:

4. Iterative development solution: The major deliverable produced from the project IDPs—construction, integration, and deployment. It is an integrated representation of business management, IT management, and project management deliverables, and major iterative solutions–iterations for construction, integration, and deployment subphases.

- Projects shall be managed and delivered based on the IT PDLC.
- Projects shall be sponsored by an executive steering committee member, adequately defined and justified, and funded by the program steering committee.
- The PDM, which includes a high-level BRS and an integrated project plan, is the proposal to the program steering committee to determine whether the funding for the project shall be reserved for eventual approval or disapproval.
- Funds shall be approved and released by the program steering committee to the project team in increments, based on the deliverables of the previous phase and detailed project plans for the next phase, according to the IT PDLC.

4.2.3 Roles and responsibilities

Executive business manager responsibilities include the following:

- Provide executive-level business support to the program business manager to ensure that the phases of the IT PDLC are enforced to better manage and control the project schedules, costs, and resources to achieve the expected result.
- Approve and release business funding for each phase of the IT PDLC.
- Approve all business changes to the approved integrated project plan for each phase of the IT PDLC.
- Approve business support resources as identified in the integrated project plan for each phase of the IT PDLC.
- Act as the overall program sponsor and chairman of the program steering committee.

Executive IT manager responsibilities include the following:

- Provide executive-level IT support to the program IT manager to ensure that the phases of the IT PDLC are enforced to control the project schedules, costs, and resources to achieve the expected result.
- Approve IT support funding for each phase of the IT PDLC.
- Approve all IT support changes to the integrated project plan for each phase of the IT PDLC.
- Approve IT support resources as identified in the integrated project plan for each phase of the IT PDLC.
- Act as the program sponsor from the IT perspective.

Executive project management manager responsibilities include the following:

- Provide executive-level PM support to the program delivery manager to ensure that the phases of the IT PDLC are enforced in order to control the project schedules, costs, and resources to achieve the expected results.

- Integrate approved business and IT funds for each phase of IT PDLC.
- Integrate all business and IT support changes in the approved integrated project plan for each phase of the IT PDLC.
- Recommend PM resources as identified in the integrated project plan for each phase of the IT PDLC.
- Act as the program sponsor with integrated business and IT responsibility and PM accountability; champion PM processes, tools, and techniques.

Program business manager responsibilities include the following:

- Submit details on the business processes and phases of the IT PDLC to the program delivery manager to manage and control the project schedules, costs, and resources to achieve the expected results.
- Report on the progress of approved business funds to the program steering committee for each phase of the IT PDLC.
- Report on all business changes to the approved integrated project plan for each phase of the IT PDLC.
- Assign business resources to business support activities for each phase of the IT PDLC.
- Report to the program steering committee with direct accountability to the business executive, and keep the program working committee informed of the current direction of the projects.

Program IT manager responsibilities include the following:

- Submit details on IT processes and phases of the IT PDLC to the program delivery manager to control the project schedules, costs, and resources to achieve the expected results.
- Report on the progress of approved IT funds to the program steering committee for each phase of the IT PDLC.
- Report on all IT changes to the approved integrated project plan for each phase of the IT PDLC.
- Assign IT resources to IT support activities for each phase of the IT PDLC.
- Report to the program steering committee with direct accountability to the IT executive, and keep the program working committee informed of the current direction of the projects.

Program delivery manager responsibilities include the following:

- Submit details on project management processes and phases of the IT PDLC to the project managers to control the project schedules, costs, and resources to achieve the expected result.
- Report on the progress of integrated business and IT funds to the program steering committee for each phase of the IT PDLC.
- Report on integrated business and IT changes to the approved integrated project plan for each phase of the IT PDLC.

- Provide integrated business and IT resource plan for business, IT, and project management support activities.
- Report to the program steering committee with direct accountability to the PM executive, and keep project managers informed of the current direction of the projects.

Project team (project manager and team) responsibilities include the following:

- Project managers to manage the projects according to the company's IT PDLC methodology;
- Project managers to develop and maintain detailed project plans for projects, to include the following:
  - Project charter;
  - Project implementation plan[5]—master WBS; project dependency diagram; project schedule; project organizational chart; project staffing profile; project budget; project payment schedule;
  - Business and IT transition plan;
  - Project communications plan;
  - PRM management plan;
  - Project issues and change management plan;
  - Project contract management plan;
  - PQM management plan;
  - Project human resource management plan.
- Project managers to integrate, assemble, and report to the program working committee on project status, issues, change requests, and risks, with direct accountability to the program delivery manager;
- Project team to develop phased deliverables according to the company's IT PDLC methodology.

4.2.4 Procedures: IT PDLC

No one was ever lost on a straight road.
—Indian proverb

The roles and responsibilities discussed in Section 4.2.3 are brief because they are intended to serve as a reference as to who does what during the execution of the IT PDLC. Procedures are now introduced to demonstrate how this IT PDLC process is executed using the major standard deliverables

5. Project implementation plan: A plan that specifies project implementation components (what), as opposed to project management components (how and when). It includes details such as WBS, project dependency diagrams, schedule, organizational chart, staffing requirements and plan, budget, payment schedule, risks, contracts, and the like.

(what), process flow (how), and checklist or measurement criteria (why) templates. The process is intended to serve as a reference based on the English-language interrogatives: who does what, how, and why. The deliverables, process flow, and checklist templates, based on real-world practical implementation, will serve as excellent references for practicing project managers during execution of the IT PDLC phases.

The deliverables template in Table 4.1 highlights the key deliverables that the practicing project manager manages during the execution of the IT PDLC. The model or guideline in Figure 4.2, based on real-world practical implementations, will provide useful references for project managers. The IT PDLC model shows the foundation representation to effectively communicate the integrated, iterative, and incremental representation of this IT PDLC to business, IT, and project management staff.

The process flow template in Figure 4.3 highlights the major processes to produce the key deliverables during the IT PDLC. These processes produce the deliverables for each phase to show the integrated, iterative, and incremental nature and dependencies of the phases and the deliverables within the IT PDLC process.

The checklist template in Table 4.2 highlights the major measurement criteria to ensure delivery of quality deliverables identified in the deliverables template and adherence to the process flows identified in the process flow template. The checklist is used to objectively determine success criteria during execution of the IT PDLC.

Table 4.1 Deliverables Template: IT PDLC Report

Section 1—Introduction
This section includes an executive summary and phased approach and introduces the report.
Section 2—Enterprise or Project Definition Phase
This section highlights the major deliverables—models, reports, and plans—produced during this phase. These deliverables show the alignment with the BSA and establish the initial baseline for further developments.
Section 3—Requirements Analysis Phase
This section highlights the major deliverables—demonstration prototypes, models, reports, and plans—produced during this phase. These deliverables show the alignment with the deliverables documented in Section 2 and establish the initial baseline for further developments.
Section 4—Architecture Design Phase
This section highlights the major deliverables—working evolutionary prototypes, models, reports, and plans—produced during this phase. These deliverables show the alignment with the deliverables documented in Section 3 and establish the initial baseline for further developments.
Section 5—IDPs
This section highlights the major deliverables—development iterations, models, reports, and plans—produced during the construction, integration, and deployment subphases. These deliverables show the alignment with the deliverables documented in Section 4 and establish the initial baseline for further iterative developments and deployments.
Appendixes—Models, Architectures, and Plans
Appendixes provide further details on the prototypes, iterations, models, architectures, reports, and plans.

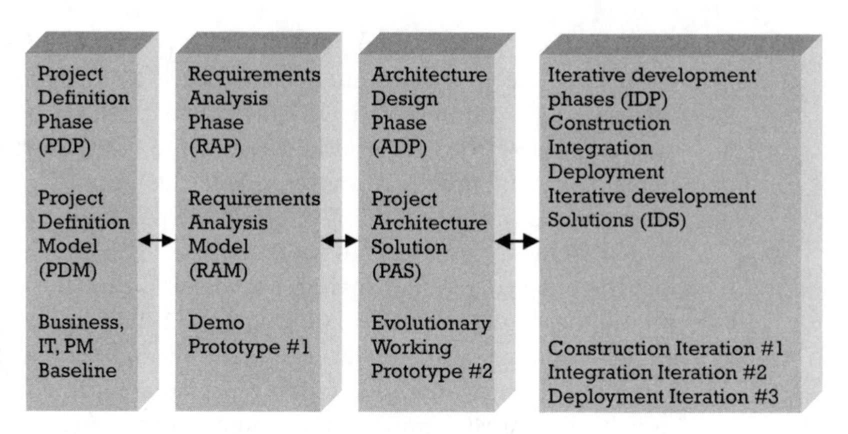

Figure 4.2 IT project delivery life-cycle model.

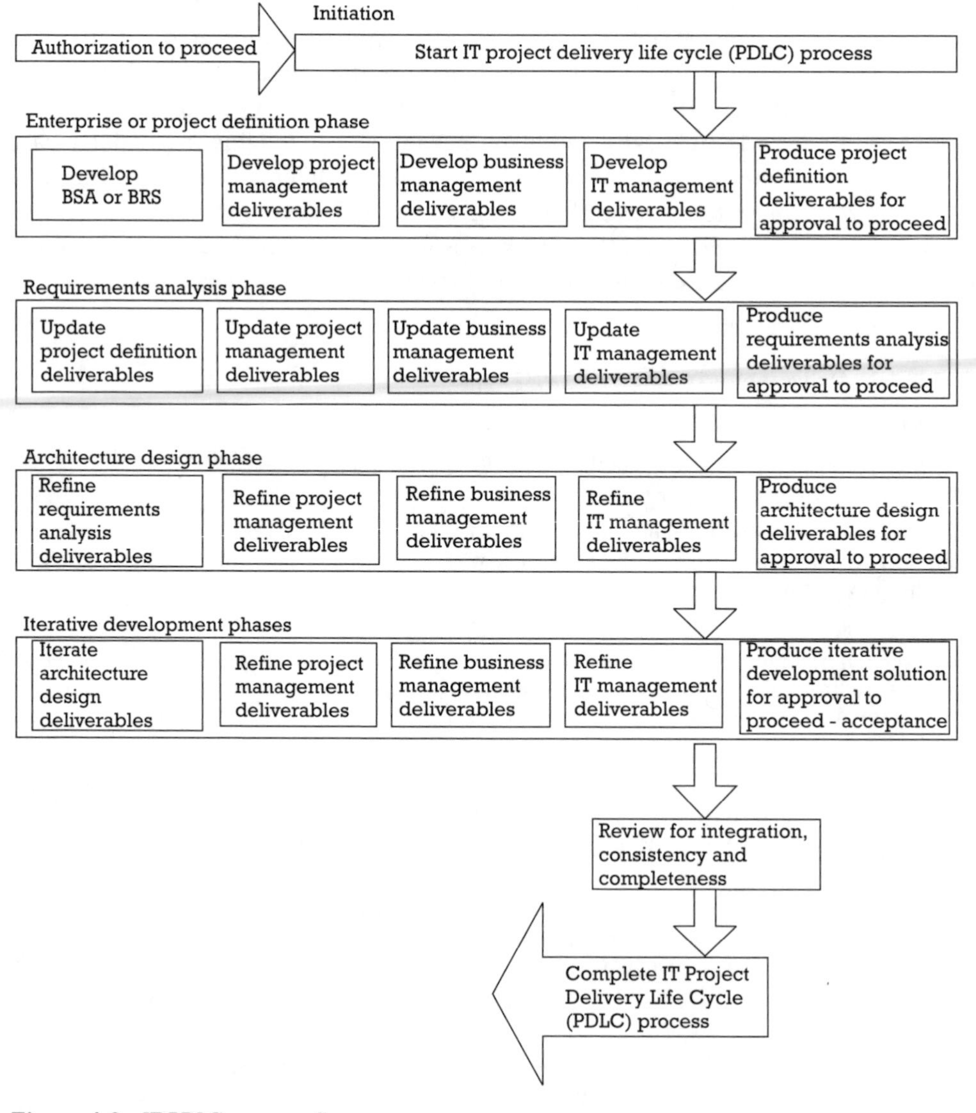

Figure 4.3 IT PDLC process flow.

Table 4.2 IT PDLC Checklist

IT PDLC Criteria	*Yes*	*No*	*IT PDLC Criteria*	*Yes*	*No*
Initiation			*PDP*		
Is management committed?	❑	❑	Does a BSA or BRS exists?	❑	❑
Have a PM sponsor and steering committee been established?	❑	❑	Are PM deliverables identified?	❑	❑
Is there a software development methodology?	❑	❑	Are BM deliverables identified?	❑	❑
Is the software development methodology deliverables based?	❑	❑	Are IT management deliverables identified?	❑	❑
Has a program delivery manager been assigned?	❑	❑	Are deliverables approval criteria established?	❑	❑
Does the assigned program delivery manager understand the PDLC phases?	❑	❑	Are there sponsor approvals to proceed to next phase?	❑	❑
RAP			*ADP*		
Are the PDP deliverables approved?	❑	❑	Are the RAP deliverables approved?	❑	❑
Are PM deliverables updated?	❑	❑	Are PM deliverables refined?	❑	❑
Are BM deliverables updated?	❑	❑	Are BM deliverables refined?	❑	❑
Are IT management deliverables updated?	❑	❑	Are IT management deliverables refined?	❑	❑
Are deliverables approval criteria established?	❑	❑	Are deliverables approval criteria established?	❑	❑
Are there sponsor approvals to proceed to next phase?	❑	❑	Are there sponsor approvals to proceed to next phase?	❑	❑

4.3 IT project management delivery processes

At all times it is better to have a method.
—Mark Caine

The IT project management delivery processes discussed in this section are very similar to the processes or knowledge areas defined in the 2000 edition of the PMBOK guide. The extended issue management and change management processes presented here should enhance the materials presented in this section to provide a clearer understanding of the IPM-IT process presented as the theme of this book. The IT project management delivery processes presented in this section are organized based on a logical implementation sequence of the core processes—scope, time, cost, and quality management. However, the facilitating processes—risk, issue, change, contract, communication and human resources management—that support the delivery of the core process do not imply any implementation sequence.

The intent of this section is not to provide detailed processes or procedures on the implementation of PMI-PMBOK project management process or similar IT project management processes. There are many excellent books on IT project management delivery processes, concepts, theory, and applicability, some of which are mentioned in the selected bibliography at the end of this chapter. The main objective of this section is to provide an overview of the contents (what), purpose (why), roles and responsibilities (who, what), and procedures (how) of the components of a typical IT project management delivery process and to demonstrate how IT project management fits within the context of the overall IPM-IT framework based on PMI-PMBOK standards. The procedure sections for each of the component processes provide baseline templates for the deliverables (what), process flows (how), and checklists (measurement criteria) for practicing project managers during the management, delivery, execution, and integration of the IT project management delivery processes.

4.3.1 Project scope management

Executive management shall use the project scope document–project charter,[6] to understand the work required to deliver the requirements of the software product and to successfully complete the project. The processes of scope management, discussed in this section, establish a scope baseline that consists of business, IT, and project management components, supporting scope management plan and project scope performance for the following:

- Scope definition and verification: project requirements management and project repository;
- Scope planning: project planning;
- Scope change control: project progress tracking.

The major components of the scope baseline document that formally authorize the project and align with the deliverables from business management, IT management, and project management are:

- *Project value justifications:* This deliverable consists of the project requirements/business needs, objectives/benefits, strategy/approach/principles, assumptions/constraints, deliverables, and critical success factors—business management alignment.
- *Product scope:* This deliverable consists of the following:
  - *Business requirements:* Details produced from BSA or BRS deliverables—business management alignment;
  - *Data requirements:* Details produced from DA deliverables—IT management alignment;
  - *Application requirements:* Details produced from AA deliverables—IT management alignment;
  - *Technology requirements:* Details derived from TA deliverables—IT management alignment;

6. Project charter: A document issued by senior management that gives the project manager authority to apply organizational resources to project activities and formally recognizes the existence of a project. It includes a description of the business need that the project was undertaken to address and a description of the product or service to be delivered by the project–PMBOK.

  - *Project management objectives:* Details on time, costs, schedule, effort, and quality objectives—project management alignment.
- *Scope management plan:* This document describes how the scope will be managed and how the scope changes; scope change requests, will be integrated into the project.
- *Project scope performance:* The success of IT projects can be attributed to the quality of the project scope document, and the quality of the project charter can negatively or positively affect project performance in terms of cost, schedule, and quality constraints. This hypothesis can be expressed as $S = f(C, T, Q)$ where S equals scope, and C, T, and Q represent cost, time, and quality, respectively. Any changes to scope management processes will impact the performance of the core project management processes (time, cost, and quality). As a result, I have included additional policies, roles, and responsibilities for the practicing project mangers to ensure that a clearer understanding is obtained of the need to develop complete and consistent project charters.

4.3.1.1 Purpose

The purpose of this process is to develop and deliver a project scope baseline or project charter, consisting of both the deliverables or product scope and the work needed to produce them, or the work scope. This document will form the scope baseline to manage the scope, cost, time, efforts, and quality objectives of the development projects. It will form an integral component of the overall integrated project plan. The program and working and steering committees will use this document as the scope baseline during the approval or disapproval of deliverables.

Requirements management

The PSM process establishes a common understanding between the executive sponsors, program managers, and IT project manager for the requirements that will be addressed by the project in the project charter. PSM establishes and maintains an agreement with the executive sponsors—business, IT, and project management—for the business, technology, and project management requirements of the project.

Requirements are initially identified at a high level in the project charter with additional detail added as the project progresses through the PDLC. PSM addresses specific project requirements during the project initiation and definition phase and manages these requirements during the phases of the IT PDLC.

After these requirements have been approved, a change control process must manage all changes. The effect of the changes on the project scope, effort, cost, schedule, and quality must be documented, and all plans and deliverables must be modified to meet the updated requirements.

Policy

The policy is the following:

- Requirements shall be documented in the project charter to establish a baseline for business management, IT management, and project management.
- Each group affected by the project requirements shall review and approve the requirements before the integrated project plan is finalized and funding approved.
- The functional groups shall review the requirements to determine whether they are feasible and appropriate.
- Project managers shall use the requirements as the basis for managing detailed project plans and product deliverables.

Program/project manager responsibilities

Program delivery manager responsibilities include the following:

- Provide sufficient resources to develop the project deliverables.
- Provide adequate tools and techniques to manage and develop project deliverables.
- Verify that project managers consistently follow PMO guidelines.
- Review the project charter for integration, completeness, and consistency, and provide recommendations for contents and structure improvements.
- Review the detailed project plan before it is accepted and funding approved.

Project manager responsibilities include the following:

- Manage and document the deliverables and their allocations to functional groups throughout the IT PDLC.
- Verify that all project management deliverables, such as delivery dates, milestones, costs, schedule, efforts, and the like, that affect the product deliverables are documented.
- Verify that all product technical requirements such as performance requirements, design constraints, and interface requirements are documented.
- Ensure that all deliverables are reflected in the detailed project plan with costs and schedule estimates.
- Develop acceptance criteria that will be used to validate the deliverables.
- Ensure that all functional groups provide relevant input on the components of the detailed project plan that they are responsible for reviewing and approving.
- Allocate approved funding to project team members to achieve the project deliverables.

Project master file/repository

The project repository contains all project documentation. It shall be utilized to do the following:

- Provide transition information when there is a change in project managers.
- Bring project team members and informed individuals up to date on the status of the project.
- Provide a history of project decisions made and the alternatives that were evaluated.
- Provide information for QA reviews.
- Provide a history of project estimates that can be used in future projects.

Policy

The policy is the following:

- Each project manager shall develop and maintain a project repository.
- The program delivery manager shall develop and maintain a master project repository.

Program and project manager responsibilities

The program delivery manager shall maintain a master project repository for all IT projects with the following structure:

- IT program management:
  - Program management plan[7]:

    Program charters;

    Program implementation plan—master WBS; projects dependency diagram; program schedule; program organization chart; program staffing plan; program budget; program payment schedule;

    QA/acceptance procedures;

    Quality management plan;

    Risk management plan;

    Change requests;

    Issue and resolution;

    Program/project close-down procedures.
  - Other program/project management plans and correspondences:

    Other program/project management plans—communications, human resources, contract, issue, change, scope, cost, time;

    Other correspondences.

The project manager shall maintain a project repository with the following structure:

- Project deliverables:

7. Program management plan: A document that describes by whom, when, and how the program/project will be managed and how scope, effort, schedule, cost, and quality objectives will be controlled. It consists of the program delivery plan, integrated project plan, program/project implementation plan, and various project management plans and detailed project plans.

- RAP;
- ADP;
- IDPs:
  - Construction phase;
  - Integration phase;
  - Deployment phase.
- Project management deliverables:
  - Project charter;
  - Project detailed project plan;
  - Project status reports;
  - Project change requests;
  - Project issues and resolutions;
  - Project risks;
  - Project approvals and budget authorizations;
  - Project QA and acceptance criteria.
- Project management plans and other correspondence:
  - Project management plans;
  - Other correspondences.

Project planning

Good results without good planning come from good luck, not good management.
—David Jaquith

Project planning provides the basis for performing and managing the project activities and represents the project manager's commitments to the business for the resources, constraints, and capabilities of the project. The project plan is based on the established project requirements baseline with support for the plans from the functional groups, business, and consultants.

Policy

The policy is the following:

- Project deliverables/activities, estimates, and commitments shall be planned and documented. The estimates shall be used in project planning and tracking.
- The project plan shall be developed in accordance with the company's IT PDLC and shall include all project deliverables. The plan shall be managed and updated to reflect approved changes to the deliverables.
- The company's IT standard project management tool shall be the database of record for the project. All resources and cost data for the project shall be incorporated into the planning tool in such a manner that it can be tracked at the deliverable level.
- The project plan shall provide an integrated view of the project to address the following:

- What is to be accomplished?
- How it is to be achieved?
- Who will perform the work, and who is responsible for each deliverable?
- When the work will take place?
- How much the project will cost?

- Estimates for the size of the deliverables shall be developed according to the company's IT standards and procedures.
- Estimates for the project cost shall be derived according to company's IT standards and procedures and should include the following: development labor expenses, overhead expenses (meals; training), travel expenses, and hardware (servers and workstations), network, and software expenses. Development labor costs shall be separated into internal IT staff, business staff, and external consultant staff.

Program/project manager responsibilities

Program delivery manager responsibilities include the following:

- Provide tools and techniques to support the project planning activities.
- Verify that budgeting and all PMO administration are complete for project tracking.
- Ensure that the project plans contain sufficient details and adhere to PMO guidelines.

Project manager responsibilities include the following:

- Negotiate project commitments with the functional managers.
- Document all project commitments from the functional managers.
- Coordinate the development of the project plan based upon the established requirements baseline and the documented commitments.
- Assign responsibilities for the project activities in a traceable and accountable manner.
- Manage and control the project plan.

Project progress tracking

The work is planned; now the plan must be worked. After the project has been properly planned in the PDP, the project manager enters the project management execution and control cycle where the key questions to answer include the following:

- Where are we currently? An assessment of the current status of the project.
- Where do we want to be? A comparison of the actual progress versus baseline project plan.
- How do we get there? A consideration of possible corrective actions to place the project back on track, if necessary, or keep it on track.

- Are we getting there? An analysis of the impact that these corrective actions will have or are having on the project

Here are six key steps to the executing and controlling process:

1. Update the plans including scope, effort, cost, schedule, and quality.
2. Update the status.
3. Analyze the risk impact.
4. Monitor and act on variances—scope, effort, cost, schedule, quality.
5. Publish the corrective actions, revisions, or changes: PM score card.
6. Communicate and inform stakeholders.

During the project management execution and control cycle, the approved baseline integrated project plan will be used as the basis for tracking the project, communicating status, and revising plans. Progress is primarily determined by comparing the actual scope, effort, cost, and schedule objectives to the baseline integrated project plan at selected milestones/deliverables or at month end. When it is determined that the integrated project plan is not being met, corrective actions are taken. These actions may include revising the plan to reflect actual accomplishments and replanning the remaining work or taking corrective actions to improve the performance objectives.

Policy

The policy is the following:

- Each resource assigned to the plan shall report actual efforts and estimates to complete the assigned tasks for the approved project plan.
- Actual results and performances shall be tracked against the approved baseline project plan at the deliverable level. The schedule performance index SPI and CPI, or percent ratio of baseline against actual, are key accomplishment indicators, which shall be evaluated on a regularly scheduled basis. CPI is budgeted cost of activities performed divided by the actual cost of the activities performed; a value of less than one indicates that the project is overbudget, and a value greater than one indicates the project is underbudget. SPI is the budgeted cost of activities performed divided by the budgeted cost of activities scheduled; a value of one indicates that the actual performance and the planned performance are the same, a value of less than one indicates that the project is behind schedule, and a value greater than one indicates that the project is ahead of schedule.
- Corrective actions shall be taken and managed to closure when actual performance deviates significantly from the plan.
- All affected groups shall agree to the committed changes, and appropriate changes shall be applied to the approved plan.

- All approved changes to the plan shall be communicated to the affected groups. Once established, baselines can only be changed with formal authorization from the program steering committee. (The baseline is a point-in-time version of a deliverable that establishes scope, effort, cost, schedule, and quality objectives that meet sponsors' expectations.)

Program/project manager responsibilities include the following:

- Review the performance of the project at appropriate milestones for schedule, cost, effort, scope, and quality constraints, and ensure adherence to PMO guidelines.
- Verify that the score card represents objective results.
- Verify that all IT PMO project tracking mechanisms have been completed.

Project manager responsibilities include the following:

- Conduct regular reviews to track actual performance against the project plan. Formal reviews of the results shall be conducted with business and IT management at milestones according to the quality assessment procedure.
- Take responsibility for the performance of the project's activities and results.
- Track project actual scope, effort, cost, and schedule on a weekly basis.
- Track the deliverables developed or services provided.
- Track the scope, effort, cost, and schedule of the planned activities.
- Take corrective action when the plan is not being achieved.
- Gain agreement from all affected groups on the project results and performance.
- Select project activities to be identified, baseline established, and controlled in order to maintain the integrity of the project deliverables.

The roles and responsibilities of executive management must be effectively communicated to all stakeholders to ensure that all stakeholders work in a cohesive manner towards an integrated solution. This section has provided some basic responsibilities guidelines of executive management, program management, and the project team that can be used as a baseline during the establishment and execution of the scope management processes.

4.3.1.2 Policy

The policy of is that the executive management, program management, and the project team shall use the scope management–requirements management, project file/repository, project planning and project progress tracking policies as specified in this section as the guiding principles in making PSM decisions.

4.3.1.3 Roles and responsibilities

Executive business manager responsibilities include the following:

- Provide executive-level business direction support to the program business manager to ensure that the project requirements, project plans and commitments, project progress, and project repository are communicated to manage and control the project scope and resources in order to achieve the expected results.
- Provide business and overall commitments to the project requirements, project plans, project progress, and project repository.
- Approve business changes to project requirements, project plans, project progress, and project repository.
- Provide business resources for delivering project requirements, project plans, project progress, and project repository scope deliverables.
- Act as the overall program sponsor and chairman of the program steering committee.

Executive IT manager responsibilities include the following:

- Provide executive-level IT direction support to the program IT manager to ensure that the project requirements, project plans and commitments, project progress, and project repository are communicated to manage and control the project scope and resources to achieve the desired goals.
- Provide IT commitments to the project requirements, project plans, project progress, and project repository.
- Approve IT changes to project requirements, project plans, project progress, and project repository.
- Provide IT resources for delivering project requirements, project plans, project progress, and project repository scope deliverables.
- Act as the program sponsor from the IT perspective.

Executive project management manager responsibilities include the following:

- Provide executive-level PM guidance, direction, and advice to the program delivery manager to ensure that the project requirements, project plans and commitments, project progress, and project repository are effectively communicated to manage and control the project scope and resources to achieve the desired goals.
- Integrate approved business and IT project requirements, project plans, project progress, and project repository.
- Integrate changes to project requirements, project plans, project progress, and project repository for scope management deliverables.
- Provide PM resources for delivering scope deliverables—project requirements, project plans, project progress, and project repository.

- Act as the program sponsor with integrated business and IT responsibility and project management accountability and champion PM processes, tools, and techniques.

Program business manager responsibilities include the following:

- Submit scope management details to program delivery manager on business guidance, direction, and advice to ensure the project requirements, project plans, project progress, and project repository delivers maximum overall business benefits to the company.
- Report on business commitment to scope management deliverables—project requirements, project plans, project progress, and project repository.
- Report on all business changes to the approved project requirements, project plans, project progress reports, and project repository.
- Assign business resources to support the delivery of project requirements, project plans, project progress reports, and project repository deliverables.
- Report to the program steering committee with direct accountability to the business executive, and keep the program working committee informed of the current scope of the projects.

Program IT manager responsibilities include the following:

- Submit scope details to the program delivery manager on IT guidance, direction, and advice to ensure the project requirements, project plans, project progress, and project repository delivers maximum overall business benefits to the company.
- Report on IT commitments to project requirements, project plans, project progress, and project repository.
- Report on all IT changes to the approved project requirements, project plans, project progress, and project repository.
- Assign IT resources to support the delivery of project requirements, project plans, project progress, and project repository deliverables.
- Report to the program steering committee with direct accountability to the IT executive, and keep the program working committee informed of the current scope of the projects.

Program delivery manager responsibilities include the following:

- Provide tools and techniques to support the project planning activities.
- Verify that budgeting and all PMO administration are complete for project tracking.
- Ensure that the project plan contains sufficient details and adheres to PMO guidelines.
- Provide sufficient resources to achieve the project deliverables.

- Provide sufficient tools and resources to support the activities of managing the deliverables.
- Verify that PMO guidelines are followed.
- Review the project charter for completeness, consistency, and approach, and provide recommendations.
- Maintain a project file for all IT and business projects.
- Provide tools and techniques to support the project planning activities.
- Ensure that the project plan contains sufficient details and adheres to PMO guidelines.
- Review the performance of the project at appropriate milestones for schedule, cost, effort, scope, and quality constraints to ensure adherence to PMO guidelines.
- Verify that the score card represents objective results.
- Verify that all IT PMO project tracking mechanisms have been completed.
- Report to the program steering committee with direct accountability to the PM executive, and keep project managers and stakeholders informed of the current scope of the projects.

Project team (project manager and team) responsibilities include the following:

- Project managers to manage the projects according to the PMO-PSM guidelines.
- Project managers to develop and maintain project plans for project development that include the following:
  - Project charter;
  - Project implementation plan—WBS; project dependency diagram; project schedule; project organizational chart; project staffing profile; project budget;
  - Business and IT transition plan;
  - Project communications plan;
  - PRM management plan;
  - Project issues and change management plan;
  - Project contract management plan;
  - PQM management plan;
  - Project human resource management plan.
- Project managers to integrate, assemble, and report to the program working committee on project status, issues, change requests, and risks, with direct accountability to the program delivery manager.
- Project team to develop phased deliverables according to the company's IT PDLC methodology.
- Project team to record actual efforts and estimates to complete deliverables described in the approved project plan on a weekly basis and to provide PMO with biweekly and monthly project tracking reports.

4.3.1.4 Procedures: scope management

Setting a goal is not the main thing. It is deciding how you will go about achieving it and staying with that plan.
—Tom Landry

The roles and responsibilities discussed in Section 4.3.1.3 are brief because they are intended to serve as a reference as to who does what during the execution of the scope management processes. Procedures are now introduced to demonstrate how this scope management process is executed using the major standard deliverables (what), process flow (how), and checklist or measurement criteria (why) templates. The process is intended to serve as a reference based on the English-language interrogatives: who does what, how, and why. The deliverables, process flow, and checklist templates, based on real-world practical implementation, will serve as excellent references for practicing project managers during execution of the scope management processes.

The deliverables template in Table 4.3 highlights the key deliverables that the practicing project manager manages during the execution of scope

Table 4.3 Deliverables Template: Scope Management Report

Section 1—Introduction
This section includes an executive summary, describes the approach, and introduces the report.
Section 2—Project Justification
This section highlights the major business need or requirements, objectives, and benefits, approach/strategy and assumptions/constraints, deliverables, and critical success factors. These deliverables show the alignment with the BSA or BRS and establish the initial justification baseline for further developments.
Section 3—Product Scope
This section elaborates on the major deliverables and requirements by providing a conceptual solution representation and supporting descriptions for business, data, applications, and technology requirements. These deliverables show the alignment with the deliverables documented in Section 2 and establish the initial product scope baseline for further developments.
Section 4—Project Scope
This section highlights the major project/work deliverables by providing the WBS, deliverables, schedule, cost, and quality objectives and supporting project repository to manage the delivery of the product scope. These deliverables show the alignment with the deliverables documented in Section 3 and establish the initial project scope baseline for further developments.
Section 5—Project Scope Management Plan
This section highlights the plan for how the scope will be managed and how scope changes and scope change requests will be integrated into the project delivery, using the change management process. These deliverables show the alignment with the deliverables documented in Section 4 and establish the initial scope management baseline plan for further developments and integration.
Section 6—Project Scope Performance Report
This section provides the scope performance reports that will be used to control the changes to the project scope. These reports show variations to the scope baseline for corrective actions and updates to the scope baseline. Scope variations and corrective actions will be documented in the project repository to provide lessons learned.
Appendixes—Conceptual Solution, WBS, Scope Performance
Appendixes provide further details—conceptual solutions, WBS, and scope performance.

management. The project charter report model in Table 4.4, based on real-world practical implementations, will provide useful references for project managers. The scope management–project charter model in Table 4.3 shows the structure of the information required to effectively communicate the scope baseline to business, IT, and project management staff.

The process flow template in Figure 4.4 highlights the major processes to produce the scope management deliverables. These processes reference the deliverables to show the integrated nature and dependencies of the deliverables.

The checklist template in Table 4.5 highlights the major measurement criteria to ensure delivery of quality deliverables identified in the deliverables template and adherence to the process flows identified in the process flow template. The checklist is used to objectively determine success criteria during execution of the scope management process.

Table 4.4 Deliverables Template: Project Charter Report Model

Scope Management–Project Charter
Background
This section summarizes the organization business—what—and identifies the purpose—why—for which the project was initiated.
Objective
This section describes the project objectives and the critical business functions that this project should achieve, specifies linkages to the business plan, and identifies the operational impacts to the business.
Scope
This section specifies the in-scope and out-of-scope requirements based on business, data, applications, and technology deliverables.
Approach or Strategy
This section defines the approach and the project management methodology that will be used to achieve the deliverables.
Assumptions and Constraints
This section specifies the assumptions and expectations that form the basis for the decision and the constraints and limitation associated with the project.
Risks
This section describes the key project risks, their probability of occurrence, their impact on scope, schedule, cost, and quality, and the risk response/mitigation strategy.
Project Completion or Success Criteria
This section identifies the major deliverables, acceptance criteria, and strategy for measuring success, which must be accepted before the project, is deemed to be successfully completed.
Project Team Responsibilities and Efforts
This section identifies the project team responsibilities and highlights the efforts involved for each team member.
Cost Requirements (Labor, Technology, and Facilities)
This section identifies the quantities needed and costs for labor, technology, and facilities.
Approvals
This section provides management signatures for project approval.

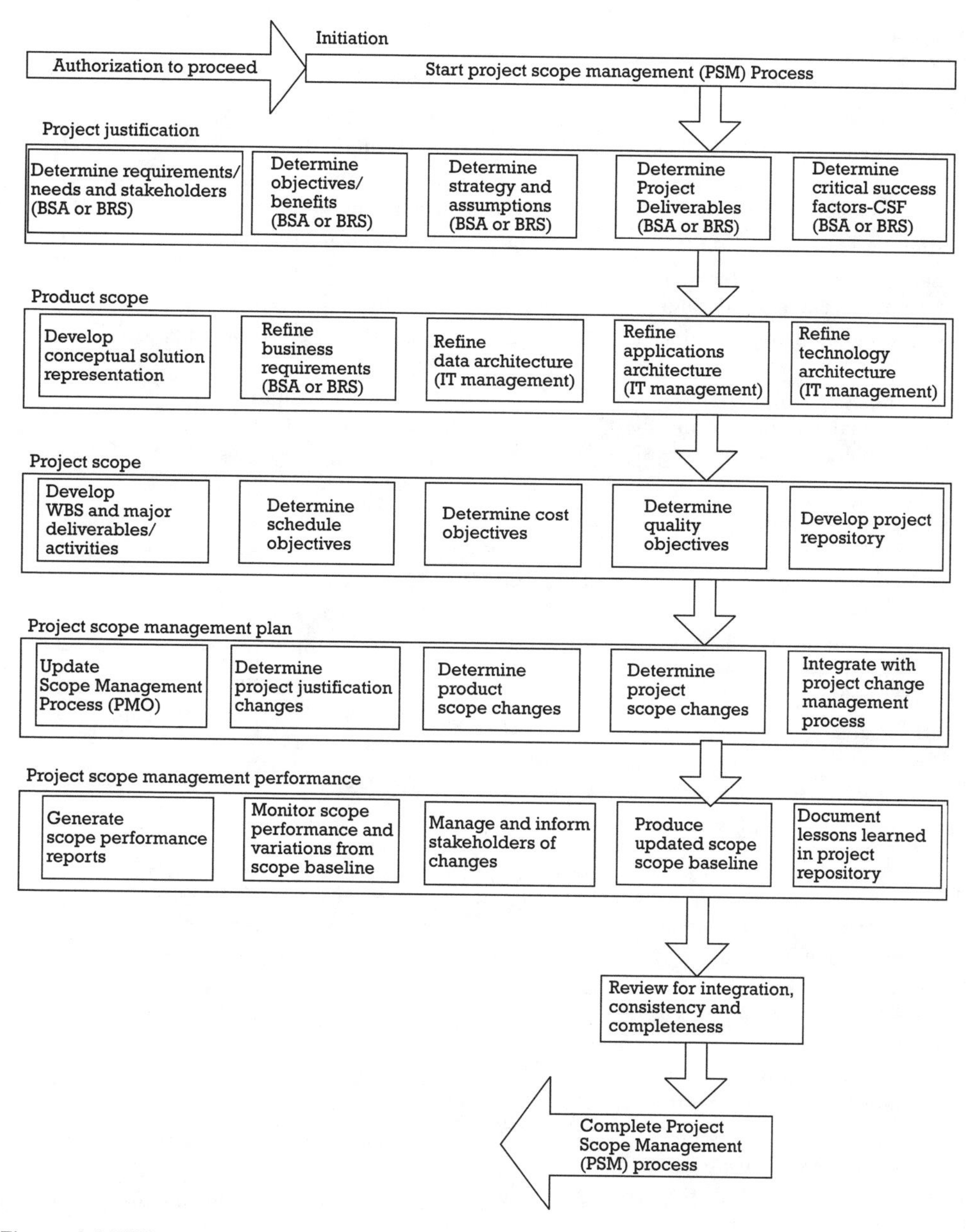

Figure 4.4 PSM process flow.

4.3.2 Project time management

Executive management shall use the integrated project schedule[8] to understand the time dimensions of the project. It provides the framework for managing and integrating the project and identifies the overall schedule that

8. Integrated project schedule: A schedule that integrates the deliverables, activities, duration, and resources for business management, IT management, and project management components.

Table 4.5 PSM Checklist

Project Scope Baseline Criteria	*Yes*	*No*	*Project Scope Baseline Criteria*	*Yes*	*No*
Initiation			*Project justifications*		
Is management committed?	❏	❏	Does a BSA/BRS exist?	❏	❏
Have a PM sponsor and steering committee been established?	❏	❏	Is the Project Justifications document approved?	❏	❏
Is there a PM methodology?	❏	❏	Are the scope components identified?	❏	❏
Does the PM methodology have scope management guidelines?	❏	❏	Are the deliverables identified?	❏	❏
Has a program delivery manager been assigned?	❏	❏	Are CSFs for deliverables documented?	❏	❏
Does the assigned program delivery manager understand the PSM deliverables and processes?	❏	❏	Is there an alignment with project/product scope?	❏	❏
Product scope			*Project scope*		
Is there a conceptual solution representation?	❏	❏	Is there a WBS?	❏	❏
Are business requirements updated?	❏	❏	Are WBS guidelines available?	❏	❏
Are IT requirements (data, applications, and technology) requirements updated?	❏	❏	Are there time, cost, and quality objectives acceptable?	❏	❏
Is there an alignment with the BSA or BRS?	❏	❏	Is there a project repository?	❏	❏
Is the product scope approved?	❏	❏	Is project scope approved?	❏	❏
Is there an alignment with the project scope?	❏	❏	Is there an alignment with scope management plan?	❏	❏
Scope management plan			*Project scope performance*		
Is the PMO–scope management process updated?	❏	❏	Is there a scope control system?	❏	❏
Are changes to project justifications documented as change requests?	❏	❏	Are scope performance reporting guidelines available?	❏	❏
Are changes to product scope documented as change requests?	❏	❏	Do variations from scope baseline exist?	❏	❏
Are changes to project scope documented as change requests?	❏	❏	Are stakeholders informed about scope changes?	❏	❏
Are change requests integrated with change management?	❏	❏	Is scope baseline document updated?	❏	❏
Are scope changes documented in project repository?	❏	❏	Are lessons learned documented in a project file?	❏	❏

is critical to the success of the project. It also serves as a baseline for measuring progress and reporting status, and it drives other project schedules. If the integrated project schedule changes, the subordinate detailed project schedules must change accordingly. The project charter and the integrated project schedule form the contract between the project team and the major stakeholders. The time management process presented in this book is similar to the processes outlined in PMBOK. However, the process presented here demonstrates a more deliverables-based approach, rather than the traditional procedural-oriented approach. The deliverables-based focus on time management presented in this section is better suited to implementing IPM for IT projects because of the dynamic and changing nature of the IT industry.

The major deliverables of time management–project schedule as presented in this chapter align with business management and IT management and consist of the following components:

- Product scope: Defined in PMBOK as resource planning;
- Project schedule estimates: resource allocations;
- Project schedule;
- Project schedule management plan;
- Project schedule control.

4.3.2.1 Purpose

The purpose of this process is to develop an integrated project schedule for managing and integrating the project schedule. It includes the overall schedule, labor resources, and effort estimates that are critical to the success of the project.

4.3.2.2 Policy

The policy of project time management is the following:

- All project schedules–integrated project schedules shall be documented to establish a baseline for measuring progress and reporting status on schedule performance. These schedules shall be included within the integrated project plan and detailed project plans.
- Every project shall provide an integrated project schedule, as part of the integrated project plan in the PDP. The integrated project plan should include sufficient detail to provide budget-level estimates for the project:
    - WBS;
    - Project dependency diagram;
    - Detailed schedule estimates;
    - Project organization;
    - Project team effort estimates.
- Schedule estimates from the BSA: The program delivery plan shall form the basis for preliminary forecast schedule estimates. The integrated project schedule within the integrated project plan shall be integrated with all detailed project schedules and shall form the basis for all baseline schedule estimates.

4.3.2.3 Roles and responsibilities

Executive business manager responsibilities include the following:

- Provide executive-level business approval to the program business manager to ensure that the integrated project plan is managed adequately and to control the integrated schedule and budget and expense authorization[9] within budgetary and schedule constraints.

- Allocate approved business budget and expense authorization and business resource scheduling within the integrated project plan.
- Approve integrated schedule–business component and business budget and expense authorization for the integrated project plan.
- Approve business resources as identified in the integrated project plan.
- Act as the overall program sponsor and chairman of the program steering committee.

Executive IT manager responsibilities include the following:

- Provide executive-level IT approval to the program IT manager to ensure that the integrated project plan is managed adequately and to control the integrated schedule and budget and expense authorization within budgetary and schedule constraints.
- Allocate approved IT support budget and expense authorization and IT resource scheduling within the integrated project plan.
- Approve integrated schedule–IT component and IT support budget and expense authorization for the integrated project plan.
- Approve IT resources as identified in the integrated project plan.
- Act as the program sponsor from the IT perspective.

Executive project management manager responsibilities include the following:

- Provide executive-level PM guidance, direction, and advice to the program delivery manager to ensure that the integrated project plan is managed adequately and to control the integrated schedule and budget and expense authorization within budgetary and schedule constraints.
- Integrate approved business and IT support budget and expense authorization and business and IT resource scheduling within the integrated project plan.
- Integrate project schedule–IT component and IT support budget and expense authorization changes in the integrated project plan.
- Approve PM resources as identified in the integrated project plan.
- Act as the program sponsor with integrated business and IT responsibility and PM accountability; champion PM processes, tools, and techniques.

Program business manager responsibilities include the following:

- Submit details on business guidance, direction, and advice to the program delivery manager to ensure the integrated program plan delivers

9. Budget and expense authorization: Authorization to spend the approved budget, sometimes referred to as authorization for expenditures (AfE).

maximum overall business benefits to the company in support of business budget and resource schedule.

- Report approved business funding to the program steering committee for each phase of the project in accordance with the integrated project plan.
- Report on all business changes to the approved integrated project plan.
- Assign business resources for business-support activities and project benefit estimates.
- Report to the program steering committee with direct accountability to the business executive, and keep the program working committee informed of the current direction of the projects.

Program IT manager responsibilities include the following:

- Submit details on IT guidance, direction, and advice to the program delivery manager to ensure the program delivers maximum overall business benefits to the company.
- Report approved IT support funding to the program steering committee for each phase of the project in accordance with the integrated program plan.
- Report on all IT changes to the approved project plan.
- Assign IT resources for IT support activities.
- Report to the program steering committee with direct accountability to the IT executive, and keep the program working committee informed of the current direction of the projects.

Program delivery manager responsibilities include the following:

- Manage program delivery plan (forecast) from BSA using project management tools (Primavera-SureTrak, Microsoft Project) as the project planning and tracking tool and integrate with integrated project plan detailed project plans.
- Develop master WBS, dependency diagram, and integrated schedule in accordance with PMO guidelines and ensure that project teams understand/follow guidelines. (The WBS provides a uniform structure for collecting resource expenditures in a consistent manner across projects. This allows for estimates for new projects to be compared with actual expenditures of previous projects.)
- Report on integrated forecasts and current program plans to the program steering committee for each of the projects within the program plan.
- Report on integrated business and IT changes to the approved integrated project plan.
- Provide integrated forecasts and a current business and IT resource plan for business, IT, and PM support.

- Report to the program steering committee with direct accountability to the PM executive, and keep project managers informed of the current status of the integrated project plan.

Project team (project manager and team) responsibilities include the following:

- Project managers to manage the projects according to the PMO-PTM guidelines.
- Project managers to develop and maintain project schedules for projects development that include the following:
  - Project charter;
  - Project implementation plan—WBS; project dependency diagram; project schedule; project organizational chart; project schedule estimates, project team effort estimates;
  - PTM management plan.
- Project managers to integrate, assemble, and report to the program working committee on project schedule status, issues, change requests, and risks, with direct accountability to the program delivery manager.
- Project team to use the integrated project plan in the project charter as the baseline during the development of the detailed project schedule.
- Project team to record actual efforts and estimates to complete deliverables described in the approved project plan on a weekly basis and to provide PMO with biweekly and monthly project tracking reports.

4.3.2.4 Procedures: time management

Never plan in more detail than you can control. The more important the project deadline, the more important the schedule becomes.
—James P. Lewis

The roles and responsibilities discussed in Section 4.3.2.3 are brief because they are intended to serve as a reference as to who does what during the execution of the time management processes. Procedures are now introduced to demonstrate how this time management process is executed using the major standard deliverables (what), process flow (how), and checklist or measurement criteria (why) templates. The process is intended to serve as a reference based on the English-language interrogatives: who does what, how, and why. The deliverables, process flow, and checklist templates, based on real-world practical implementation, will serve as excellent references for practicing project managers during execution of the time management processes.

The deliverables template in Table 4.6 highlights the key deliverable that the practicing project manager manages during the execution of time management. The time management WBS model in Figure 4.5, based on real-world practical implementations, will provide useful references for project managers. The time management model–WBS shows the structure of the

Table 4.6 Deliverables Template: Time Management Report

Section 1—Introduction
(This section includes an executive summary, describes the approach, and introduces the report.)
Section 2—Project Scope
(This section highlights the major project/work deliverables by providing the WBS, deliverables, schedule, cost, and quality objectives and supporting project repository to manage the delivery of the product scope. These deliverables show the alignment with the deliverables documented in the scope management report and establish the initial project baseline for further developments.)
Section 3—Project Schedule Estimates
(This section highlights the project activity/deliverables duration estimates for the allocated resources to be used in developing the project schedule. These schedule effort estimates for each phase are detailed in Chapter 5 and establish the initial baseline for schedule estimating.)
Section 4—Project Schedule
(This section provides the project schedule based on the resource allocation and schedule estimates to be used in managing and controlling the project schedule. The schedule for each project, phase, and deliverable is detailed in this section and the initial schedule baseline for schedule control is established.)
Section 5—Project Schedule Management Plan
(This section highlights the plan for how the schedule will be managed and schedule changes and schedule change requests will be integrated into the project delivery using the change management process. These deliverables show the alignment with the deliverables documented in Section 4 and establish the initial schedule management baseline plan for further developments and integration.)
Section 6—Project Schedule Control
(This section provides the schedule performance reports that will be used to control the changes to the project schedule. These reports show variations to the schedule baseline for corrective actions and updates to the schedule baseline. Schedule variations and corrective actions will be documented in the project repository to provide lessons learned.
Appendixes—Project Management Software, WBS, Schedule Performance
(Appendixes provide further detail on project management software, WBS, and schedule performance reports generated from software.)

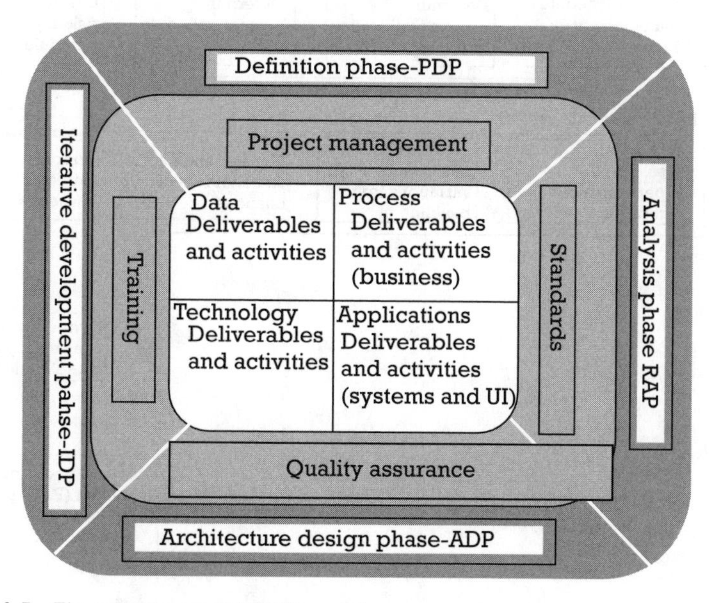

Figure 4.5 Time management WBS model.

project scope that is used to effectively communicate the deliverables and activities schedule to the business, IT, and project management staff. This WBS structure forms the basis for implementation of the integrated project schedule, using project management tools.

The process flow template in Figure 4.6 highlights the major processes to produce the time management deliverables. These processes reference the deliverables to show the integrated nature and dependencies of the deliverables.

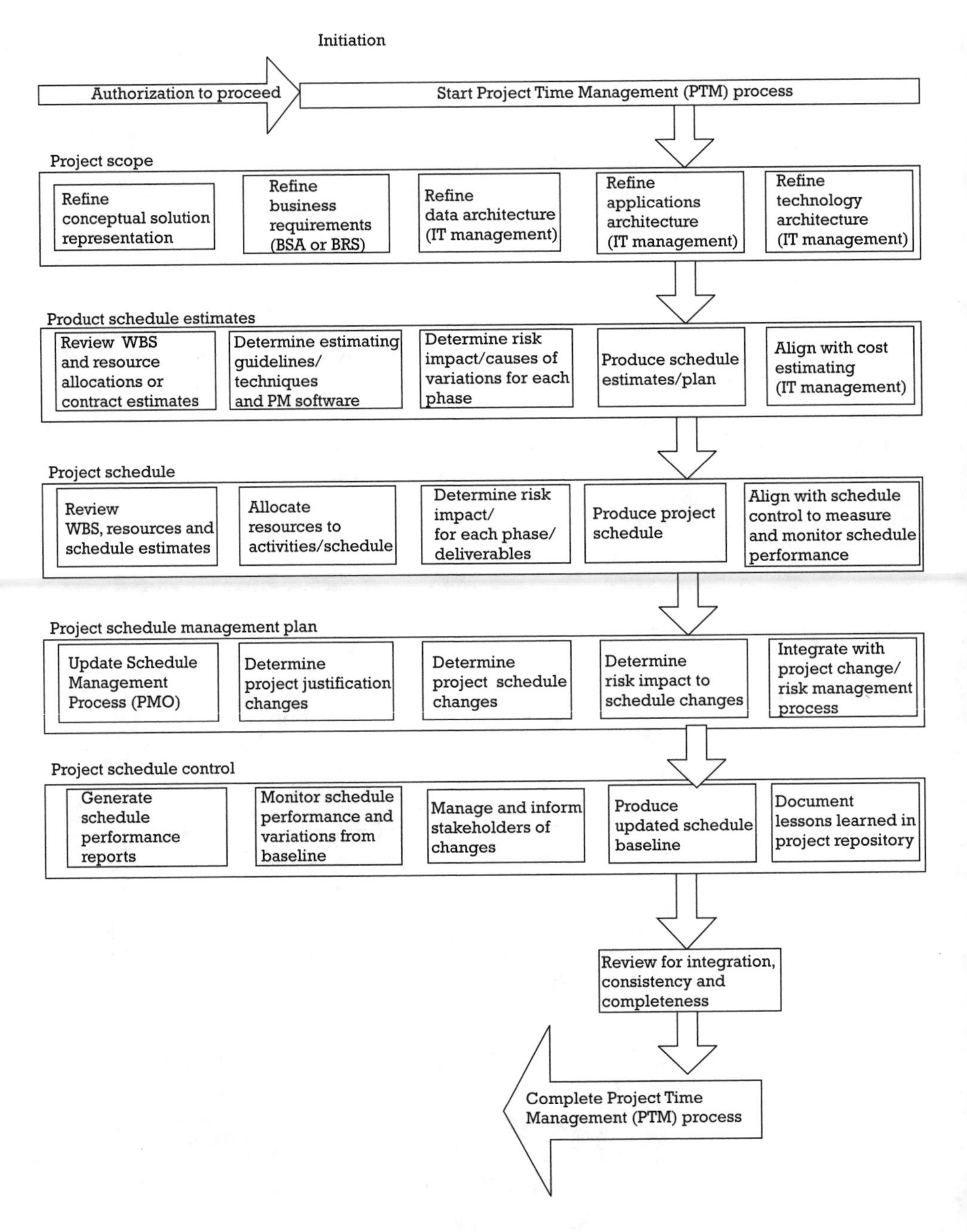

Figure 4.6 PTM process flow.

The checklist template in Table 4.7 highlights the major measurement criteria to ensure delivery of quality deliverables identified in the deliverables template and adherence to the process flows identified in the process flow template. The checklist is used to objectively determine success criteria during execution of the time management process.

Deliverables template: time management WBS model

Figure 4.5 is a master WBS to support the core data, business process, applications (systems processes and user interfaces) and technology deliverables,

Table 4.7 Time Management WBS Model

Project Schedule Baseline Criteria	*Yes*	*No*	*Project Schedule Baseline Criteria*	*Yes*	*No*
Initiation			*Project scope*		
Is management committed?	❑	❑	Does a WBS exist?	❑	❑
Have a PM sponsor and steering committee been established?	❑	❑	Is the project scope document approved?	❑	❑
Is there a PM methodology?	❑	❑	Are the scope objectives updated?	❑	❑
Does the PM methodology have schedule management guidelines?	❑	❑	Are the time objectives updated?	❑	❑
Has a program delivery manager been assigned?	❑	❑	Are the cost objectives updated?	❑	❑
Does the assigned program delivery manager understand the PTM deliverables and processes?	❑	❑	Are the quality objectives updated?	❑	❑
Project schedule estimates			*Project schedule*		
Are the WBS and resource allocations updated?	❑	❑	Is there a WBS?	❑	❑
Are there estimating guidelines?	❑	❑	Are WBS guidelines available?	❑	❑
Is there a risk impact assessment?	❑	❑	Are the time, cost, and quality objectives acceptable?	❑	❑
Are the schedule estimates acceptable?	❑	❑	Is there a project repository?	❑	❑
Is the product scope approved?	❑	❑	Is the project schedule approved?	❑	❑
Is there an alignment with the project scope?	❑	❑	Is there an alignment with the schedule management plan?	❑	❑
Schedule management plan			*Project schedule control*		
Is the PMO–time management process updated?	❑	❑	Is there a schedule control system?	❑	❑
Are changes to the project schedule documented as change requests?	❑	❑	Are schedule performance reporting guidelines available?	❑	❑
Are change requests integrated with project change management process?	❑	❑	Do variations from the schedule baseline exist?	❑	❑
Are schedule changes documented in project repository?	❑	❑	Are stakeholders informed about schedule changes?	❑	❑
			Is the schedule baseline document updated?	❑	❑
			Are lessons learned documented in a project file?	❑	❑

and supporting project management, training, standards, and QA deliverables for each of the PDLC phases: PDP, RAP, ADP, and IDP.

4.3.3 Project cost management

Cutting costs without improvements in quality is futile.
—W. Edwards Deming

Executive management shall use project cost management—budget and expense authorization and earned-value management (EVM)[10]—a major component of the integrated project plan, as the baseline in determining the cost status and progress of the projects. This process provides the framework for managing, integrating and controlling the projects cost, and identifies the overall budget and expense authorization that is critical to the success of the project. It also serves as a baseline for measuring progress and reporting status for cost performance. If the budget and expense authorization changes, the integrated project plan and the subordinate detail project plans must change accordingly. The project charter–budget and expense authorization, a major component of the integrated project plan, forms the cost contract between the project teams and the major stakeholders. The cost management process presented in this book, is similar to the cost management knowledge area of PMBOK, but focuses more on budget and expense authorization. It consists of the following components:

- Resource allocations (defined in PMBOK as resource planning);
- Project cost estimates–cost estimating;
- Project budget and expense authorization;
- Project cost control–EVM.

4.3.3.1 Purpose

The purpose of this process is to provide budgeting and expense authorization details that align with the integrated project schedule for monitoring, measuring, and reporting on the status and progress of the IT projects for time and costs performance. This information will form the baseline to manage and integrate the cost and time objectives of the development projects. The program steering committee will use this budget and expense authorization report as a baseline in determining the budget status and progress of the overall program and of each project.

4.3.3.2 Policy

The policy of project cost management is the following:

10. Earned-value management: Analysis of a project's schedule and financial progress as compared to the original plan. This analysis is based on actual cost (AC), planned value (PV), earned value (EV), budget at completion (BAC), cost variance (CV), cost performance index (CPI), schedule variance (SV), and schedule performance index (SPI).

- All project costs and budget and expense authorization shall be documented to establish the baseline for measuring progress and reporting the status of cost performance.
- Every project shall produce project budgeting and expense authorization (PBEA) details within the integrated project plan during the PDP. The integrated project plan shall include sufficient details to provide budget-level baseline estimates for the project:
  - Master WBS;
  - Project dependency diagram;
  - Integrated schedule;
  - Project organization;
  - Staffing profile;
  - Project budget and expense authorization;
  - Payment schedule.
- Cost estimates from the BSA–program delivery plan shall be used as preliminary forecast estimates for all IT delivery projects and shall be integrated with the cost estimates in the integrated project plan and detailed project plans.

4.3.3.3 Roles and responsibilities

Executive business manager responsibilities include the following:

- Provide executive-level business approval to the program business manager to ensure that the business budget and expense authorization is managed adequately and to control the overall program budget and expense authorization within budgetary constraints.
- Allocate approved business budget and expense authorization within the program.
- Approve all business expense authorization to the approved budget.
- Approve business resources as identified in the budget.
- Act as the overall program sponsor and chairman of the program steering committee.

Executive IT manager responsibilities include the following:

- Provide IT executive-level approval to the program IT manager to ensure that the IT support budget and expense authorization is managed adequately and to control the overall program budget and expense authorization within budgetary constraints.
- Allocate approved IT budget and expense authorization for each project.
- Approve all IT support expense authorization for the approved budget.
- Approve IT resources as identified in the budget.
- Act as the program sponsor from the IT perspective.

Executive project management manager responsibilities include the following:

- Attend program steering committee meetings and provide PM executive-level approval to the program delivery manager to ensure that the budget and expense authorization processes are executed adequately and to control the overall program budget and expense authorization within budgetary constraints.
- Integrate approved business and IT budget and expense authorization for each project within the program.
- Integrate business and IT budget and expense authorization with approved budget.
- Recommend PM resources as identified in the program budget.
- Act as the program sponsor with integrated business and IT responsibility and PM accountability; champion PM processes, tools, and techniques.

Program business manager responsibilities include the following:

- Attend program steering committee meetings and present details on business budget and expense authorization to the program sponsors to control the overall program budget and expense authorization within budgetary constraints.
- Report on the progress of approved business budget and expense authorization to the program steering committee for each of the projects within the program plan.
- Report on all changes to the approved business budget and expense authorization.
- Assign business resources for business budget and expense authorization support.
- Report to the program steering committee with direct accountability to the business executive, and keep the program working committee informed of the current status of the program budget and expense authorization.

Program IT manager responsibilities include the following:

- Attend program steering committee meetings and present details on IT budget and expense authorization to the program sponsors to control the overall program budget and expense authorization within budgetary constraints.
- Report on the progress of approved IT support budget and expense authorization to the program steering committee for each of the projects within the program budget.
- Report on all IT changes to the approved program budget and expense authorization.
- Assign IT resources for IT budget and expense authorization support.

- Report to the program steering committee with direct accountability to IT executive, and keep the program working committee informed of the current status of the program budget and expense authorization.

Program delivery manager responsibilities include the following:

- Attend program steering committee meetings and present details on integrated business and IT budget and expense authorization to the program sponsors to control the overall program budget and expense authorization within budgetary constraints.
- Report on integrated approved business and IT budget and expense authorization to the program steering committee for each of the projects within the program plan.
- Report on integrated business and IT changes to the approved program budget.
- Provide integrated business and IT resource plans for business, IT, and PM budget and expense authorization support.
- Report to the program steering committee with direct accountability to the PM executive, and keep project managers informed of the current status of the program budget and expense authorization.
- Integrate the program implementation plan–forecast from the BSA with the current integrated program plan and detailed project plans.
- Integrate the program implementation plan–BSA using project management tools (SureTrack or Microsoft Project) as the project planning and tracking tool and integrate with IT integrated project plan and detailed project plans.
- Develop master WBS, dependency diagram, and integrated project schedule in accordance with company's guidelines and ensure that project team understands and follows guidelines. (The master WBS provides a uniform structure for collecting resource expenditures in a consistent manner across projects. This allows estimates for new projects to be compared with actual expenditures of previous projects.)

Project team (project manager and team) responsibilities include the following:

- Project managers to manage the projects according to budget and expense authorization guidelines documented in PMO–cost management guidelines.
- Project managers to develop and maintain detailed project plans for projects' development using the integrated project plan as the baseline:
  - Master WBS;
  - Project dependency diagram;
  - Integrated schedule;
  - Project organization;
  - Staffing profile;

- Project budget;
- Project schedule.
- Project managers to integrate, assemble, and report to the program working committee on project budget and expense authorization.
- Project team to develop deliverables within the constraints of project budget and expense authorization.

4.3.3.4 Procedures: cost management

The roles and responsibilities discussed in Section 4.3.3.3 are brief because they are intended to serve as a reference as to who does what during the execution of the cost management processes. Procedures are now introduced to demonstrate how this cost management process is executed using the major standard deliverables (what), process flow (how), and checklist or measurement criteria (why) templates. The process is intended to serve as a reference based on the English-language interrogatives: who does what, how, and why. The deliverables, process flow, and checklist templates, based on real-world practical implementation, will serve as excellent references for practicing project managers during execution of the cost management processes.

The deliverables template in Table 4.8 highlights the key deliverables that the practicing project manager manages during the execution of cost management. The budget authorization model in Table 4.9, based on real-world practical implementations, will provide useful references for project

Table 4.8 Cost Management Report—Budget and Expense Authorization

Section 1—Introduction
This section includes an executive summary, cost baseline and introduces the report.
Section 2—Resource Allocation
This section highlights the major cost components—labor, technology infrastructure, hardware, software, network, and facilities, training, and administration—and quantities to be used in developing the cost estimates. These resource allocations for each phase are detailed in Chapter 5 and establish the initial baseline for resource planning.
Section 3—Cost Estimating
This section highlights the various project cost estimates—macro, construction, and steward able—for the allocated resources to be used in allocating budget and expense authorization. These cost estimates for each phase are detailed in Chapter 5 and establish the initial baseline for cost estimating.
Section 4—Budget and Expense Authorization (AfE)
This section provides the budget and expense authorization, or AfE, based on resource allocation and cost estimates, to be used in managing and controlling the project's budget. The budgets for each project, phase, and deliverable are detailed in this section, and the initial cost baseline for cost control is established.
Section 5—Cost Control
This section highlights the EVM deliverables, in terms of planned value, earned value, and actual costs to accomplish the earned value. The cost control reports for each project, phase, and deliverable are detailed in this section.
Appendixes—Models, Architectures, and Plans
Appendixes provide further details on EVM deliverables.

Table 4.9 PBEA Report Model

Projects	*Phases*	*AfE # Approve*	*Budget*	*Current*	*Cumulative*	*Forecast*	*Variance*
Project-A	PDP						
	RAP						
	ADP						
	IDP-I1						
	IDP-I2						
	IDP-I3						
Project-B	PDP						
	RAP						
	ADP						
	IDP-I1						
	IDP-I2						
	IDP-I3						
Project-C	PDP						
	RAP						
	ADP						
	IDP-I1						
	IDP-I2						
	IDP-I3						
Project-D	PDP						
	RAP						
	ADP						
	IDP-I1						
	IDP-I2						
	IDP-I3						
Project-E	PDP						
	RAP						
	ADP						
	IDP-I1						
	IDP-I2						
	IDP-I3						
Project-F	PDP						
	RAP						
	ADP						
	IDP-I1						
	IDP-I2						
	IDP-I3						

managers. The cost management model–budget and expense authorization shows a structure to record and track costing information to effectively communicate the budget and expense authorization to the business, IT, and project management staff.

The process flow template in Figure 4.7 highlights the major processes to produce the cost management deliverables. These processes reference the deliverables to show the integrated nature and dependencies of the deliverables.

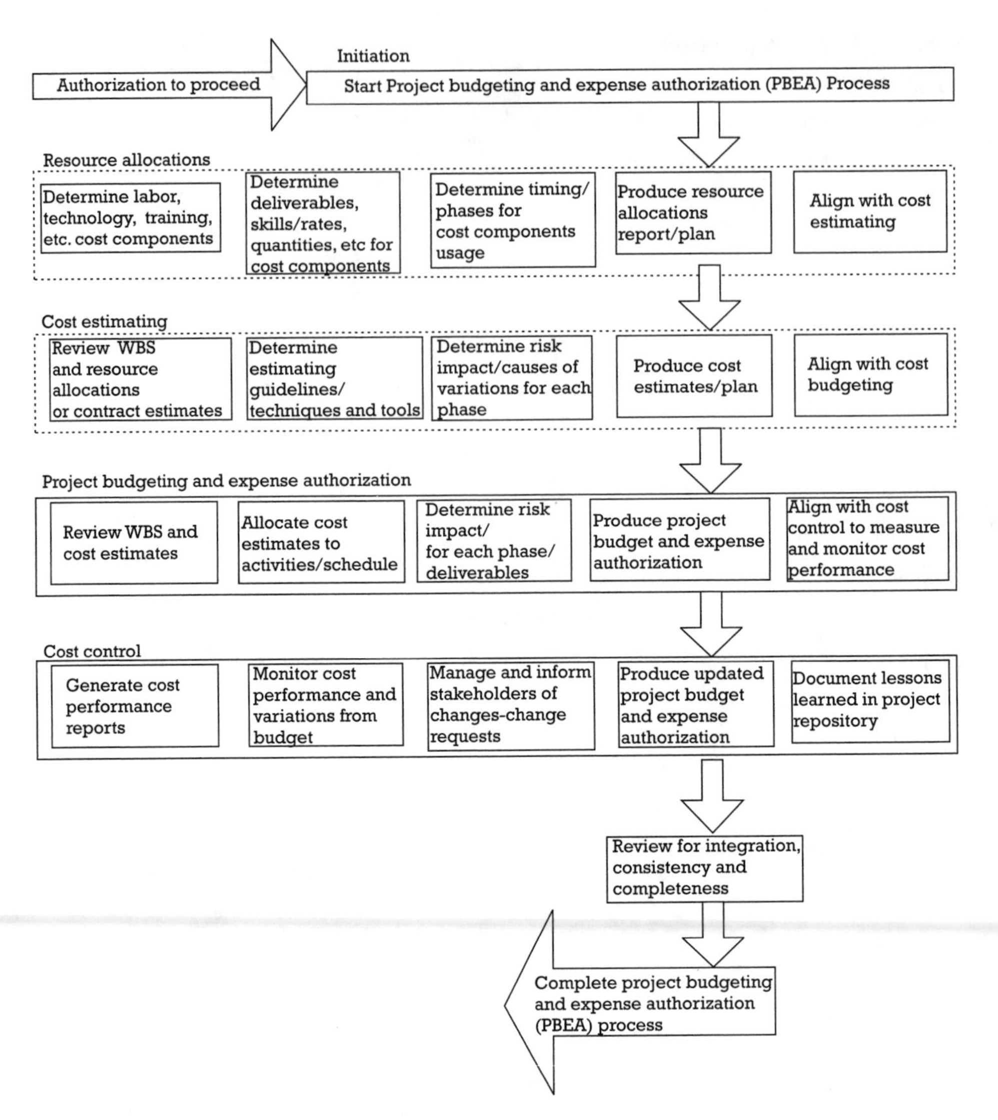

Figure 4.7 Project PBEA process—flow.

The checklist template in Table 4.10 highlights the major measurement criteria to ensure delivery of quality deliverables identified in the deliverables template and adherence to the process flows identified in the process flow template. The checklist is used to objectively determine success criteria during execution of the cost management process.

4.3.4 Project quality management

There is never enough time to do it right, but there is always time to do it over.
—Murphy's Law

Executive management shall use the quality management report to ensure that the appropriate level of authority has been applied to the project and

Table 4.10 Budgeting and Expense Authorization Checklist

PBEA Criteria	*Yes*	*No*	*PBEA Criteria*	*Yes*	*No*
Initiation			*Resource allocations*		
Is management committed?	❑	❑	Does a resource allocation plan exist?	❑	❑
Have a PM sponsor and steering committee been established?	❑	❑	Is the resource allocation plan approved?	❑	❑
Is there a PM methodology?	❑	❑	Are the cost components identified?	❑	❑
Does the PM methodology have PBEA component?	❑	❑	Are the detailed cost components identified?	❑	❑
Has a program delivery manager been assigned?	❑	❑	Are the timing/phases for cost components usage identified?	❑	❑
Does the assigned program delivery manager understand the PBEA process?	❑	❑	Is there an alignment with cost estimating?	❑	❑
Cost estimating			*Budgeting and expense authorization*		
Is there a WBS?	❑	❑	Is there a WBS?	❑	❑
Are estimating guidelines available?	❑	❑	Are budgeting guidelines available?	❑	❑
Is there risk impact analysis?	❑	❑	Is there risk impact analysis?	❑	❑
Are cost estimates acceptable?	❑	❑	Are budget and expense authorizations acceptable?	❑	❑
Are cost estimates time based?	❑	❑	Are budget and expense authorizations time based?	❑	❑
Is there an alignment with cost budgeting?	❑	❑	Is there an alignment with cost control?	❑	❑
Cost control			*Appendixes*		
Is there a cost-control system?	❑	❑	Do the appendixes include details on EVM performance reports?	❑	❑
Are cost performance reporting guidelines available?	❑	❑			
Do variations from the cost baseline exist?	❑	❑			
Are stakeholders informed about changes?	❑	❑			
Is the budget and expense authorization updated?	❑	❑			
Are lessons learned documented in a project file?	❑	❑			

that project quality has not been reduced to inappropriate levels due to political or other project pressures. The quality management process presented in this book is similar to the quality management knowledge area of PMBOK, but focuses more on QA. QA verifies that the project adheres to the company's IT policies, standards, and procedures to ensure the following:

- Compliance with organizational policies;
- Compliance with guidelines and requirements imposed by the PDLC guidelines;
- Compliance with PMO guidelines that are appropriate for use by the project.

The components of quality management are:

- Quality deliverables and standards;
- Quality management plan;
- QA and quality control.

4.3.4.1 Purpose

The purpose of this process is to provide quality standards to ensure that an appropriate level of quality has been applied to the project and that project quality has not been reduced to some inappropriate level because of political or other project pressures. This information will form the baseline to manage and integrate the quality objectives of the development projects. The program steering committee will use these quality standards as a baseline in determining the quality status and progress of the overall program and of the individual projects.

4.3.4.2 Policy

The policy of project quality management is the following:

- The project quality shall be maintained at the level that is appropriate for the project. Adherence to PMO guidelines and policies will be verified objectively.
- The minimum level of quality shall be established by PMO policies, guidelines, and procedures. Specific project requirements may require higher quality levels and shall be documented in the PQM management plan.[11]
- The company's policies, guidelines, and procedures shall be published and readily available. Project managers shall develop project-specific variances based upon the quality requirements of the project. Variances shall be documented in the QA and quality control document.
- Checklists shall be developed and made available as guidelines in performing quality checks on major project deliverables for each phase of the PDLC. PMO checklists shall also be made available during the development of the integrated project plan.
- QA checkpoints and reviews with the appropriate management approval shall be conducted at major milestones of the PDLC.
- The QA review team shall provide recommendations and corrective action plans for each quality review.

4.3.4.3 Roles and responsibilities

Executive business manager responsibilities include the following:

- Provide executive-level business approval to the program business manager to ensure that the quality management deliverables are

11. Quality management plan: A document that describes by whom, when, and how deliverables will be managed and controlled to support project quality objectives.

managed adequately and to control the QA process within budgetary and schedule constraints.

- Provide executive-level business commitment for quality management deliverables.
- Approve business support changes to quality management deliverables.
- Approve business resources to support delivery of quality management deliverables.
- Act as the overall program sponsor and chairman of the program steering committee.

Executive IT manager responsibilities include the following:

- Provide executive-level IT approval to the program IT manager to ensure that the quality management plan is managed adequately and to control the QA process within budgetary and schedule constraints.
- Provide executive-level IT commitment to the quality management deliverables.
- Approve all IT support changes to quality management deliverables.
- Provide IT resources to support delivery of quality management deliverables.
- Act as the program sponsor from the IT perspective.

Executive project management manager responsibilities include the following:

- Provide executive-level PM guidance, direction, and advice to the program delivery manager to ensure that the required levels of the quality management plan and QA process are effectively communicated in order to manage and control the QA process.
- Integrate approved business and IT QA requirements.
- Integrate changes to business and IT QA processes and plans and changes to quality deliverables.
- Provide PM resources for delivering quality deliverables.
- Act as the program sponsor with integrated business and IT responsibility and PM accountability; champion PM processes, tools, and techniques.
- Provide quality management guidance, direction, and advice to the program managers to ensure the program delivers the required quality level to provide overall business benefits to the company.

Program business manager responsibilities include the following:

- Submit details on business guidance, direction, and advice to the program delivery manager to ensure the required quality management plan and process delivers maximum overall business benefits to the company, based on project scope, cost, and time constraints.
- Report on business commitment to quality deliverables.
- Report on business changes to the approved quality deliverables.

- Assign business resources for supporting the delivery of the quality deliverables.
- Report to the program steering committee with direct accountability to the business executive, and keep the program working committee informed of the current quality of the projects.

Program IT manager responsibilities include the following:

- Submit quality details on IT guidance, direction, and advice to the program delivery manager to ensure the required quality management deliverables and process delivers maximum overall business and IT benefits to the company, based on project scope, cost, and time constraints.
- Report on IT commitment to approved quality management deliverables.
- Report on IT support changes to the approved quality management deliverables.
- Assign IT resources for supporting the delivery of quality management deliverables.
- Report to the program steering committee with direct accountability to the business executive, and keep the program working committee informed of the current quality of the project deliverables.

Program delivery manager responsibilities include the following:

- Conduct QA reviews in accordance with the approved quality management plan.
- Verify that the project meets PMO quality policies, guidelines, and procedures.
- Verify that the project meets the project-specific quality requirements by utilizing the checklists.
- Report variances to the appropriate management team and recommend solutions.
- Change the QA process as needed to keep it effective.
- Report to the program steering committee with direct accountability to the PM executive, and keep project managers and stakeholders informed of the current quality of the project deliverables.

Project team (project manager and team) responsibilities include the following:

- Develop a QA plan.
- Document project-specific quality requirements in the project charter.
- Adhere to PMO quality policies, guidelines, and procedures.
- Ensure that a program is in place to achieve the expected quality.

- Monitor the project to ensure the quality management plan is being followed.
- Establish measurements to gauge the effectiveness of the quality management plan.
- Schedule QA reviews according to the approved quality management plan.
- Change the quality management plan as needed to keep it effective.
- Project managers to manage the projects according to quality requirements and standards documented in PMO–quality management guidelines.
- Project managers to integrate, assemble, and report to the program working committee on the quality of the deliverables.
- Project team to develop deliverables within the constraints of project scope, time, and costs constraints.

4.3.4.4 Procedures: quality management

The roles and responsibilities discussed in Section 4.3.4.3 are brief because they are intended to serve as a reference as to who does what during the implementation of the quality management process. Procedures are now introduced to demonstrate how this quality management process is executed using the major standard deliverables (what), process flow (how), and checklist or measurement criteria (why) templates. The process is intended to serve as a reference based on the English-language interrogatives: who does what, how, and why. The deliverables, process flow, and checklist templates, based on real-world practical implementation, will serve as excellent references for practicing project managers during implementation of the quality management processes.

The deliverables template in Table 4.11 highlights the key deliverables that the practicing project manager manages during the implementation of quality management. The project planning checklist model in Table 4.12, based on real-world practical implementations, will provide useful references for project managers. The detailed project planning checklist shows a model to measure the quality of the detailed project plan in order to establish and communicate successful completion criteria.

The process flow template in Figure 4.8 highlights the major processes to produce the quality management deliverables. These processes reference the deliverables to show the integrated nature and dependencies of the deliverables.

The checklist template in Table 4.13 highlights the major measurement criteria to ensure delivery of quality deliverables identified in the deliverables template and adherence to the process flows identified in the process flow template. The checklist is used to objectively determine success criteria during execution of the quality management process.

Table 4.11 Deliverables Template: Quality Management Report

Section 1—Introduction
This section includes an executive summary, quality requirements, standards, and approach and introduces the report.
Section 2—Quality Deliverables and Standards
This section highlights the major business management, IT management, and project management milestone deliverables produced during each phase of IT delivery life cycle. It also highlights the standards to measure the quality of the deliverables. The deliverables for each phase presented in Chapter 6 and the supporting standards documented within the company's PMO standards and processes establish the quality baseline for quality planning.
Section 3—Quality Management Plan
This section highlights the quality requirements, organizational structure, responsibilities, measurement criteria and checklists, review processes, and resources needed to measure the quality of the deliverables for each phase based on the established standards. This quality management plan establishes the project quality system for performing QA and quality control.
Section 4—Quality Assurance and Control
This section provides the results of the review processes by providing quality measurement reports containing defect statistics and suggested corrective actions on the quality of the deliverables. The corrective actions are documented as change requests during the change management process. The QA and quality control reports for each project, phase, and deliverable are detailed in this section.
Appendixes—Quality Measurement Reports
Appendixes provide further details on defect statistics and corrective actions.

4.3.5 Project risk management

Risk is like fire: If controlled, it will help you; if uncontrolled, it will rise up and destroy you.
—Theodore Roosevelt

Executive management shall use the risk management report to ensure that the appropriate level of authority has been applied to the project risks and that appropriate categories of risks are identified and quantified during each phase of the project. They shall ensure that appropriate mitigation action plans are developed to maximize the results of positive events and minimize the consequences of adverse events. Each project will have a different tolerance to risk associated with it. The project manager will understand the risk tolerance level associated with the project. Risk tolerances are influenced and ranked based on the following project characteristics:

- Is it a mission critical project?
- Does it require operational capabilities (24 hours a day, 7 days a week)?
- Is optimum business process and technology performance necessary?
- What is the business impact of not being operational?
- What is the adverse effect on scope, schedule, cost, and quality?

The risk management process presented in this book is similar to the risk management knowledge area of PMBOK, but focuses more on risk identifications and assessments. Risk response strategies[12] or mitigation strategies are developed and prioritized based on established risk categories to ensure consistency and integrity among the identified risks.

Table 4.12 Detailed Project Planning Checklist Model

Project Planning Criteria	*Yes*	*No*	*Project Planning Criteria*	*Yes*	*No*
Are a maximum of three tasks assigned per person per week?	❏	❏	Is the duration of each phase limited to six months or less?	❏	❏
Are tasks limited to 40 to 80 hours?	❏	❏	Is overtime scheduled?	❏	❏
Is there a deliverable for each task?	❏	❏	Does a network diagram exist?	❏	❏
Do tasks relate to deliverables?	❏	❏	Does every task have a person assigned to it?	❏	❏
Have completion criteria been established for each deliverable?	❏	❏	Will team members receive a weekly work schedule?	❏	❏
Have deliverables/tasks been reviewed with project team members?	❏	❏	Is the project plan to be maintained? If yes, who will maintain the plan?	❏	❏
Do project team members agree on deliverables?	❏	❏	Are the following plans complete?	❏	❏
Estimated effort			Staffing plan		
Planned start date			Project budget		
Planned completion date			Hardware/software requirements		
Dependency			Test plan		
			Data conversion plan		
			Training plan		
Is one person responsible for each deliverable?	❏	❏	Have QA deliverables/tasks been included?	❏	❏
Have all deliverables and tasks needed to accomplish objectives been included in plan?	❏	❏	Have naming conventions been followed?	❏	❏
			Project/phase		
			WBS		
			Resource name?		
Did you use the PMO template?	❏	❏	Does the PM project schedule include time for efforts such as issue, change request, quality, scope, and contract management	❏	❏
Is project schedule documented in PM tools?	❏	❏	Has PMO been informed? Is linkage to general ledger (GL) established?	❏	❏
Do project team members agree to record the following data weekly?	❏	❏	Has the original/revised plan been approved/reviewed by the following individuals?	❏	❏
Actual effort			Infrastructure group		
Forecasted effort			Project/program manager		
Actual completion			Business project manager		
			Business and IT project sponsor		

The components of risk management are as follows:

- Project charter and risk management policies;

12. Risk response strategies: A document that describes a strategy for avoiding, transferring, mitigating, and accepting potential risks.

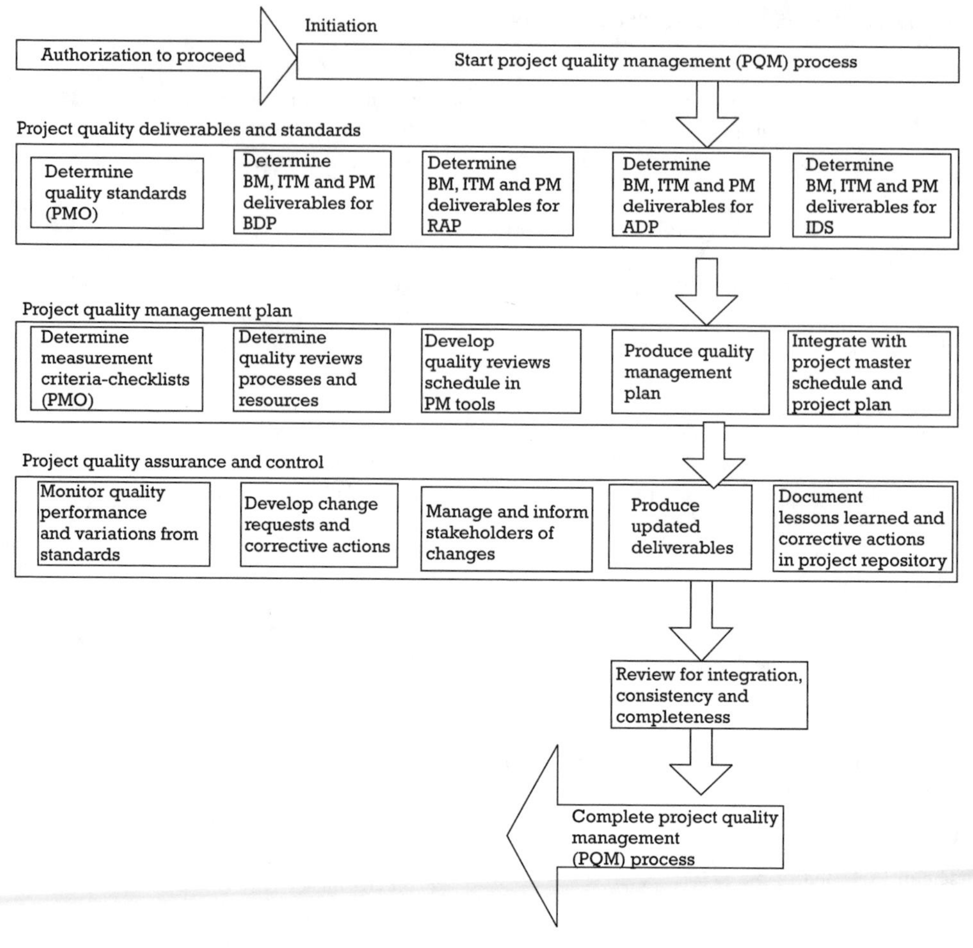

Figure 4.8 PQM process flow.

- PRM management plan;
- Project risk;
- Project risk response plan.

4.3.5.1 Purpose

The purpose of this process is to provide a risk response plan to ensure that an appropriate degree of risk strategies has been applied to the project risks and that appropriate category of risks are identified and quantified during each phase of the project. This information will form the baseline to manage, integrate, and prioritize the potential risks of the development projects. The program steering committee will use this risk response plan as a baseline in approving the risk response strategies.

4.3.5.2 Policy

The policy of project risk management is the following:

Table 4.13 Quality Management Checklist

PQM Criteria	*Yes*	*No*
Initiation		
Is management committed?	❑	❑
Have a PM sponsor and steering committee been established?	❑	❑
Is there a PM methodology?	❑	❑
Does the PM methodology have a PQM component?	❑	❑
Has a program delivery manager been assigned?	❑	❑
Does the assigned program delivery manager understand the PQM process?	❑	❑
Quality deliverables and standards		
Do quality standards exist?	❑	❑
Are the quality standards approved?	❑	❑
Are the deliverables for PDP completed?	❑	❑
Are the deliverables for RAP completed?	❑	❑
Are the deliverables for ADP completed?	❑	❑
Are the deliverables for IDP completed?	❑	❑
Quality management plan		
Are quality requirements determined?	❑	❑
Are quality checklists available–PMO?	❑	❑
Is there a quality management plan?	❑	❑
Are quality reviews scheduled?	❑	❑
Are quality review deliverables and activities included in project plan?	❑	❑
Is there an alignment with quality control?	❑	❑
QA and quality control		
Is there a quality control system?	❑	❑
Are quality performance reporting guidelines available?	❑	❑
Do variations from the quality baseline exist?	❑	❑
Are stakeholders informed about quality changes?	❑	❑
Is quality management plan and deliverables updated?	❑	❑
Are lessons learned documented in the project file	❑	❑
Appendixes		
Do the appendixes include details on defect statistics and corrective actions?	❑	❑

- Project risks shall be identified, quantified, and managed within the stakeholders' risk tolerance levels established for the project.
- Project risks, priorities, and response strategies shall be included in the project charter.
  - Project risks shall be identified and assessed based on established risk categories.
- The minimum level of risk tolerance shall be established by PMO policies, guidelines, and procedures. Specific project requirements may require higher risk tolerance levels, which shall be documented in the project risks management report.

4.3.5.3 Roles and responsibilities

Executive business manager responsibilities include the following:

- Provide executive-level business approval to the program business manager to ensure that the risks and risk response plan are managed

adequately in order to control the risk management process within budgetary and schedule constraints.
- Provide executive-level business commitment for the risk response plan.
- Approve all business changes to the approved risk response plan.
- Provide business resources to support development of the risk response plan.
- Act as the overall program sponsor and chairman of the program steering committee.

Executive IT manager responsibilities include the following:

- Provide executive-level IT approval to the program IT manager to ensure that the risks and risk response plan is managed adequately in order to control the risk management process within budgetary and schedule constraints.
- Provide executive-level IT commitment for the risk response plan.
- Approve all IT changes to the approved risk response plan.
- Provide IT resources to support development of the risk response plan.
- Act as the program sponsor from the IT perspective.

Executive project management manager responsibilities include the following:

- Provide executive-level PM guidance, direction, and advice to the program delivery manager to ensure that the required tolerance levels for the risk response plan and risk management process are effectively communicated in order to manage and control the risk management process.
- Integrate approved business and IT risk response strategies.
- Integrate changes to business and IT risk management processes and plans and changes to risk management deliverables.
- Provide PM resources for delivering risk response strategies and plan.
- Act as the program sponsor with integrated business and IT responsibility and PM accountability; champion PM processes, tools, and techniques.
- Provide risk management guidance, direction, and advice to the program delivery manager to ensure that the project risks are managed properly.

Program business manager responsibilities include the following:

- Submit details on business guidance, direction, and advice to the program delivery manager to ensure the required risk response plan and process delivers maximum overall business benefits to the company within project scope, cost, and time constraints.
- Report on business commitment to the approved risk response plan.
- Report on all business changes to the approved risk response plan.
- Assign business resources for supporting the delivery of the risk response plan.

- Report to the program steering committee with direct accountability to the business executive, and keep the program working committee informed of the current status of the risks response strategies.

Program IT manager responsibilities include the following:

- Submit details on IT guidance, direction, and advice to the program delivery manager to ensure the required risk response plan delivers maximum overall business and IT benefits to the company within project scope, cost, time, and quality constraints.
- Report on IT commitment to the approved risks response plan.
- Report on all IT deviations from the identified risk response plan.
- Assign IT resources for supporting the delivery of the risk response plan.
- Report to the program steering committee with direct accountability to the business executive, and keep the program working committee informed of the current status of the risks response strategies.

Program delivery manager responsibilities include the following:

- Define what the project managers can and cannot do when a risk occurs.
- Establish an agreed-upon process for submitting the risk and evaluating its impact on the current scope, schedule, cost, and quality baseline.
- Conduct risk reviews and facilitate risk response strategies based on issues and change requests documented in the issue and change requests logs.
- Verify that the project team adheres to PMO risk management policies and procedures by utilizing the checklists included in the risk management plan.
- Manage the potential risks by reporting on the contents and corrective actions to the appropriate management team.
- Update the risk management process as needed to keep it effective.
- Report to the program steering committee with direct accountability to the PM executive, and keep project managers and stakeholders informed of the current status of the risks.

Project team (project manager and team) responsibilities include the following:

- Develop the initial risk response plan during the project initiation and definition phase, and manage the risk plans through each phase of the PDLC. (The risk response plan will consist of identifying, analyzing, and mitigating the project risk. The risk checklist is available to help identify potential risks in three categories: known, predictable, and unpredictable. A risk identification template is also available for identifying potential risk areas.)
- Analyze and estimate the probability of the risk becoming a reality and its potential cost and priority so that focus can be directed to the most critical items.

- Develop risk mitigation plans to identify the action to be taken to eliminate, reduce, transfer, or control the project risk.
- Communicate the risks to program managers—business, IT, and project management.

4.3.5.4 Procedures: risk management

The roles and responsibilities discussed in Section 4.3.5.3 are brief because they are intended to serve as a reference as to who does what during the implementation of the risk management processes. Procedures are now introduced to demonstrate how this risk management process is executed using the major standard deliverables (what), process flow (how), and checklist or measurement criteria (why) templates. The process is intended to serve as a reference based on the English-language interrogatives: who does what, how, and why. The deliverables, process flow, and checklist templates, based on real-world practical implementation, will serve as excellent references for practicing project managers during the implementation of risk management processes.

The deliverables template in Table 4.14 highlights the key deliverables that the practicing project manager manages during the implementation of risk management. The risk management report model in Table 4.15, based on real-world practical implementations, will provide useful references for project managers. The risk category and risk assessment templates model in Table 4.15 shows how to identify and assess risks in order to ensure consistency and integration of potential risks during the development and prioritization of risk response strategies.

The process flow template in Figure 4.9 highlights the major processes to produce the risk management deliverables. These processes reference the deliverables to show the integrated nature and dependencies of the deliverables.

The checklist template in Table 4.16 highlights the major measurement criteria to ensure delivery of quality risk management deliverables identified in the deliverables template and adherence to the process flows identified in the process flow template. The checklist is used to objectively determine success criteria during execution of the risk management process.

4.3.6 Project communications management

There is a profound difference between information and meaning.
—Warren Bennis

When all is said and done, a lot more is said than done in reporting project progress, so be alert.

Executive management shall use the communications management report to ensure that the appropriate level of project status reporting has been communicated to the stakeholders. PCM provides formal written communication between the project team and senior management. The project

Table 4.14 Deliverables Template: Risk Management Report

Section 1—Introduction
This section includes an executive summary, risk management policies, and approach and introduces the report.
Section 2—Project Charter and Risk Management Policies
This section highlights the major deliverables of the PMBOK risk planning process by providing the scope baseline—project justification, project scope, product scope, scope management plan, scope performance reports—described in the project charter and supporting PMO risk management policies. This scope baseline document produced during the scope management process is used in Section 3 to establish the baseline for the development of the risk management plan and in Section 4 in the identification of the risks.
Section 3—Risk Management Plan
This section highlights the major deliverables of the PMBOK risk planning and risk analysis or assessment processes by providing processes, guidelines, and schedules for how and when the risk management processes—from risk identification to risk response planning—will be managed. These risk management processes and schedules integrate with the project plan produced during the PDP. This risk management plan establishes the baseline project risk processes and procedures for identifying and prioritizing, assessing, and managing risks.
Section 4—Risk
This section highlights the major deliverables of the PMBOK risk identification and risk analysis processes by providing the risk assessment worksheet, which consists of identified risks, assessment results, and resulting risk strategies. This worksheet is developed using the project charter deliverables outlined in Section 1 in accordance with the processes, guidelines, and schedules outlined in Section 3.
Section 5—Risk Response Plan
This section highlights the major deliverables of the PMBOK risk response planning, risk monitoring, and risk control processes by providing risk response plans and risk progress reports containing risks, risk response strategies, and suggested corrective actions for risks. These risk progress reports integrate with performance reports produced during the communications management process and are used to measure the progress of the risk based on their alignment with the risk worksheet. The corrective actions are documented as risk change requests and integrate with the change management process.
Appendixes—Risk Responses
Appendixes provide further details on risk analysis and response plans.

managers and the program delivery manager will compile the project biweekly and monthly status reports that represent the project team status, progress, and forecasts. The reports for each project will be compiled by the PMO into a portfolio and will be distributed each month to the program steering committee meetings and to other stakeholders, as appropriate.

The communications management process presented in this book is similar to the communications management knowledge area of PMBOK, but focuses more on program/project status reporting. The communications management plan[13] is established based on the stakeholders' communications requirements to ensure the completeness, consistency, and integrity of project status reports.

The components of communications management are as follows:

13. Communications management plan: A document that describes by whom, when, and how project information will be gathered, distributed or disseminated, and stored. It is part of the integrated project plan.

Table 4.15 Deliverables Template: Risk Management Report Model

Risks Category	*Risk: Uncertainty Probability*		
	High	*Medium*	*Low*
Technical			
Hardware			
System software/database			
Application software			
Network			
Facilities			
Requirements			
Scope			
Deliverables			
Contracts			
Methodology			
Environment			
Organization			
Policies			
Space			
Politics			
Culture			
Staff			
Management skills			
Technical skills			
Project management skills			
Consultant skills			

Risk Template–Risk Assessment

Risk #	*Risk Category (T, R, E, S)*	*Risk Title (Text)*	*Risk Priority (H, M, L)*	*Risk Impact (C, T, Q, S)*	*Risk Probability (H, M, L)*	*Risk Cost (H, M, L)*	*Risk Strategy (Text)*	*Risk Response Strategy (V, T, M, A)*

(T, R, E, S): technical, requirements, environment, staff;
(V, T, M, A): avoid, transfer, mitigate, accept;
(C, T, Q, S): cost, time, quality, scope;
(H, M, L): high, medium, low.

- Project communication requirements;
- PCM plan;
- Project repository and PMIS;
- Project performance reports.

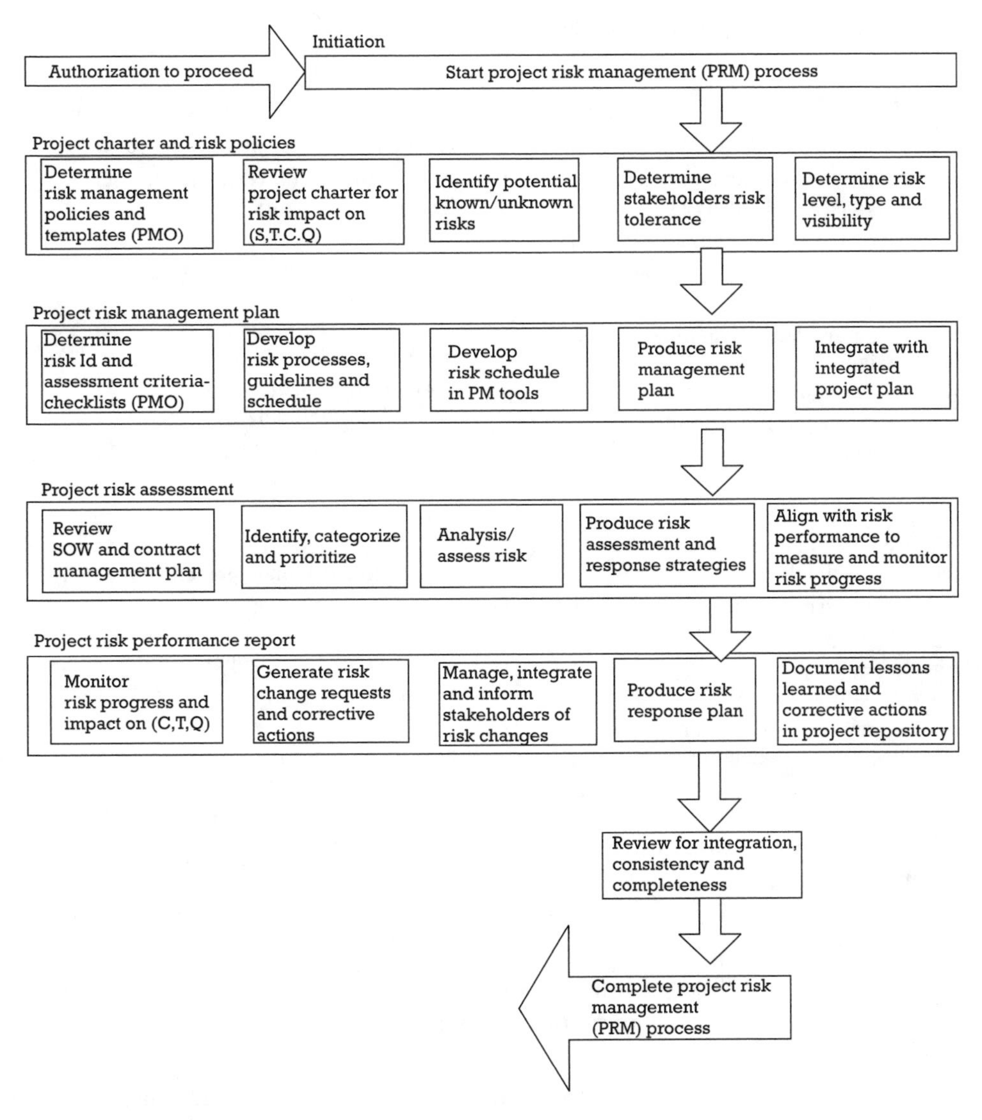

Figure 4.9 PRM process flow.

4.3.6.1 Purpose

The purpose of this process is to provide a communications management plan to ensure that the appropriate level of project status reporting has been communicated to the stakeholders. This information will form the baseline to communicate, integrate, and prioritize the project progress. The program steering committee will use this communications management plan as a baseline in determining the project progress for approvals to proceed with further developments.

4.3.6.2 Policy

The policy of project communications management is the following:

Table 4.16 Risk Management Checklist

PRM Criteria	*Yes*	*No*
Initiation		
Is management committed?	❑	❑
Have a PM sponsor and steering committee been established?	❑	❑
Is there a PM methodology?	❑	❑
Does the PM methodology have PRM component?	❑	❑
Has a program delivery manager been assigned?	❑	❑
Does the assigned program delivery manager understand the PRM process?	❑	❑
Project charter and risk policies		
Do risk policies exist?	❑	❑
Is the impact on cost, time, scope, and quality determined?	❑	❑
Are the risks documented in the project charter?	❑	❑
Are the stakeholders' risk tolerance levels identified?	❑	❑
Are risks prioritized and risks responses defined?	❑	❑
Are risk identified based on established categories?	❑	❑
Risk management plan		
Is there a risk management template available?	❑	❑
Are risk assessment guidelines available?	❑	❑
Does schedule include risk management activities?	❑	❑
Does integrated project plan include a risk management plan?	❑	❑
Is the risk management plan approved?	❑	❑
Is there an alignment with potential risks?	❑	❑
Risk assessment		
Does contractor's SOW include a risk management plan?	❑	❑
Are risk assessments based on risk response strategies?	❑	❑
Is there a risk impact analysis?	❑	❑
Is the budget for risk management acceptable?	❑	❑
Are risk response strategies approved by the stakeholders?	❑	❑
Is there an alignment with risk performance?	❑	❑

- Every project manager shall provide biweekly and monthly status reports in accordance with the schedule and format provided by the PMO.
- Project managers and program delivery manager shall develop a project communications plan that will form an integral component of the integrated project plan.
- A project repository and PMIS[14] shall be used to produce project status reports to ensure completeness, consistency, and integrity of project progress.
- Project performance reports shall communicate the impact to scope, schedule, cost, and quality baselines.

4.3.6.3 Roles and responsibilities

Executive business manager responsibilities include the following:

14. Project management information system (PMIS): An automated system that is used to access, manipulate, integrate, and update the project repository. It integrates with various subsystems such as the project management forecasting system (PMFS) and the project management tracking system (PMTS), and various project repositories.

- Provide executive-level business approval to the program business manager to ensure that the communications management plan is managed adequately in order to control the communications management process within budget and schedule constraints.
- Provide executive-level business commitment to the communications management plan.
- Approve all business changes to the approved communications management plan.
- Approve business resources to support the development of the communications management plan.
- Act as the overall program sponsor and chairman of the program steering committee.

Executive IT manager responsibilities include the following:

- Provide executive-level IT approval to the program IT manager to ensure that the communications management plan is managed adequately in order to control the communications management process within budgetary and schedule constraints.
- Provide executive-level IT commitment for the communications management plan.
- Approve all IT changes to the approved communications management plan.
- Provide IT resources to support development of communications management plan.
- Act as the program sponsor from the IT perspective.

Executive project management manager responsibilities include the following:

- Provide executive-level PM guidance, direction, and advice to the program delivery manager to ensure that the required levels of communications management plan and communications management process are effectively communicated in order to manage and control the communications management process.
- Integrate business and IT communications management requirements.
- Integrate changes to business and IT communications management processes and plans with changes to communications management deliverables.
- Provide PM resources for delivering the communications management plan.
- Act as the program sponsor with integrated business and IT responsibility and PM accountability; champion PM processes, tools, and techniques.
- Provide communications management guidance, direction, and advice to the program delivery manager to ensure the project progress

reporting provides communications value to the stakeholders in order to optimize scope, schedule, cost, and quality performance.

Program business manager responsibilities include the following:

- Submit details on business guidance, direction, and advice to the program delivery manager to ensure the required communications management plan and process delivers maximum overall business benefits to the company in support of project scope, cost, and time constraints.
- Report on business commitment to the approved communications management plan.
- Report on all business changes to the approved communications management plan.
- Assign business resources for supporting the delivery of the communications management plan.
- Report to the program steering committee with direct accountability to the business executive, and keep the program working committee informed of the current status of the project on a monthly basis.

Program IT manager responsibilities include the following:

- Submit details on IT guidance, direction, and advice to the program delivery manager to ensure the required communications process delivers maximum overall business and IT benefits to the company in support of project scope, cost, time, and quality constraints.
- Report on IT commitment to the approved communications management plan.
- Report on all IT changes to the communications plan.
- Assign IT resources for supporting the development of the communications plan.
- Report to the program steering committee with direct accountability to the business executive, and keep the program working committee informed of the current status of the projects.

Program delivery manager responsibilities include the following:

- Define what the project managers can and cannot do when producing status reports.
- Establish an agreed-upon process for submitting a status report and evaluating its impact on the current baseline.
- Conduct project biweekly and monthly reviews of the status reports for the approved project plan.
- Verify that the project team adheres to PMO communications management policies, guidelines, and procedures.
- Verify that the project meets the project-specific communication requirements by utilizing the checklists and are included in the project plan.

- Manage the communications management plan by reporting on the contents and recommended solutions to the appropriate management team.
- Update communications management process as needed to keep it effective.
- Compile status reports into a portfolio and distribute each month to the program steering committee.
- Report to the program steering committee with direct accountability to the PM executive, and keep project managers and stakeholders informed of the current status of the project within the program.

Project team (project managers and team) responsibilities include the following:

- Project managers to compile biweekly and monthly status reports according to PMO guidelines.
- Project managers to submit status reports for the project team to PMO. The program delivery manager will present the progress of projects at the monthly program steering committee meetings.
- Project managers to attend biweekly status meetings called by the program working committee chairperson–program delivery manager.
- Project managers to integrate, assemble, and report to the program working committee, on the progress of the project.
- Project team to communicate progress during development of deliverables.

4.3.6.4 Procedures: communications management

The roles and responsibilities discussed in Section 4.3.6.3 are brief because they are intended to serve as a reference as to who does what during the implementation of the communications management processes. Procedures are now introduced to demonstrate how this communications management process is executed using the major standard deliverables (what), process flow (how), and checklist or measurement criteria (why) templates. The process is intended to serve as a reference based on the English-language interrogatives: who does what, how, and why. The deliverables, process flow, and checklist templates, based on real-world practical implementation, will serve as excellent references for practicing project managers during implementation of the communications management processes.

The deliverables template in Table 4.17 highlights the key deliverables that the practicing project manager manages during the implementation of communications management. The PMIS reporting model in Figure 4.10, based on real-world practical implementations, will provide useful references for project managers. The PMIS model is a representation of an IPM reporting process in order to ensure completeness, consistency, and integration of project progress reports during the IT PDLC.

Table 4.17 Deliverables Template: Communications Management Report

Section 1—Introduction
This section includes an executive summary, communications management policies, and approach and introduces the report.
Section 2—Project Communications Requirements
This section highlights the major source of the PMBOK communications planning process by providing the communications requirements of the project stakeholders. These communications requirements establish the baseline for the development of the communications management plan in Section 3 and the development of the project repository and PMIS in Section 4.
Section 3—PCM Plan
This section highlights the major deliverables of the PMBOK communications planning process by providing a plan for the plan's structure and contents (what), methods (how), schedule (when), organization or staff (who), and location (where) in which the project status reports will be generated, collected, disseminated, stored, and disposed of. This communications management plan integrates with the integrated project plan produced during the PDP. It establishes the baseline information directory for managing the project documentation and distributing the information to the stakeholders.
Section 4—Project Communications–Reporting Repository and PMIS
This section highlights the major deliverables of PMBOK information distribution processes by providing the repository contents and information retrieval system for storing, retrieving, accessing, disseminating, and deleting project performance reports. This PMIS supports the communications requirements outlined in Section 1, based on the processes, guidelines, and schedule outlined in Section 3.
Section 5—Project Communications–Performance Reports
This section highlights the major deliverables of the PMBOK performance reporting by providing project status, progress, forecasting performance reports, and suggested corrective actions for the variations to scope, schedule, cost, and quality. These performance reports represent an integration of the performance reports produced the scope, time, cost, quality, risk, and contract project management processes, based on the integration features of the repository and PMIS outlined in Section 4. The corrective actions are documented as communications change requests and integrate with the integrated change management process.
Appendixes—Project Communications–Performance Reports
Appendixes provide further details on the PMIS and performance reports.

The process flow template in Figure 4.11 highlights the major processes to produce the communications management deliverables. These processes reference the deliverables to show the integrated nature and dependencies of the deliverables.

The checklist template in Table 4.18 highlights the major measurement criteria to ensure delivery of quality deliverables identified in the deliverables template and adherence to the process flows identified in the process flow template. The checklist is used to objectively determine success criteria during execution of the communications management process.

4.3.7 Project human resources management

An expert is one who knows more and more about less and less until he knows absolutely everything about nothing.
—Murphy's Law

Executive management shall use the human resources management report to ensure that the human resource utilizations are optimized to an acceptable level of staff knowledge, skills, and experiences. Optimizing human

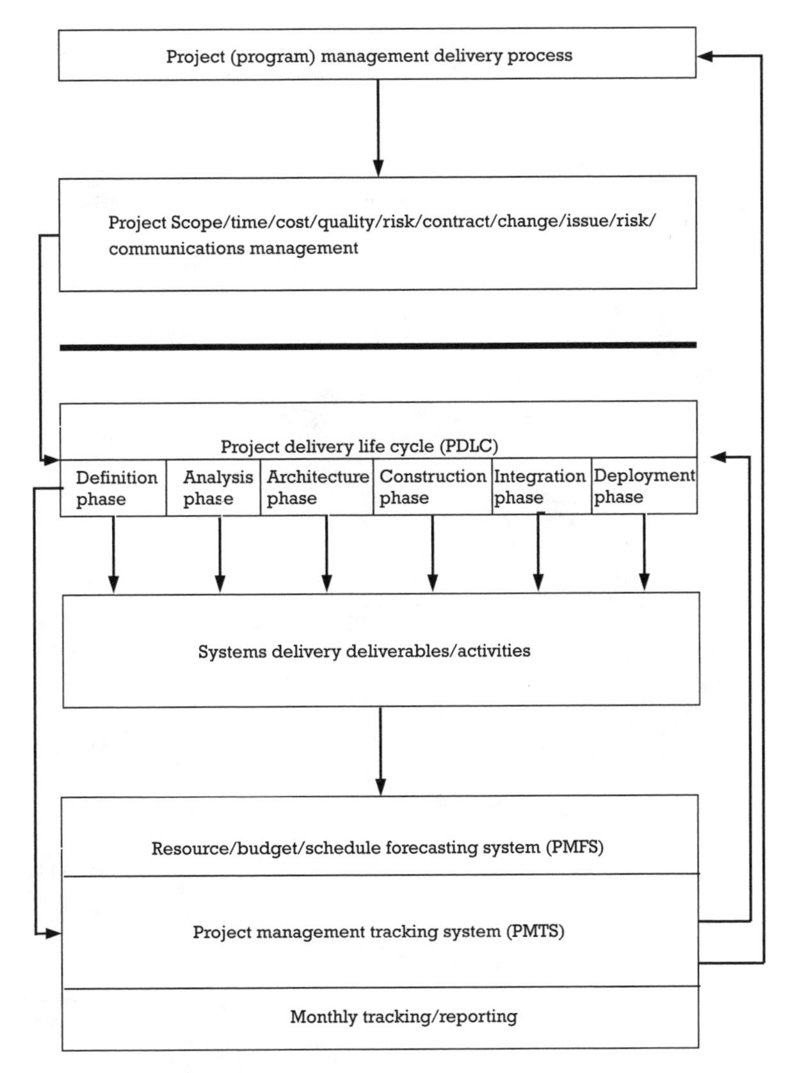

Figure 4.10 PMIS reporting model.

resources efforts is often the key to minimizing project costs. Human resources management ensures that the right staff members with the right skills are assigned the right tasks at the right time to optimize the utilization of people resources. The project managers and the program delivery manager will establish a responsibility assignment matrix (RAM)[15] that represents the project team's current and forecasted skills requirements. The RAM for each project and the overall program will be compiled by the PMO into a portfolio and will be distributed each month at the program steering committee meetings and to other stakeholders, as appropriate.

The human resources management process presented in this book is similar to the human resources management knowledge area of PMBOK,

15. RAM: A structure used to relate the WBS to individual resources to ensure that each element of the project's scope of work is assigned to an individual PMBOK.

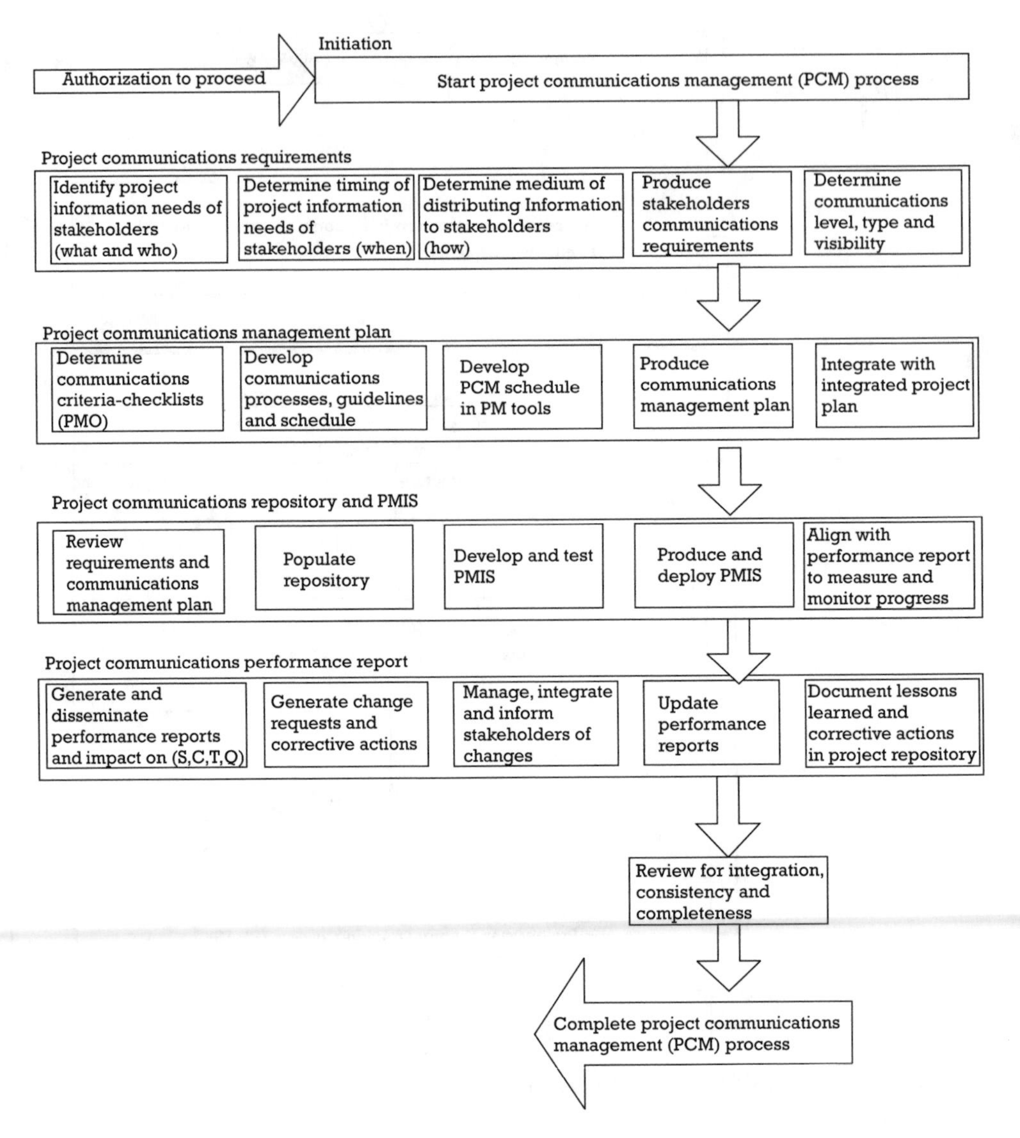

Figure 4.11 PCM process flow.

but focuses more on program/project staff requirements–RAM and staff management plan.[16] The project staff management plan is established based on the project staff requirements to ensure optimum utilization of human resources during each phase of the IT PDLC.

The components of human resource management are as follows:

- Project staff requirements;
- Project staff management plan;
- Project staff;

16. Staff management plan: Document that describes when and how human resources will become part of the project team and when they will return to their organizational units; it may be a part of the overall project plan PMBOK.

Table 4.18 Communications Management Checklist

PCM Criteria	*Yes*	*No*	*PCM Criteria*	*Yes*	*No*
Initiation			*Communications requirements*		
Is management committed?	❑	❑	Are stakeholders identified?	❑	❑
Have a PM sponsor and steering committee been established?	❑	❑	Are stakeholders' information needs determined?	❑	❑
Is there a PM methodology?	❑	❑	Are the media for report distribution determined?	❑	❑
Does the PM methodology have PCM component?	❑	❑	Is the timing for report distribution determined?	❑	❑
Has a program delivery manager been assigned?	❑	❑	Are communications requirements documented?	❑	❑
Does the assigned program delivery manager understand the PCM process?	❑	❑	Is there a communications management strategy?	❑	❑
Communications management plan			*Communications repository and PMIS*		
Is there a communications management plan?	❑	❑	Is there a PM repository?	❑	❑
Are guidelines available?	❑	❑	Is there a PMIS?	❑	❑
Is the communications plan accepted?	❑	❑	Does the PMO manage the PM repository and PMIS?	❑	❑
Are communications activities included in project plan?	❑	❑	Is the PM repository populated?	❑	❑
Is the communications plan part of the integrated project plan?	❑	❑	Are the PM repository and PMIS communicated to project team?	❑	❑
Is the communications plan documented in the repository?	❑	❑	Is there an alignment with communications performance reports?	❑	❑
Communications performance reports			*Appendixes*		
Is there a communications control process in PMIS?	❑	❑	Do the appendixes include further details on PMIS and performance reporting?	❑	❑
Are project performance reporting guidelines available?	❑	❑			
Do variations from the communications plan exist?	❑	❑			
Are stakeholders informed about changes?	❑	❑			
Are performances updated?	❑	❑			
Are lessons learned documented in a project file?	❑	❑			
	❑	❑			

- Project staff performance report.

4.3.7.1 Purpose

The purpose of this process is to provide a RAM and staff management plan to ensure that the right staff members with the right skills are assigned the right tasks at the right time to optimize the utilization of people resources. This information will form the baseline to communicate, integrate, forecast, and optimize utilization of people resources during each phase of the project. The program steering committee will use the RAM and staff management plan as a baseline in determining the current and forecasted resource requirements for approvals to proceed with further developments.

4.3.7.2 Policy

The policy of project human resources management is the following:

- Every project manager shall provide biweekly and monthly staff utilization reports for each phase of the project, according to the schedule and format provided by the PMO.
- Project managers/program delivery managers shall develop RAM and a human resources management plan that will form an integral part of the integrated project plan.
- The project repository and PMIS shall be used to produce staff utilization reports to ensure completeness, consistency, integrity, and optimization of people resources.
- Project staff performance reports shall communicate the impact to scope, schedule, cost, and quality baselines.

4.3.7.3 Roles and responsibilities

Executive business manager responsibilities include the following:

- Provide executive-level business approval to the program business manager to ensure that the human resource management plan is executed effectively in order to optimize human resource utilization within budgetary and schedule constraints.
- Provide executive-level business commitment for human resource assignments.
- Approve all business changes to the approved human resource management plan.
- Approve business resources to support development of the human resource management plan.
- Act as the overall program sponsor and chairman of the program steering committee.

Executive IT manager responsibilities include the following:

- Provide executive-level IT approval to the program IT manager to ensure that the human resource management plan is executed effectively in order to optimize human resource utilization within budgetary and schedule constraints.
- Provide executive-level IT commitment for human resource assignments.
- Approve all IT changes to the approved human resource management plan.
- Approve IT support resources to support development of the human resource management plan.
- Act as the program sponsor from the IT perspective.

Executive project management manager responsibilities include the following:

- Provide executive-level PM guidance, direction, and advice to the program delivery manager to ensure that the required levels of human resource skills are effectively communicated and to manage and control the human resource management process.
- Integrate approved business and IT human resource management requirements.
- Integrate changes to business and IT human resource management processes and plans and changes to staff assignments.
- Provide PM resources for developing the human resource management plan.
- Act as the program sponsor with integrated business and IT responsibility and PM accountability; champion PM processes, tools, and techniques.
- Provide human resource management guidance, direction, and advice to the program delivery manager to ensure resource utilization is optimized within scope, schedule, cost, and quality constraints.

Program business manager responsibilities include the following:

- Submit details on business guidance, direction, and advice to the program delivery manager to ensure the required human resource management plan and staff utilization delivers maximum overall business benefits to the company within the project scope, cost, time, and quality constraints.
- Report on business commitment to the approved human resource management plan.
- Report on all business changes to the approved human resource management plan.
- Assign business resources for supporting development of the human resource management plan.
- Report to the program steering committee with direct accountability to the business executive, and keep the program working committee informed of the current status of staff utilization for each phase of the project on a monthly basis.

Program IT manager responsibilities include the following:

- Submit details on IT guidance, direction, and advice to the program delivery manager to ensure the required human resource management plan and staff utilization delivers maximum overall business and IT benefits to the company within the project scope, cost, time, and quality constraints.
- Report on IT commitment to the approved human resources management plan.
- Report on all IT changes to the approved human resource management plan.

- Assign IT resources for supporting development of the human resource management plan.
- Report to the program steering committee with direct accountability to the business executive, and keep the program working committee informed of the current status of staff utilization for each phase of the project on a monthly basis.

Program delivery manager responsibilities include the following:

- Define what project managers can and cannot do when assigning staff to a project
- Establish an agreed-upon process for submitting resource utilization and evaluating its impact on the current baseline.
- Conduct project biweekly and monthly reviews for resource utilization based on the approved project plan.
- Verify that the project team adheres to PMO human resource management policies, guidelines, and procedures.
- Verify that the project meets the project-specific staff requirements by utilizing the checklists included in the project plan.
- Manage the human resource management plan by reporting on staff utilization and recommend solutions to the appropriate management team.
- Update the human resource management process as needed to keep it effective.
- Report to the program steering committee with direct accountability to the PM executive, and keep project managers and stakeholders informed of the current status of the staff utilization for each phase of the project on a monthly basis.

Project team (project managers and team) responsibilities include the following:

- Project managers to compile biweekly and monthly staff utilization reports according to PMO guidelines.
- Project managers to submit staff utilization reports for the project team to PMO and program delivery manager to present the staff utilization progress at the monthly program steering committee meetings.
- Project managers to attend biweekly status meetings called by the program working committee chairperson–program delivery manager and to report on staff progress.
- Project managers to integrate, assemble, and report to the program working committee on the progress of effort utilization for each phase of the project.
- Project team to produce deliverables based on the effort baseline.

4.3.7.4 Procedures: human resources management

The roles and responsibilities discussed in Section 4.3.7.3 are brief because they are intended to serve as a reference as to who does what during the implementation of the human resources management processes. Procedures are now introduced to demonstrate how this human resources management process is executed using the major standard deliverables (what), process flow (how), and checklist or measurement criteria (why) templates. The process is intended to serve as a reference based on the English-language interrogatives: who does what, how, and why. The deliverables, process flow, and checklist templates, based on real-world practical implementation, will serve as excellent references for practicing project managers during implementation of the human resources management process.

The deliverables template in Table 4.19 highlights the key deliverables that the practicing project manager manages during the implementation of human resources management. The responsibility assignment model in Table 4.20, based on real-world practical implementations, will provide useful references for project managers. The RAM model is a representation of staff

Table 4.19 Deliverables Template: Human Resources Management Report

Section 1—Introduction
This section includes an executive summary, human resource management policies, organizational chart, and approach and introduces the report.
Section 2—Project Staffing Requirements
This section highlights the major source of the PMBOK organizational planning process by providing the staffing requirements of the project stakeholders. These staffing requirements establish the baseline for development of the PSM management plan in Section 3 and the assignment of individuals to the project team in Section 4.
Section 3—Project Staffing Management Plan
This section highlights the major deliverables of PMBOK organizational planning process by providing a plan on the staff responsibilities (who, what—responsibility assignment matrix), schedule (when), and methods (how)—efforts/costs—for hiring/releasing human resources in/out of the project team. This PSM management plan integrates with the project plan produced during the PDP. It establishes the baseline staffing plan for managing the project team and distributing the human resource assignment information to the stakeholders.
Section 4—Project Staff
This section highlights the major deliverables of the PMBOK staff acquisition processes by providing the project team directory, current and forecasted project staff and staff assignment matrixes that support the PSM management plan in Section 3, and the company's hiring policies. The staff assigned in the project team directory must integrate and be consistent with the human resources assigned during project schedule or integrated project plan development and updates.
Section 5—Project Staff Performance Reports
This section highlights the major deliverables of the PMBOK team development process by providing individual and project team performance reports and suggested corrective actions for improvements to skills and attitudes in producing the deliverables within the scope, schedule, cost, and quality constraints. These human resource performance reports must be aligned with the individual and project team progress status reports produced during the PCM process. The corrective actions or improvements form the basis for input to project staff performance appraisals.
Appendixes—Project Team Performance Reports
Appendixes provide further details on the RAM and project staff performance reports.

Table 4.20 RAM Report Model

Roles / *Phases/ Deliverables*	*Executive Business Manager*	*Executive IT Manager*	*Executive PM Manager*	*Program Business Manager*	*Program IT Manager*	*Program Delivery Manager*	*Project Manager and Team*
Project definition							
PDM	RA	R	R	S	S	M/I	C
Requirements analysis							
RAM demonstration prototype	RA	R	R	S	S	M/I	C
Architecture design							
PAS working prototype	R	RA	R	S	S	M/I	C
Iterative developments–construction							
Iteration #1	R	RA	R	S	S	M/I	C
Iterative developments–integration							
Iteration #2	R	RA	R	S	S	M/I	C
Iterative developments–deployments							
Iteration #3	RA	R	R	S	S	M/I	C

Note: C = create; R = review; A = approve; M/I = manage/integrate; S = support
The shaded areas represent the major project management deliverables produced during the PDLC phase.

assignments in order to ensure completeness, consistency, integration, and optimization of resource utilization during each phase of the IT PDLC.

The process flow template in Figure 4.12 highlights the major processes to produce the human resources management deliverables. These processes reference the deliverables to show the integrated nature and dependencies of the deliverables.

The checklist template in Table 4.21 highlights the major measurement criteria to ensure delivery of quality deliverables identified in the deliverables template and adherence to the process flows identified in the process flow template. The checklist is used to objectively determine success criteria during execution of the human resources management process.

4.3.8 Project contract management

Some contractors and/or consultants are like the bottom half of a double boiler; they get all heated up, but don't know what's cooking.—Unknown

Executive management shall use the contract management report to ensure that the company's contract guidelines and procedures are used to select

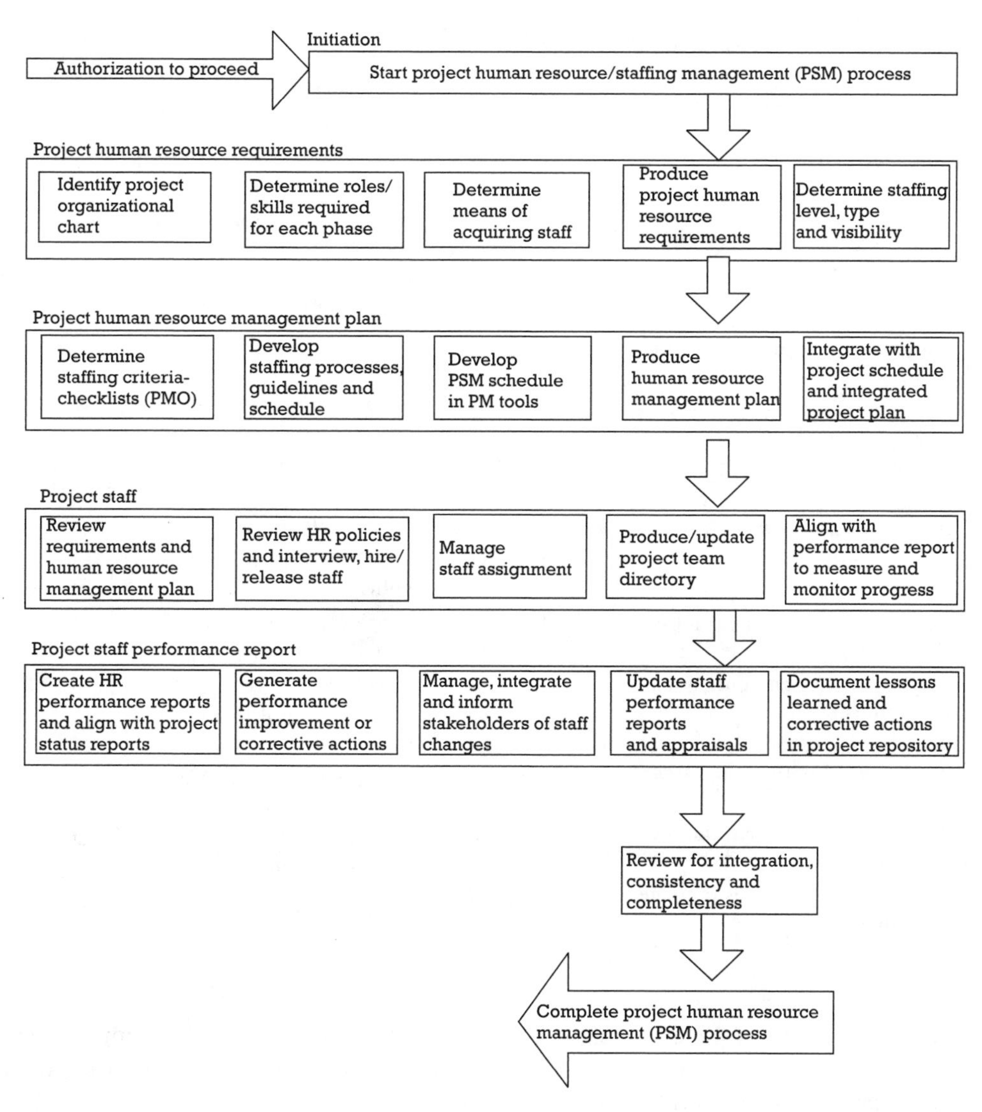

Figure 4.12 Project PSM process flow.

and manage contractors or consultants for fixed-priced or time-and-materials contracts. Optimizing contractors' efforts is the key to minimizing project costs. Contract management ensures that the right contracting staffs with the right skills are assigned the right tasks at the right time to optimize the utilization of contract resources. The project manager and program delivery manager will develop a SOW[17] that describes the stakeholders' requirements. The SOW for each project will be compiled by the PMO into a

17. Statement of work: Narrative description of products or services to be supplied under contract that states the specifications or other minimum requirements; quantities; performance dates, times, and locations, if applicable; and quality requirements. It serves as the basis for the contractor's response and as a baseline against which the progress and subsequent contractual changes are measured during contract performance.

Table 4.21 Human Resources–PSM Checklist

PSM Criteria	*Yes*	*No*	*PSM Criteria*	*Yes*	*No*
Initiation			*Human resources requirements*		
Is management committed?	❑	❑	Does a project organizational chart exist?	❑	❑
Have a PM sponsor and steering committee been established?	❑	❑	Is the project organizational chart approved?	❑	❑
Is there a PM methodology?	❑	❑	Are the roles and skills requirements determined?	❑	❑
Does the PM methodology have PSM component?	❑	❑	Are the means of acquiring staff determined?	❑	❑
Has a program delivery manager been assigned?	❑	❑	Are the human resources requirements documented?	❑	❑
Does the assigned program delivery manager understand the PSM process?	❑	❑	Is there a project staffing strategy?	❑	❑
Human resources management plan			*Project staff*		
Is there a human resources management plan?	❑	❑	Are there human resources policies?	❑	❑
Are human resources guidelines available?	❑	❑	Is project staff assigned?	❑	❑
Are the staff assignments documented in PM tools?	❑	❑	Is there a project staff directory?	❑	❑
Is the human resources management plan included within the integrated project plan?	❑	❑	Is the project directory updated?	❑	❑
Is the human resources management plan approved?	❑	❑	Are deliverables and tasks assigned to each member?	❑	❑
Is the human resources management plan documented in repository?	❑	❑	Is there an alignment with staff performance?	❑	❑

portfolio and will be distributed at the program steering committee meetings and to other stakeholders, as appropriate, for approval to proceed with contract negotiations.

The contract management process presented in this book is similar to the procurement management knowledge area of PMBOK, but focuses more on SOW–client requirements and contract management plan.[18] The project contract management plan is established based on the project contract requirements to ensure optimum utilization of contract resources during each phase of the IT PDLC.

The components of contract management are as follows:

- Project SOW and contract standards;
- Project contract management plan;
- Project contract;
- Project contract performance reports.

18. Contract management plan: A document that describes by whom, when, and how the project contract will be administered. It is part of the integrated project plan.

4.3.8.1 Purpose

The purpose of this process is to provide a SOW and contract management plan to ensure that the right contract resources with the right skills are assigned to the right tasks at the right time to optimize the utilization of people resources. Optimizing the contractor's effort is the key to minimizing project costs. This information will form the baseline to communicate, integrate, forecast, and optimize utilization of contract resources during each phase of the project. The program steering committee will use this SOW and project contract as a baseline in determining the current and forecasted contract resource requirements for approvals to proceed with further developments using contract resources.

4.3.8.2 Policy

The policy of project contract management is the following:

- Contractors shall be selected based upon established selection criteria with planning, tracking, and oversight activities to be appropriately performed.
- The contractor shall be selected based upon a balanced assessment of both technical and nontechnical criteria for the project. The project requirements shall be determined based the client SOW and the IT PDLC with the appropriate partitioning of the requirements into release(s).
- The contractor SOW shall be prepared, reviewed, agreed to, revised when necessary, and managed and controlled.
- The contractor's development plan to include the contractor's plan for tracking and communicating status shall be reviewed and approved by the project manager and program delivery manager.
- Changes to the contractor SOW, contract terms and conditions, and other commitments shall be made according to change control process.
- The project manager and the contractor management shall conduct periodic status and technical reviews.
- Formal reviews shall be conducted to address the contractor's accomplishments and results at selected milestones.
- QA reviews shall be conducted as appropriate.
- Acceptance testing shall be conducted as part of the delivery of the contractor's deliverable based on established acceptance criteria.

4.3.8.3 Roles and responsibilities

Executive business manager responsibilities include the following:

- Provide executive-level business approval to the program business manager to ensure that the contract is managed properly in order to control the contract management process within budgetary and schedule constraints.
- Provide executive-level business commitment for the contract.

- Approve all business support changes to the approved contract.
- Approve business resources to support development of the contract management plan.
- Act as the overall program sponsor and chairman of the program steering committee.

Executive IT manager responsibilities include the following:

- Provide executive-level IT approval to the program IT manager to ensure that the contract is managed properly; in order to control the contract management process within budgetary and schedule constraints.
- Provide executive-level IT commitment for the contract.
- Approve all IT changes to the approved contract.
- Approve IT resources to support development of the contract management plan.
- Act as the program sponsor from the IT perspective.

Executive project management manager responsibilities include the following:

- Provide executive-level PM guidance, direction, and advice to the program delivery manager to ensure that the required levels of contract negotiations are effectively communicated in order to manage and control the contract management process.
- Integrate approved business and IT requirements into contract.
- Integrate approved changes to business and IT requirements into contract and approve changes to meet quality requirements.
- Provide PM resources for development of contract management plan.
- Act as the program sponsor with integrated business and IT responsibility and PM accountability; champion PM processes, tools, and techniques.
- Provide contract management guidance, direction, and advice to the project managers to ensure the contractor delivers the SOW to provide overall business benefits to the company.

Program business manager responsibilities include the following:

- Submit details on business guidance, direction, and advice to the program delivery manager to ensure the required contract delivers maximum overall business benefits to the company within the project scope, cost, time, and quality constraints.
- Report on business commitment to the approved contract.
- Report on all business changes to the approved contract.
- Assign business resources for supporting the development of the contract management plan.
- Report to the program steering committee with direct accountability to the business executive, and keep the program working committee informed of the current status of the contract

Program IT manager responsibilities include the following:

- Submit details on IT guidance, direction, and advice to the program delivery manager to ensure the required contract delivers maximum overall business and IT benefits to the company within project scope, cost, time, and quality constraints.
- Report on IT commitment to the approved contract.
- Report on all IT changes to the approved contract.
- Assign IT resources for supporting the development of the contract management plan.
- Report to the program steering committee with direct accountability to the business executive, and keep the program working committee informed of the current status of the contracts.

Program delivery manager responsibilities include the following:

- Conduct contract reviews according to the approved contract management plan.
- Verify that the project meets business and IT contract policies, guidelines, and procedures.
- Verify that the project meets the project-specific contract goals by utilizing the checklists.
- Report variances to the appropriate management team and recommend solutions.
- Change the contract process as needed to keep it effective.
- Report to the program steering committee with direct accountability to the PM executive, and keep project managers and stakeholders informed of the current status of the contracts.

Project team (project managers and team) responsibilities include the following:

- Project manager to select the contractors or contracting firm;
- Project manager to negotiate the contract terms with the external contractor;
- Project manager to manage the contractors' work;
- Project manager to review and approve the contractor's development plan;
- Project manager to coordinate the technical scope to be contracted and the terms and conditions of the contract with the affected parties;
- Project manager to arrange for development and support of the accepted deliverables based on the contract agreement;
- Contract project manager to compile biweekly and monthly progress reports according to PMO guidelines;
- Contract project managers to submit contract progress reports for the project team to PMO and program delivery manager to present the progress at the monthly program steering committee meetings;

- Contract project manager to attend biweekly status meetings called by the program working committee chairperson–program delivery manager and to report on progress;
- Contract project manager to integrate, assemble, and report to the program working committee on the progress of the project for each phase of the project;
- Contract project team to produce deliverables based on the approved contract agreement.

4.3.8.4 Procedures: contract management

The roles and responsibilities discussed in Section 4.3.8.3 are brief because they are intended to serve as a reference as to who does what during the execution of the contract management processes. Procedures are now introduced to demonstrate how this contract management process is executed using the major standard deliverables (what), process flow (how), and checklist or measurement criteria (why) templates. The process is intended to serve as a reference based on the English-language interrogatives: who does what, how, and why. The deliverables, process flow, and checklist templates, based on real-world practical implementation, will serve as excellent references for practicing project managers during execution of the contract management process.

The deliverables template in Table 4.22 highlights the key deliverables that the practicing project manager manages during the implementation of contract management. The software evaluation/selection process (SES) model or guideline in Figure 4.13, based on real-world practical implementations, will provide useful references for project managers. This SES model shows a typical process for selecting a vendor or software and for contract assignments in order to ensure completeness, consistency, and integration during the selection of the vendor or software.

The process flow template in Figure 4.14 highlights the major processes to produce the contract management deliverables. These processes reference the deliverables to show the integrated nature and dependencies of the deliverables.

The checklist template in Table 4.23 highlights the major measurement criteria to ensure delivery of quality deliverables identified in the deliverables template and adherence to the process flows identified in the process flow template. The checklist is used to objectively determine success criteria during execution of the contract management process.

4.3.9 Project issue management

What happens is not as important as how you react to what happens.
—Thaddeus Golas

Executive management shall use the issue management report to ensure that project issues are documented, assessed, and prioritized based on

Table 4.22 Deliverables Template: Contract Management Report

Section 1—Introduction
This section includes an executive summary, contract overview, contract standards, and approach and introduces the report.
Section 2—Contract SOW Deliverables and Standards
This section highlights the major deliverables of the PMBOK procurement planning and solicitation planning processes by providing the client requirements—business, technical, and project management—described in the SOW and supporting company's and PMO standards. This SOW supports the scope baseline document–project charter produced during the scope management process, which is used to select the contractor. This SOW establishes the baseline for development of the contract management plan in Section 3 and the award and establishment of the contract agreement in Section 4.
Section 3—Contract Management Plan
This section highlights the major deliverables of the PMBOK procurement planning process by providing processes and schedule on how and when the contract management processes—from SOW to contract closure—will be managed. These contract processes and schedule integrate with the project plan produced during the PDP. This contract management plan establishes the baseline project contract processes and procedures for managing the contract.
Section 4—Contract
This section highlights the major deliverables of the PMBOK solicitation process by providing the agreement with the contractor and the client, consisting of the contractor's statement of work, or CSOW. This CSOW will support the client business, technical, and project management requirements, described in the CSOW document in Section 2 and according to the processes and schedule outlined in Section 3.
Section 5—Contract Performance Report
This section highlights the major deliverables of the PMBOK procurement administrative and procurement closure processes by providing contract progress reports containing contractors' performance and suggested corrective actions for the contract agreement. These contract performance reports integrate with performance reports produced during the communications management process and are used to measure the progress of the contract agreement based on alignment with the deliverables outlined in Sections 2 and 5. The corrective actions are documented as contract change requests and integrate with the change management process.
Appendixes—Contract Performance Reports
Appendixes provide further details on contract agreement and performance.

critical business needs. Managing and controlling issues is a major element to controlling project budget. Issue management maintains a focus on resolving issues by documenting, assessing, prioritizing, and developing action plans and reporting status on a periodic basis. Issues logs shall be reviewed at team status meetings and unresolved issues shall be presented at the program working committee meetings. Issues will be prioritized, action plans developed, due dates established, and responsibilities assigned. Action items shall also be managed using the same tracking system as the issues log. The project manager and program delivery manager will populate the issues log that describes the project issues. The issues for each project will be compiled by the PMO into the project issues management system (PIMS),[19] and critical issues will be presented to the program steering

19. PIM system (PMIS): An automated system that is used to access, manipulate, integrate, and update project issues stored in the issues log. It is an integrated component of the PMIS.

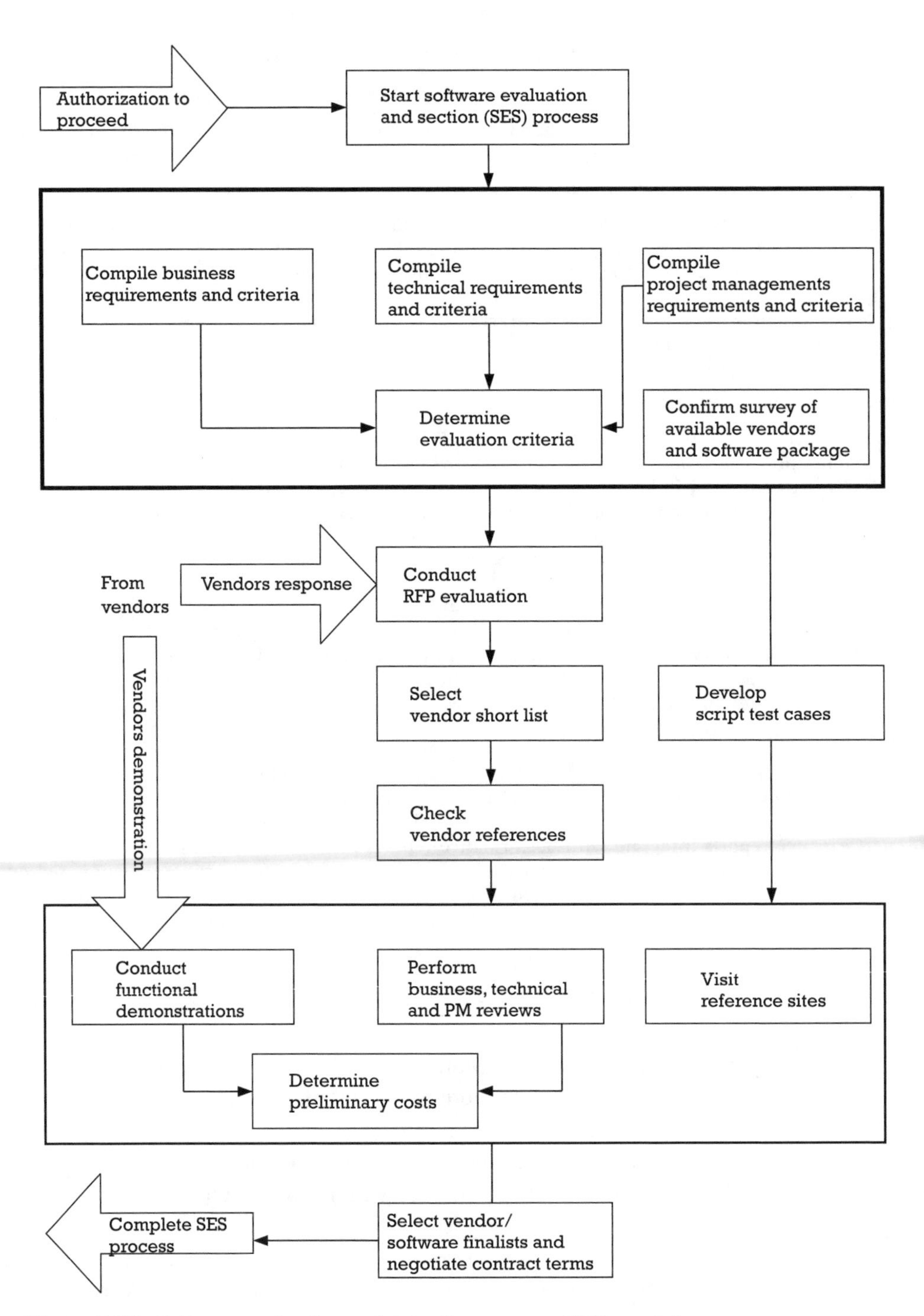

Figure 4.13 Software evaluation and selection process (SES) model.

committee meetings and to other stakeholders, as appropriate, for approval to proceed with appropriate action plans.

The issue management process presented in this book is an extension of the knowledge area of PMBOK. PIMS is an integrated component of the

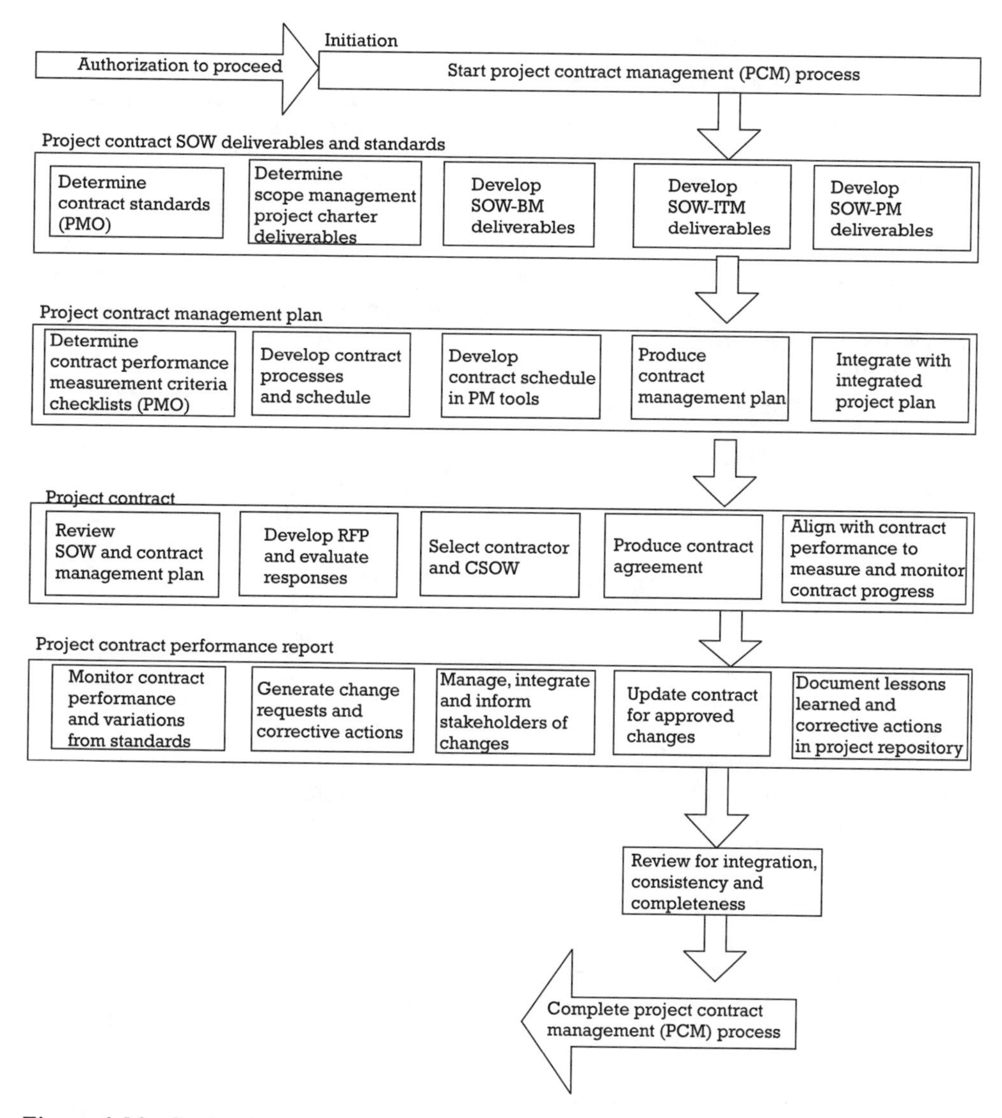

Figure 4.14 Contract management process flow.

PMIS. The PIM management plan[20] outlines by whom, when, and how issues are addressed and resolved based on the impact to the scope, time, cost, effort, and quality performance objectives. Unresolved issues form the basis for the creation of change requests.

The components of PIM include the following:

- Project issues;
- PIM management plan;
- Project issues log and PIMS;
- Project issue resolution reports.

20. PIM management plan: A document that describes by whom, when, and how project issues will be managed and controlled. It is part of the integrated project plan.

Table 4.23 Contract Management Checklist

Project Contract Management Criteria	*Yes*	*No*	*Project Contract Management Criteria*	*Yes*	*No*
Initiation			*Contract requirements–SOW*		
Is management committed?	❑	❑	Do contract standards exists?	❑	❑
Have a PM sponsor and steering committee been established?	❑	❑	Is the SOW completed?	❑	❑
Is there a PM methodology?	❑	❑	Is the SOW approved?	❑	❑
Does the PM methodology have project contract management component?	❑	❑	Does the SOW include business deliverables?	❑	❑
Has a program delivery manager been assigned?	❑	❑	Does the SOW include IT deliverables?	❑	❑
Does the program delivery manager understand the project contract management process?	❑	❑	Does the SOW include PM deliverables?	❑	❑
Contract management plan:			*Project contract*		
Is there a checklist for measuring contract progress?	❑	❑	Is there a contract work breakdown structure (CWBS)?	❑	❑
Is the contract management plan approved?	❑	❑	Is there an RFP and contract agreement?	❑	❑
Does the contract management plan include processes, roles and responsibilities, and schedule?	❑	❑	Does the contract include acceptance criteria?	❑	❑
Is the contract management plan included within integrated project plan?	❑	❑	Does the contract include deliverables milestones?	❑	❑
Is the contract management plan schedule integrated with the project schedule?	❑	❑	Is the contract signed?	❑	❑
Is the contract management plan developed by the PMO?	❑	❑	Is there alignment with contract performance?	❑	❑

4.3.9.1 Purpose

The purpose of this process is to provide an issue management log and PIMS to ensure that project issues are documented, assessed, and prioritized based on critical business needs to minimize the impact to project scope, time, cost, effort, and quality performance objectives. This issues log will form the baseline to communicate, prioritize, and integrate issues for potential resolutions. The program steering committee will use this issues log as a baseline in determining the project progress and approvals to proceed with further developments or recommend the need for a change request.

4.3.9.2 Policy

The policy of project issue management is the following:

- Every project manager shall document all issues in the issues log and track and report the current status and/or resolution in the PIMS.

- The unresolved issues shall form the basis for determining the need for a change request.
- Issues shall be documented in the issues log in a format that describes the impact to the scope, time, cost, effort, and quality performance objectives.

4.3.9.3 Roles and responsibilities

Executive business manager responsibilities include the following:

- Provide executive-level business approval to the program business manager to ensure that the issues log is managed effectively, in order to control the issue management process within budgetary and schedule constraints.
- Provide executive-level business commitment for addressing critical issues.
- Approve all business changes to the approved issues management plan.
- Approve business resources to support development of the issue management plan.
- Act as the overall program sponsor and chairman of the program steering committee.

Executive IT manager responsibilities include the following:

- Provide executive-level IT approval to the program IT manager to ensure that the issues log is managed effectively in order to control the issue management process within budgetary and schedule constraints.
- Provide executive-level IT commitment for addressing critical issues.
- Approve all IT changes to the approved issue management plan.
- Provide IT resources to support development of the issue management plan.
- Act as the program sponsor from the IT perspective.

Executive project management manager responsibilities include the following:

- Provide executive-level PM guidance, direction, and advice to the program delivery manager to ensure that all critical issues and the overall issue management process are effectively communicated in order to manage and control the issues log within budgetary and schedule constraints.
- Integrate approved business and IT issues for resolution.
- Integrate changes to business and IT issue management plans.
- Provide PM resources for developing the issue management plan.
- Act as the program sponsor with integrated business and IT responsibility and PM accountability; champion PM processes, tools, and techniques.

- Provide issue management guidance, direction, and advice to the program delivery manager to ensure that critical issues are addresses within the budget and schedule constraints.

Program business manager responsibilities include the following:

- Submit details on business guidance, direction, and advice to the program delivery manager to ensure that the resolution of business-related issues and issue management process will deliver maximum overall business benefits to the company within the project scope, cost, and time constraints.
- Report on business commitment to resolving critical issues.
- Report on all business changes to the approved issue management plan.
- Assign business resources for supporting development of the issue management plan.
- Report to the program steering committee with direct accountability to the business executive, and keep the program working committee informed of the current status of the issues in the issues log.

Program IT manager responsibilities include the following:

- Submit details on IT guidance, direction, and advice to the program delivery manager to ensure that the resolution of IT-related issues and the issue management process will deliver maximum overall business benefits to the company within project scope, cost, and time constraints.
- Report on IT commitment to resolving critical IT issues.
- Report on all IT changes to the approved issue management plan.
- Assign IT resources for supporting the development of issue management plan.
- Report to the program steering committee with direct accountability to the business executive, and keep the program working committee informed of the current status of the issue requests in the issues log.

Program delivery manager responsibilities include the following:

- Define what the project managers can and cannot do when an issue of scope occurs.
- Establish an agreed-upon process for submitting the issue and evaluating its impact on the current baseline.
- Conduct issue request reviews for the issues according to the issue management plan.
- Verify that the project team adheres to PMO issue management policies, guidelines, and procedures.
- Verify that the project meets the project-specific issue request goals by utilizing the checklists, and ensure that issue management activities are included in project plan.

- Manage the issue management log by reporting on the contents and recommended action plan to the appropriate management team.
- Update the issue request management process as needed to keep it effective.
- Report to the program steering committee with direct accountability to the PM executive, and keep project managers and stakeholders informed of the current status of the issues in the issues log.

Project team (project managers and team) responsibilities include the following:

- Project managers to document and maintain an issues log using PIMS;
- Project managers to prioritize action, establish due dates, and assign responsibility for logged issues;
- Project managers to report status of issues to IT and business managers;
- Project team to review issues log at team status meetings.

4.3.9.4 Procedures: issue management

The roles and responsibilities discussed in Section 4.3.9.3 are brief because they are intended to serve as a reference as to who does what during the execution of the issue management processes. Procedures are now introduced to demonstrate how this issue management process is executed using the major standard deliverables (what), process flow (how), and checklist or measurement criteria (why) templates. The process is intended to serve as a reference based on the English-language interrogatives: who does what, how, and why. The deliverables, process flow, and checklist templates, based on real-world practical implementation, will serve as excellent references for practicing project managers during execution of the issue management process.

The deliverables template in Table 4.24 highlights the key deliverables that the practicing project manager manages during the implementation of issue management. The issues log and issue management report model in Table 4.25, based on real-world practical implementations, will provide useful references for project managers. This model shows a typical template for reporting and recording project issues in order to ensure completeness, consistency, and integration during the prioritization and resolution project issues.

The process flow template in Figure 4.15 highlights the major processes to produce the issue management deliverables. These processes reference the deliverables to show the integrated nature and dependencies of the deliverables.

The checklist template in Table 4.26 highlights the major measurement criteria to ensure delivery of quality deliverables identified in the deliverables template and adherence to the process flows identified in the process flow template. The checklist is used to objectively determine success criteria during execution of the issue management process.

Table 4.24 Deliverables Template: Issue Management Report

Section 1—Introduction
This section includes an executive summary, issue management policies, and approach and introduces the report.
Section 2—Project Issues
This section highlights the major deliverables of the issue identification process by providing the issue title, issue description, project impact—if not resolved—priority, issue status, originator, issue resolution plan strategy, date, and person responsible for resolution. These issues documented in the issues log establish the baseline for development of the PIM management plan in Section 3 and the development of the issues log and PIMS in Section 4.
Section 3—PIM Management Plan
This section highlights the major deliverables of the issue planning process by providing a plan that outlines the processes and procedures for documenting, prioritizing, resolving, and reporting the project issues. It includes guidelines or checklists on how to determine the impact of the issues—if not resolved—priority, alternatives, recommendations, and how to integrate, review, and approve issues. This issue management plan integrates with the integrated project plan and master schedule produced during the PDP. It establishes the baseline processes and procedures for managing, resolving, and reporting on issues resolution to stakeholders.
Section 4—Project Issues Log and PIMS
This section highlights the major deliverables of the issue distribution processes by providing the issues log contents and PIM system for storing, retrieving, accessing, reporting, and deleting project issues. This PIMS and issues log provides automated support for the project issues outlined in Section 1, based on the processes, guidelines, and schedule outlined in Section 3.
Section 5—Project Issue Resolution Reports
This section highlights the major deliverables of the issue resolution process by providing project issue status, progress, issue resolution, and suggested corrective actions for the impact on scope, schedule, cost, and quality. The unresolved issues in the issue resolution reports are converted to and managed as change requests during the integrated change management process. These corrective actions documented as issue change requests are integrated with the change management process.
Appendixes—Project Issue Management System
Appendixes provide further details on PIMS and performance reports.

Real-world scenario

Nothing is less productive than to make more efficient what should not be done at all.
—Peter Drucker

Situation: A major corporation that favored the outsourcing mode of operations for all IT project development initiatives decided to contract most project management services rather than using the services of in-house project managers. They experienced situations where some consultant-type project managers seemed to operate in a mode of turning clients' problems into their personal goldmines. The primary focus of project management seemed to be, first, identifying potential issues and recommending solutions with the goal of generating more billable hours and, then, secondly, addressing the needs of the corporation. In some cases, many issues were identified and packaged into unnecessary projects labeled "phase-1 to *n*" of the original project name, to hide the real goal—generating more billable hours. The original project evolved into multiple overlapping, disjointed, and redundant projects, labeled as "phase-1 to *n*" of the original project name, which

Table 4.25 Deliverables Template: Issue Management Report Model

Issue Report

Issue ID: Person who initiated issue: Date initiated:

Issue name: Issue priority: Issue status:

Issue description:

Impact (if not resolved): ❑ scope ❑ schedule ❑ budget ❑ quality ❑ effort

(Explanation)

Alternatives:

Issue resolution recommendation:

Approved by: Date approved: Date expected:

Person assigned: Date assigned: Date completed:

Issues Log

Issue ID	*Issue Name*	*Priority*	*Person Initiated*	*Date Initiated*	*Person Assigned*	*Date Assigned*	*Date Completed*	*Status*

resulted in an uncontrolled project situation. The client's budget escalated exponentially, so senior management decided to address this budget overrun concern by hiring another consulting firm whose mode of operation was similar to that of the previous consulting firm, but with a more professional style of management. The uncontrolled budget environment, which resulted in the identification of projects to address the uncontrolled issues, continued to manifest itself, and this corporation's project management environment evolved to an "issues firefighting" mode of operations.

Root cause of problem: In this situation, I believe that the root cause of these identified problems is the lack of a proper issue management process to assess, prioritize, and integrate these issues. Management must be aware of the subtle professional qualities of some consulting firms, whose hidden objective may be to turn clients' problems into their personal goldmines. These type of consulting firms usually provide auditing-type project management services, with the main focus being on the identification of project issues, rather than providing real project management services

Solution: The solution to this type of situation is the establishment and application of an issue management process, similar in content and structure to that presented in this chapter. Senior management involvement, approval, and understanding of issue management processes are absolutely

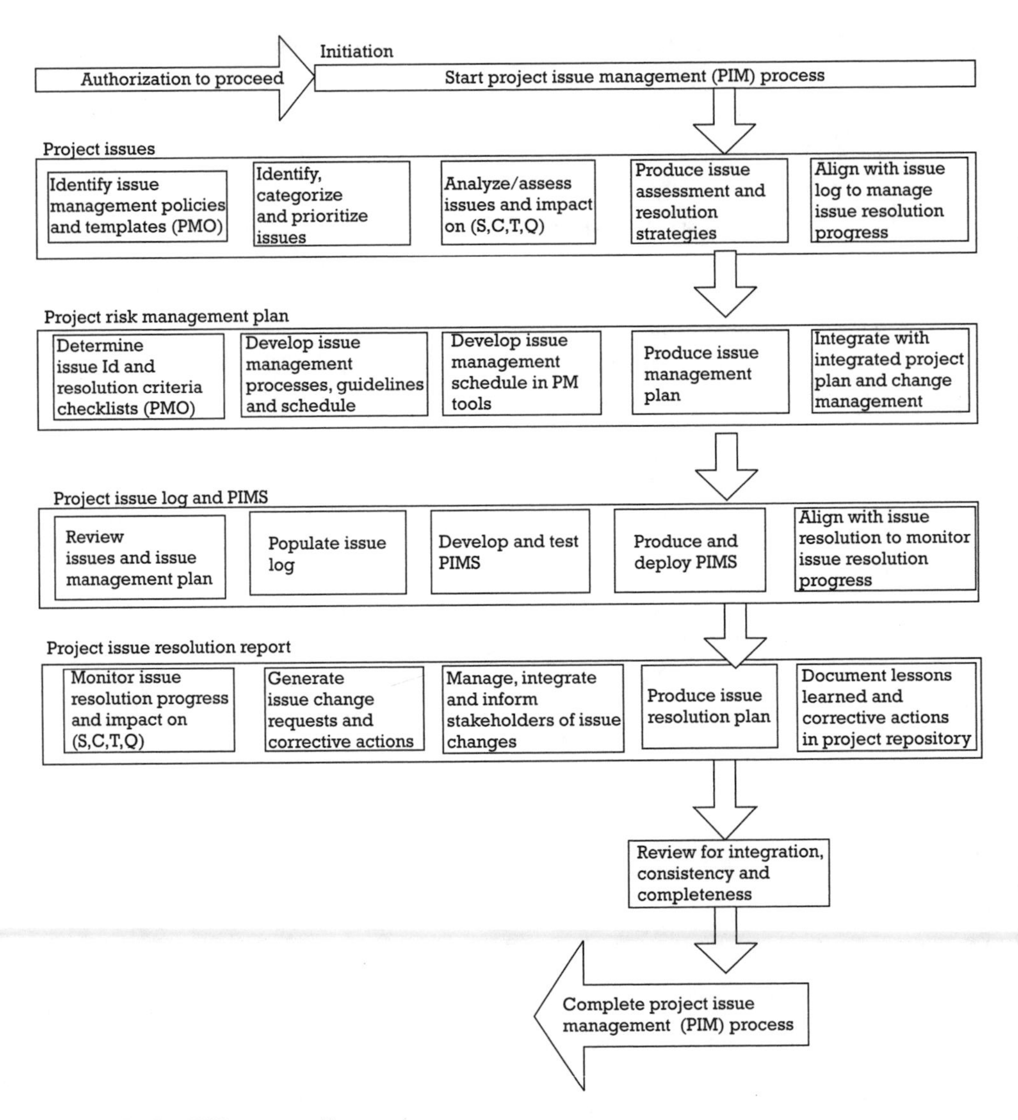

Figure 4.15 PIM process flow.

necessary for successful implementation. Managing and controlling issues are often the key to controlling project budgets.

4.3.10 Project change management

There's no "school" solution to managing change.
—Jack Welch, General Electric Corporation

Executive management shall use the change management report to ensure that project change requests are documented, assessed, prioritized, and communicated to stakeholders, based on critical issues in the issues log that need further resolution. Managing and controlling project change requests

Table 4.26 Issue Management Process Checklist

PIM Criteria	*Yes*	*No*	*PIM Criteria*	*Yes*	*No*
Initiation			*Project issues*		
Is management committed?	❑	❑	Do issues policies exist?	❑	❑
Have a PM sponsor and steering committee been established?	❑	❑	Are issues categorized and prioritized?	❑	❑
Is there a PM methodology?	❑	❑	Is the impact on issues determined?	❑	❑
Does the PM methodology have a PIM component?	❑	❑	Is the issue resolution strategy determined?	❑	❑
Has a program delivery manager been assigned?	❑	❑	Is the issue approved for resolution?	❑	❑
Does the assigned program delivery manager understand the PIM process?	❑	❑	Is an individual assigned to resolve the issue?	❑	❑
Issue management plan (IMP)			*Issues log and PIMS*		
Is there a process for measuring issues progress?	❑	❑	Is there an issues log?	❑	❑
Is the IMP approved?	❑	❑	Is there a PIMS?	❑	❑
Does the IMP include processes, roles and responsibilities, and schedule?	❑	❑	Are issues documented in an issues log?	❑	❑
Is the IMP included within Integrated Project Plan?	❑	❑	Is the issues log and PIMS communicated to team?	❑	❑
Is the IMP integrated with the change request?	❑	❑	Is the project team trained in the used of PIMS?	❑	❑
Is the IMP developed by the PMO?	❑	❑	Is there alignment with issue resolution?	❑	❑
Issue resolution report			*Appendixes*		
Is there an issue cost control process?	❑	❑	Does the appendix include details on PIMS and issue resolution reports?	❑	❑
Are issue resolution reporting guidelines available?	❑	❑			
Do changes to the issue resolution deadline exist?	❑	❑			
Are stakeholders informed about changes?	❑	❑			
Is issue resolution accepted?	❑	❑			
Are lessons learned documented in a project file?	❑	❑			

is often the key to controlling project scope. Project change management, as defined in this chapter, refers to the process of change request management and configuration management. Change requests encompasses changes made to project scope plans, standards, schedules, budgets, resources and labor, and product scope deliverables. They are:

- Identified and documented;
- Evaluated and assessed for risk impact;
- Communicated to the affected groups and individuals;
- Tracked to completion in the change request log.

Changes to product scope refer to additions, modifications, or deletions made to the product or services deliverables. These changes to product scope affect the business processes, data, applications, and technology deliverables.

Project configuration management provides adequate controls, reviews, and approvals for tracking and controlling the changes to project and product deliverables. A CRMS–change request log is established to manage the versions of the changes to the deliverables.

The project manager and program delivery manager will populate the change request log that describes the project change requests. The change requests for each project will be compiled by the PMO using the CRMS, and critical change requests will be presented at the program steering committee meetings and to other stakeholders, as appropriate, for approval to proceed with appropriate action plans. The program steering committee will review the change request, and if they decide the request has merit, they will agree to fund the changes, pending further investigations.

The change management process presented in this chapter is an extension to the knowledge areas of PMBOK that focus on the change request management process. The CRMS[21] is aligned with the PIMS, which is an integrated component of the PMIS. The project change request management (PCRM) plan[22] outlines by whom, when, and how scope changes are addressed and resolved based on the impact on time, cost, effort, and quality performance objectives.

The components of change management are as follows:

- Project change requests;
- PCRM management plan;
- Project change request log and CRMS;
- Project change request resolution reports.

4.3.10.1 Purpose

The purpose of this process is to provide a change request log and CRMS to ensure that project change requests are documented, assessed, and prioritized based on critical business needs to minimize the impact to project scope, time, cost, effort, and quality performance objectives. This change request log will form the baseline to communicate, prioritize, and integrate scope changes for implementation. The program steering committee will use this change request log as a baseline in determining the project progress and approvals to proceed with implementing the change requests. They will review the change requests, and if they decide a request has merit, they will agree to fund the changes, pending further investigations.

21. Change request management system: An automated system that is used to access, manipulate, integrate, and update project change requests stored in the change requests log. It is an integrated component of the PMIS.

22. PCRM management plan: A document that describes by whom, when, and how change requests will be managed and controlled. It is part of the integrated project plan.

4.3.10.2 Policy

The policy of project change management is the following:

- A change control process will manage all changes to the approved requirements and plans. It should be the result of unresolved issues recorded in the issues log.
- Every project manager shall document all change requests in the change request log and track and report on the current status recorded in the CRMS.
- Change requests shall be associated with one or many issues in the issues log.
- Changes requests shall be documented in the change request log in a format that describes the impact to scope and related time, cost, effort, and quality performance objectives.

4.3.10.3 Roles and responsibilities

Executive business manager responsibilities include the following:

- Provide executive-level business approval to the program business manager to ensure that the change requests are managed adequately in order to control the change management process within budgetary and schedule constraints.
- Provide executive-level business commitment for approving the change requests.
- Approve all business changes to the approved change requests.
- Approve business resources to support development of the change requests and supporting management plan.
- Act as the overall program sponsor and chairman of the program steering committee

Executive IT manager responsibilities include the following:

- Provide executive-level IT approval to the program IT manager to ensure that the change requests are managed adequately in order to control the change management process within budgetary and schedule constraints.
- Provide executive-level IT commitment for approving the change requests.
- Approve all IT changes to the approved change requests.
- Approve IT resources to support development of the change requests and supporting management plan.
- Act as the program sponsor from the IT perspective.

Executive project management manager responsibilities include the following:

- Provide executive-level PM guidance, direction, and advice to the program delivery manager to ensure that the required levels of changes are effectively communicated in order to manage and control the change management process.

- Integrate approved business and IT changes requests.
- Integrate changes to business and IT change management processes and plans.
- Provide PM resources for implementing change requests and supporting management plan.
- Act as the program sponsor with integrated business and IT responsibility and PM accountability; champion PM processes, tools, and techniques.
- Provide change management guidance, direction, and advice to the project managers to ensure that change requests are implemented within the budget and schedule constraints.

Program steering committee responsibilities include the following:

- Review change of scope request within the parameters approved by management and authorize funding to investigate the change, if appropriate.
- Evaluate the potential impact of the change and decide whether to approve the request.
- Set priorities for the approved change requests.
- Review the change request and decide whether the request has merit, then agree to funding, pending further investigation.
- Assign change request to be investigated, after which an analysis of the impact of the change requests on the project will be conducted and recommendations determined.
- Evaluate the potential impact of the change and decide whether to approve the change; if the change is approved the program steering committee will set priorities for the change, and funding and schedule will also be approved, as appropriate.

Program business manager responsibilities include the following:

- Submit details on business guidance, direction, and advice to the program delivery manager to ensure the required change request and supporting management plan delivers maximum overall business benefits to the company within the project scope, cost, and time constraints.
- Report on business commitment to approved change requests.
- Report on all business changes to implementing the approved change requests.
- Assign business resources for supporting the delivery of the change request and supporting change management plan.
- Report to the program steering committee with direct accountability to the business executive, and keep the program working committee informed of the current status of the change requests.

Program IT manager responsibilities include the following:

- Submit details on IT guidance, direction, and advice to the program IT manager to ensure the required changes and process delivers maximum overall business and IT benefits to the company within the project scope, cost, time, and quality constraints.
- Report on IT commitment to approved change requests.
- Report on all IT changes to the approved change requests.
- Assign IT resources for supporting the delivery of the change requests.
- Report to the program steering committee with direct accountability to the business executive, and keep the program working committee informed of the current status of the change requests.

Program delivery manager responsibilities include the following:

- Define what the project managers can and cannot do when a change of scope occurs.
- Establish an agreed-upon process for submitting the change requests and evaluating their impact on the current baseline.
- Conduct change request reviews for the approved change requests, according to the change management plan.
- Verify that the project team adheres to PMO change management policies, guidelines, and procedures.
- Verify that the project meets the project-specific change request goals by utilizing the checklists.
- Manage the change request log by reporting on the contents and recommended solutions to the appropriate management team.
- Update the change request management process as needed to keep it effective.
- Report to the program steering committee with direct accountability to the PM executive, and keep project managers and stakeholders informed of the current status of the change requests.

Project team (project manager and team) responsibilities include the following:

- Manage the change requests within the parameters approved by management.
- Manage the change of scope process.
- Complete the section of the change control form that describes the change and its benefits.
- Assign a change request number and record the date received, after which the change is placed on the agenda for the next program working committee meeting.
- Update all appropriate plans, deliverables, end products or services, specifications, applicable standards, schedules, and budgets to incorporate the approved change.

- Communicate the changes to the affected groups and individuals.
- Implement the change requests.

4.3.10.4 Procedures: change management

The roles and responsibilities discussed in Section 4.3.10.3 are brief because they are intended to serve as a reference as to who does what during the implementation of the change management process. Procedures are now introduced to demonstrate how the change management process is executed using the major standard deliverables (what), process flow (how), and checklist or measurement criteria (why) templates. The process is intended to serve as a reference based on the English-language interrogatives: who does what, how, and why. The deliverables, process flow, and checklist templates, based on real-world practical implementation, will serve as excellent references for practicing project managers during execution of the change management process.

The deliverables template in Table 4.27 highlights the key deliverables that the practicing project manager manages during the implementation of change management. The change requests log and change request report model in Table 4.28, based on real-world practical implementations, will provide useful references for project managers. This model shows a typical template for reporting and recording project change requests to ensure completeness, consistency, and integration in prioritizing and approving change requests.

The process flow template in Figure 4.16 highlights the major processes to produce the change management deliverables. These processes reference the deliverables to show the integrated nature and dependencies of the deliverables.

The checklist template in Table 4.29 highlights the major measurement criteria to ensure delivery of quality deliverables identified in the deliverables template and adherence to the process flows identified in the process flow template. The checklist is used to objectively determine success criteria during execution of the change request management process.

Real-world scenario

When it is not necessary to change, it is necessary not to change.
—Lucius Cary, Second Viscount Falkland (1610–1643)

Situation: The real-world scenario presented here shows the similarities in problems that occur between issues and change requests management processes. A major corporation, which preferred the outsourcing mode of operations for all IT project development initiatives, made the decision to contract most project development services rather than using the services of an in-house staff. The contract was awarded to a consulting firm whose primary mode of operations seemed to first focus more on generating change requests, and then focus on addressing the client's requirements. This firm thrived on creating change requests with unclear links to business issues and recommended solutions with the main objective of generating more billable

Table 4.27 Deliverables Template: Change Request Management Report

Section 1—Introduction
This section includes an executive summary, change request management policies, and approach and introduces the report.
Section 2—Project Change Requests
This section highlights the major deliverables of the change request identification process by providing the change request title, description, project impact—if not resolved—priority, status, originator, change request resolution plan strategy, date, and person responsible for resolution. These change requests, documented in the change request log, establish the baseline for development of the PCRM management plan in Section 3 and the development of the change request log and CRMS in Section 4.
Section 3—Project Change Request Management Plan
This section highlights the major deliverables of the change request planning process by providing a plan that outlines the procedures for documenting, prioritizing, resolving, and reporting the project change requests. It includes guidelines or checklists for how to determine the impact of the change requests, their priority, alternatives, recommendations, and how to execute the integration, review, and approvals processes. This change request management plan integrates with the project plan produced during the PDP. It establishes the baseline processes and procedures for managing and resolving the change requests and reporting the information to the stakeholders.
Section 4—Project Change Request Log and CRMS
This section highlights the major deliverables of the change request distribution processes by providing the change request log contents and PCRM system for storing, retrieving, accessing, reporting, and deleting project change requests. This CRMS and change request log provides automated support for the project change requests outlined in Section 1 to support the processes, guidelines, and schedule outlined in Section 3.
Section 5—Project Change Request Progress Reports
This section highlights the major deliverables of the change request progress process by providing project change request status, progress, change request resolution, and suggested corrective actions for the impact to scope, schedule, cost, and quality. The approved change requests documented in the change request log integrate with the issue management process and the project plan and schedule.
Appendixes—Project Change Request Management System
Appendixes provide further details on the CRMS and change request solution.

hours, rather than addressing the needs of the corporation. In some cases, these consultants spent most of their billable time searching for creative ways to generate change requests. This scenario resulted in the identification of many duplicated change requests that were eventually packaged into unnecessary projects with different labels than the original project name to hide the real objective—generating more billable hours. The original project evolved into multiple overlapping, disjointed, and redundant projects, resulting in an uncontrolled project situation. The client's budget escalated exponentially as a result of the additional change requests, and senior management was faced with uncontrolled scope expansion and budget overrun concerns. Senior management addressed these uncontrolled scope creep and budget environments by firing various consultant project managers. However, these problems continued to reoccur, and this major development project was finally completed with a budget overrun of 10 times the original contract agreement.

Root cause of the problem: In this situation, I believe that the root cause of these identified problems is the lack of a proper change request management

Table 4.28 Change Request Report Model

Change Request (CR) Report		
CR ID:	Person who initiated CR:	Date requested:
CR name:	CR priority:	CR status:
Project ID:	Project manager:	Project name:
CR description:		
Reason for change: (Explanation)		
Impact of change: Scope: ❑ Business ❑ Data ❑ Applications ❑ Technology ❑ Process Schedule: ❑ Deliverable dates Effort: ❑ Quantity ❑ Type Budget: ❑ <10% ❑ 10%–20% ❑ >20% Projects internal to program: Projects external to program: Related issues: Additional details:		
CR Approvals:		
Approved by:	Date approved:	PM assigned:
Project/program sponsor:	Date assigned:	Date completed:

Change Request Log

CR ID	*CR Name*	*CR Priority*	*Related Issues*	*Date Requested*	*PM Assigned*	*Date Assigned*	*Date Completed*	*Status*

process to assess, prioritize, and integrate these changes. Management must be aware of the motives of certain consulting firms, whose major objective may be to expand the scope of the project in order to generate more billable hours. These firms usually focus on providing auditing-type project management services, with the main focus on the identification of project change request independent of business issues, rather than on the provision of real project management and delivery services.

Solution: The key solution to this problem scenario is the establishment and application of an effective change management process similar in content and structure to that presented in this chapter. Senior management involvement, approval, and understanding of the change management processes are absolutely necessary for successful implementation. Managing

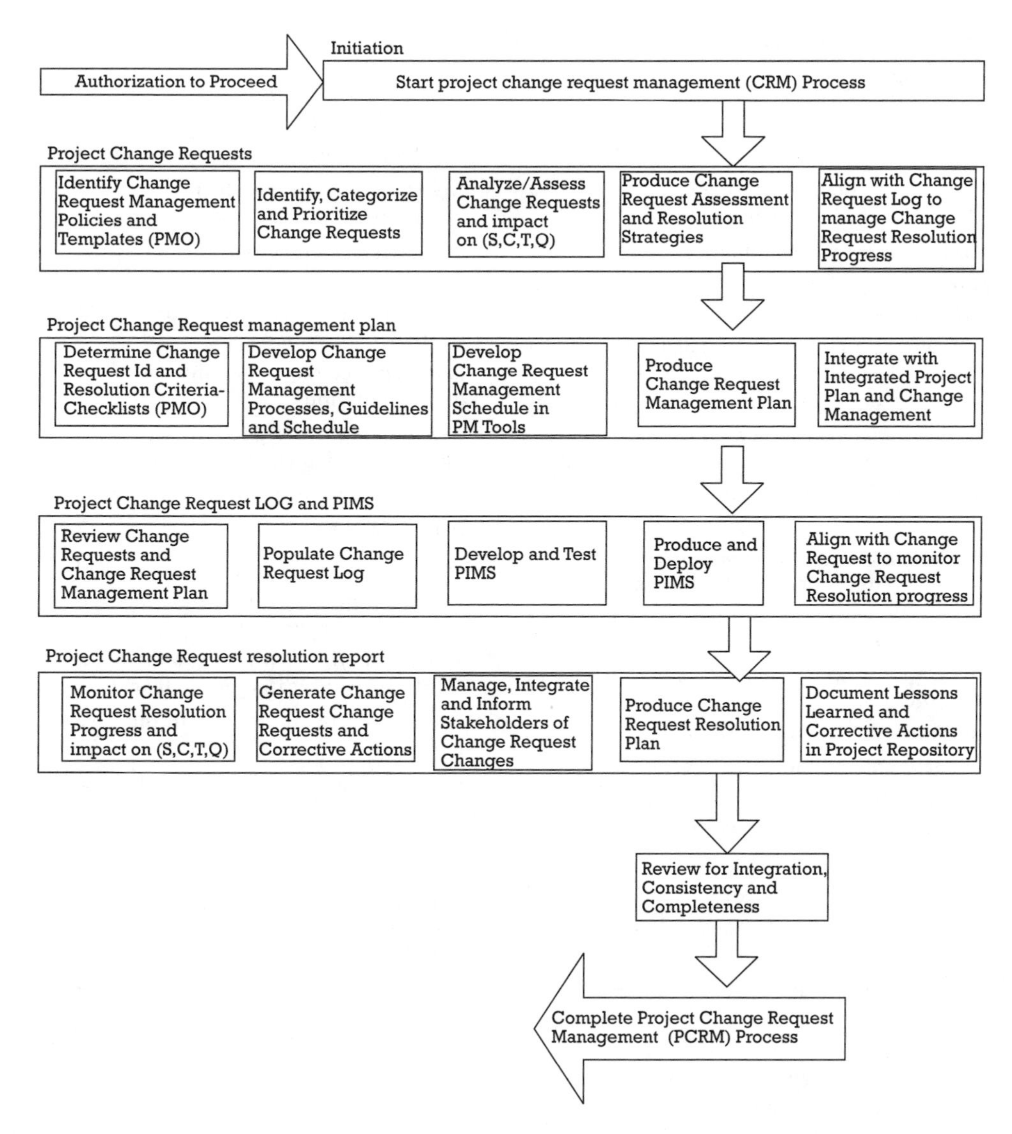

Figure 4.16 PCRM process flow.

and controlling change requests are often the key to controlling project scope and budget.

4.4 PMO support

Whoever admits that he is too busy to improve his methods has acknowledged himself to be at the end of his rope.
—J. Ogden Armour in Information Week

Executive management shall use the PMO report to ensure that all project deliverables, plans, reports, and processes are documented consistently in the

Table 4.29 Change Request Management Checklist

PCRM Criteria	*Yes*	*No*	*PCRM Criteria*	*Yes*	*No*
Initiation			*Project change requests*		
Is management committed?	❑	❑	Do change request policies exist?	❑	❑
Have a PM sponsor and steering committee been established?	❑	❑	Are change requests integrated and prioritized?	❑	❑
Is there a PM methodology?	❑	❑	Is impact to change requests determined?	❑	❑
Does the PM methodology have a PCRM component?	❑	❑	Is the change request resolution strategy determined?	❑	❑
Has a program delivery manager been assigned?	❑	❑	Is the change request approved for implementation?	❑	❑
Does the assigned program delivery manager understand the PCRM process?	❑	❑	Are individuals assigned to implement each change request?	❑	❑
Change request management plan (CRMP)			*Change request issues log and CRMS*		
Is there a process for measuring change request progress?	❑	❑	Is there a change request log?	❑	❑
Is the CR management plan approved?	❑	❑	Is there a CRMS?	❑	❑
Does the CRMP include processes, roles and responsibilities, and schedule?	❑	❑	Is the change request documented in a change request log?	❑	❑
Is the CRMP included within the integrated project plan?	❑	❑	Are the change request log and CRMS communicated to the team?	❑	❑
Is the CRMP integrated with issue requests?	❑	❑	Is the project team trained in the use of CRMS?	❑	❑
Does the CRMP involve the PMO?	❑	❑	Is there alignment with issues in the issues log?	❑	❑
Change request resolution report			*Appendixes*		
? Is there a change request cost control process?	❑	❑	Do the appendixes include details on CRMS and change request performance reports?	❑	❑
? Are change request resolution reporting guidelines available?	❑	❑			
? Do deviations from the change request delivery deadline exist?	❑	❑			
? Are stakeholders informed about changes?	❑	❑			
? Is the change request solution accepted?	❑	❑			
? Are lessons learned documented in a project file?	❑	❑			

PMO repository.[23] Managing and controlling project deliverables and reports in the project repository is often the key to ensuring the project delivers complete, consistent, and integrated deliverables. This process will enhance the communications among the stakeholders internal and external to the project.

The project team, program delivery manager, and PMO support staff will populate the project deliverables in the PMO repository according to PMO

23. PMO repository–PMIR: A database containing project deliverables, reports, and plans.

standards, policies, procedures, and guidelines. The project performance reports and metrics for each project will be compiled by the PMO using the PMO repository, and inconsistencies will be presented to the program delivery manager and to other stakeholders, as appropriate, for corrective actions. The program steering committee will review the structure and contents of the reports and will suggest changes to improve completeness, consistency, and integration of the reports in order to enhance the communications among team members and stakeholders.

The PMO infrastructure support process presented in this chapter emphasizes the need for consistency and standardization of processes, deliverables, tools, and techniques to ensure the IPM-IT environment is effectively implemented. The PMO supports the project team by providing methodology, process deployment, project management training, project metrics, project measurement criteria, tools and project management reporting support to increase project team productivity and reusability.

The components of PMO, discussed in this chapter are as follows:

- PMO infrastructure support requirements;
- PMO management plan;[24]
- PMO repository–PMIR and PMIS;
- PMO metrics and performance reporting.

The processing components of the PMO, which focuses on common responsibilities, are as follows:

- Establishing and deploying a common set of project management processes and templates, which saves each project manager or organization the trouble of having to create these on their own time. These reusable project management components help projects start up more quickly and with less effort.
- Building the methodology and updating it to account for improvements and best practices. For example, as new or revised processes and templates are made available, the PMO deploys them consistently throughout the organization.
- Facilitating improved project team communications by having common processes, deliverables, and terminology. Fewer misunderstandings and less confusion occur if everyone uses the same language and terminology for project-related work.
- Providing training (internal or outsourced) to build core project management competencies and a common set of experiences. If the training is delivered by the PMO, there is a further reduction in overall training costs paid to outside vendors.

24. PMO management plan: A document that describes by whom, when, and how PMO processes will be deployed.

- Delivering project management coaching services to keep projects from getting into trouble. Projects at risk can also be coached to ensure they don't become more risky.
- Tracking basic information on the current status of all projects in the organization and providing project visibility to management in a common and consistent manner.
- Tracking organization-wide metrics on the state of project management, project delivery, and the value being provided to the business. The PMO also assesses the general project delivery environment on an ongoing basis to determine the improvements that have been made.
- Acting as the overall advocate for project management to the organization, including actively educating and selling managers and team members on the value gained through the use of consistent IPM processes.

In summary, companies are finding that they need to standardize on how projects are managed by utilizing consistent and integrated processes. They are recognizing that the process takes much more than just training the staff. It requires a holistic approach, covering many aspects of work and the company culture.

The intent of this section is not to provide detailed processes or procedures on the establishment and implementation of the PMO. There are many excellent books on PMO processes, concepts, theory, and applicability, some of which are mentioned in the selected bibliography at the end of this chapter. The main objective of this section is to provide an overview of the contents (what), purpose (why), roles and responsibilities (who does what), and procedures (how) of the major components of the PMO and to demonstrate how this PMO infrastructure support process fits within the context of the overall IPM-IT framework. The procedure sections of the PMO process provide real-world baseline templates for the deliverables (what), process flow (how), and checklist (measurement criteria) that practicing project managers can readily retrofit and apply during the establishment, implementation, and deployment of the PMO.

4.4.1 Purpose

The purpose of this PMO support process is to support the project team with methodology, templates, process deployment, project management training, project metrics, project measurement criteria, project management tools, and project management reporting to increase the project team productivity and reusability. The PMO staffs advocate the need for consistency, standardization, and integration of processes, deliverables, tools, and techniques to ensure that the IPM-IT environment is effectively implemented.

4.4.2 Policy

The policy of PMO support is the following:

- PMO methodology and processes shall be deliverables based rather than procedure oriented. The processes or methodologies shall be SMART—specific, measurable, achievable, realistic, and timely—rather than DUMB—doubtful, unrealistic, massive, and boring. They must be flexible enough to be easily adapted to different project development approaches.
- Executive and senior management shall commit to using the company's approved methodology and processes during project delivery.
- Program/project managers shall ensure that all PMO deliverables/activities are included in the integrated project plan and the detailed project plans.
- Project teams shall adhere to PMO policies, standards, guidelines, and procedures.

4.4.3 Roles and responsibilities

Executive business manager responsibilities include the following:

- Provide executive-level business approval to the program business manager to ensure that project management deliverables conform to PMO guidelines to commit to consistency, standardization, and integration within budget and schedule constraints.
- Provide executive-level business commitment for conformance to PMO processes.
- Approve all business changes required to adapt to PMO processes.
- Approve business resources to support development of the PMO processes.
- Act as the overall program sponsor and chairman of the program steering committee.

Executive IT manager responsibilities include the following:

- Provide executive-level IT approval to program IT manager to ensure that project management deliverables conform to PMO guidelines to commit to consistency, standardization, and integration within budgetary and schedule constraints.
- Provide executive-level IT commitment for conformance to PMO processes.
- Approve all IT changes to adapt to PMO processes.
- Provide IT resources to support development of the PMO processes.
- Act as the program sponsor from the IT perspective.

Executive project management manager responsibilities include the following:

- Provide executive-level PM guidance, direction, and advice to the program delivery manager to ensure that the PMO processes are effectively and consistently applied.

- Integrate approved business and IT project management processes into PMO processes.
- Integrate changes to existing business and IT project management processes into PMO processes.
- Provide PM resources for delivering PMO processes.
- Act as the program sponsor with integrated business and IT responsibility and PM accountability; champion PM processes, tools, and techniques.
- Provide PMO processes guidance, direction, and advice to the project managers to ensure the project managers produce project management deliverables in accordance with PMO guidelines to provide overall business benefits to the company.

Program business manager responsibilities include the following:

- Submit details on business guidance, direction, and advice to the program delivery manager to ensure the required PMO processes deliver maximum overall business benefits to the company within project scope, cost, time, and quality constraints.
- Report on business commitment to approved PMO processes.
- Report on all business changes to adapt to the approved PMO processes.
- Assign business resources for supporting the delivery of the PMO processes.
- Report to the program steering committee with direct accountability to the business executive, and keep the program working committee informed of the effectiveness of the PMO processes.

Program IT manager responsibilities include the following:

- Submit details on IT guidance, direction, and advice to the program IT manager to ensure the required PMO processes deliver maximum overall business and IT benefits to the company within project scope, cost, time, and quality constraints.
- Report on IT commitment to approved PMO processes.
- Report on all IT changes to adapt to the approved PMO processes.
- Assign IT resources for supporting the delivery of the PMO processes.
- Report to the program steering committee with direct accountability to the business executive, and keep the program working committee informed of the effectiveness of the PMO processes.

Program delivery manager responsibilities include the following:

- Provide policies, guidelines, best practices, and templates to enable cost effective and efficient project delivery.
- Maintain the project repository using PMIS.
- Provide guidelines to an acceptable level of analysis for estimating and cost/benefit analysis.

- Lead and coordinate the integration of all IT standards, templates, and checklists for projects into a single consolidated repository for access by project teams.
- Provide standard templates and tools for project management tasks and project reporting.
- Provide guidelines and standards for reporting frequency.
- Assist project teams in the use of project management templates and spreadsheets.
- Provide the necessary training on the use of project management standards practices and templates.
- Provide information on past projects relevant to current projects.
- Provide guidance and counsel to the various project roles.
- Identify to the steering committee, when appropriate, trend analysis of key indicators for performance monitoring in terms of effort, cost, schedule, scope, and quality.
- Conduct or arrange for externally conducted scheduled project reviews during project execution and postimplementation.
- Conduct or arrange for externally conducted program steering and working committee and steering committee directed reviews.
- Ensure that postproject reviews are completed and findings are executed.
- Communicate PMO policies and ensure project teams adhere to IT PMO methodology, processes, tools, and techniques.
- Manage the PMO processes by reporting on the contents and recommended changes to the appropriate management team.
- Update PMO processes as needed to keep them effective.
- Report to the program steering committee with direct accountability to the PM executive, and keep project managers and stakeholders informed of the effectiveness of the PMO processes.

Project team (project manager and team) responsibilities include the following:

- Project manager determines PMO processes that are applicable to the project during the PDP and advocates the applicability to the project team in concert with the PMO staff.
- Project manager keeps program delivery manager informed of the effectiveness of the PMO process.
- Project manager communicates PMO policies and processes and ensures project teams adhere to IT PMO methodology, processes, tools, and techniques.
- Project team populates the project repository with project deliverables, using PMO standards, policies and procedures, and PMIS.
- Project team keeps project managers informed of the effectiveness of the PMO processes.

4.4.4 Procedures: PMO support

The roles and responsibilities discussed in Section 4.4.3 are brief because they are intended to serve as a reference as to who does what during the implementation of the PMO processes. Procedures are now introduced to demonstrate how the PMO processes is executed using the major standard deliverables (what), process flow (how), and checklist or measurement criteria (why) templates. The process is intended to serve as a reference based on the English-language interrogatives: who does what, how, and why. The deliverables, process flow, and checklist templates, based on real-world practical implementation, will serve as excellent references for practicing project managers during execution of the PMO processes.

The deliverables template in Table 4.30 highlights the key deliverables that the practicing project manager manages during the implementation of the PMO processes. A PMO model in Table 4.31, based on real-world practical implementations, will provide useful references for project managers. This model shows a typical template for reporting and recording the status of project progress to improve communications among the stakeholders and

Table 4.30 Deliverables Template: PMO Report

Section 1—Introduction
This section includes an executive summary, PMO policies, approach and introduces the report.
Section 2—Project PMO Infrastructure Support Requirements
This section highlights the major deliverables of the PMO requirements process by providing the methodology and process deployment, training and coaching, metrics, measurement criteria, tools support, and reporting requirements to support the development and implementation of the deliverables during each phase of the project. The deliverables produced are incorporated into the PMO repository using the PMIS in Section 4, based on the procedures established in the PMO management plan in Section 3.
Section 3—Project PMO Management Plan (Processes and Schedule)
This section highlights the major deliverables of the PMO planning process by providing a plan that outlines the PMO processes, procedures, and schedule for documenting, tracking, supporting, reporting, and managing the project deliverables. It includes guidelines or checklists on how to ensure the consistency, completeness, and integrity of the project deliverables. This PMO management plan integrates with the project plan and master schedule produced during the PDP. It establishes the baseline PMO processes, procedures, and schedule for managing and tracking the project deliverables and reporting the information to the stakeholders.
Section 4—Project PMO Repository and PMIS
This section highlights the major deliverables of the PMO information distribution process by providing the project management milestone deliverables produced during each phase of the IT delivery life cycle by providing the PMO repository contents and PMIS for storing, retrieving, accessing, reporting, and deleting project deliverables. This PMIS and PMO repository provide automated support for the PMO requirements outlined in Section 1, based on the processes, guidelines, and schedule outlined in Section 3.
Section 5—Project PMO Metrics and Performance Reports
This section highlights the major deliverables of the PMO performance reporting process by providing PMO metrics and performance reports for PMO support services and suggested corrective actions on the impact of PMO standards. The corrective actions are enhanced to improve the overall PMO processes on future projects.
Appendixes—Project PMO Methodology
Appendixes provide further details on PMO processes, tools, and techniques.

Table 4.31 Deliverables Template: PMO Report Model

Project Status Report for Time Period

Project ID: | Project Name: | Project Manager:

Overall Status:

Accomplishments

Phase	*Deliverables*	*Original Baseline Date*	*Revised Baseline Date*	*Completion Date*	*Status (G, Y, R)*
PDP	PDM				
RAP	RAM-prototype #1				
ADP	PAS-prototype #2				
IDP-C	Iteration #1				

Issues

Issue ID-Name	*Issue Description*	*Issue Status*	*Person Assigned*	*Date Initiated*	*Date Assigned*	*Date Resolved*
Issue #1		❑ New ❑ Existing				

Change Requests (CR)

CR ID-Name	*CR Description*	*CR Status*	*Person Assigned*	*Date Initiated*	*Date Assigned*	*Date Resolved*
Issue #1		❑ New ❑ Existing				

Budget Information

Budget Category	*Budget Cost*	*Approved Cost*	*Forecast Cost*	*Current Cost*	*Current Variance*
Staff					
Technology					
Facilities					
Total					

to ensure completeness, consistency, and integration of project progress status.

The process flow template in Figure 4.17 highlights the major PMO processes required to produce the PMO deliverables. These processes reference the deliverables to show the integrated nature and dependencies of the deliverables.

The checklist template in Table 4.32 highlights the major measurement criteria to ensure delivery of quality deliverables identified in the deliverables template and adherence to the process flows identified in the process flow template. The checklist is used to objectively determine success criteria during execution of the PMO support processes.

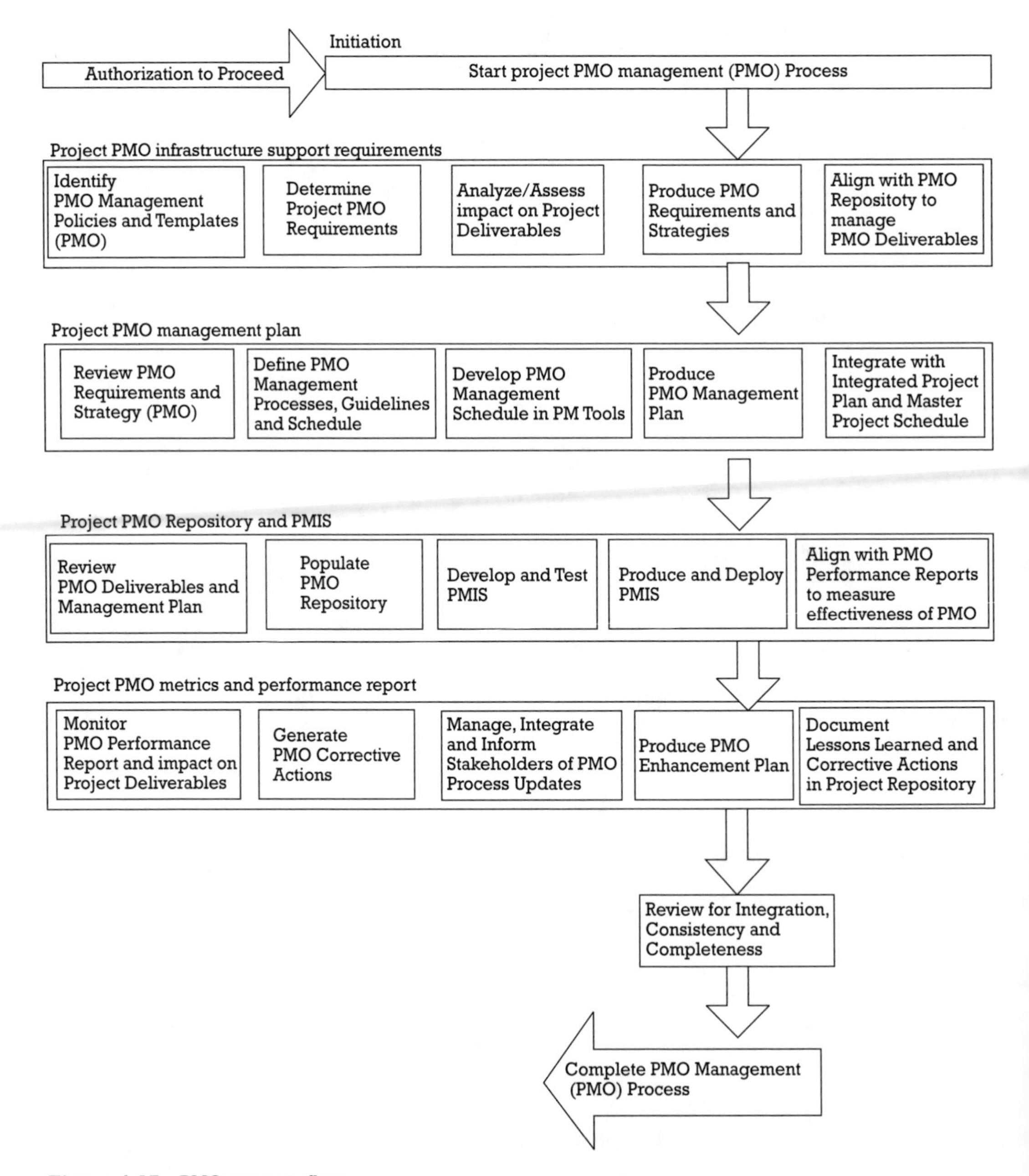

Figure 4.17 PMO process flow.

Table 4.32 PMO Process Checklist

Initiation Criteria	*Yes*	*No*	*PMO Requirements Criteria*	*Yes*	*No*
Initiation			*Project issues*		
Is management committed?	❑	❑	Is there a checklist for measuring PMO progress?	❑	❑
Have a PM sponsor and steering committee been established?	❑	❑	Are the PMO project requirements determined?	❑	❑
Is there a PM methodology?	❑	❑	Are PMO deliverables and activities included in the project schedule?	❑	❑
Does the PM methodology have a PMO component?	❑	❑	Are PMO strategies accepted by the team?	❑	❑
Has a program delivery manager been assigned?	❑	❑	Is the PMO schedule integrated with the project plan?	❑	❑
Does the assigned program delivery manager understand the PMO process?	❑	❑	Is PMO staff involved in determining requirements?	❑	❑
PMO management plan			*PMO repository and PMIS*		
Is there a checklist for measuring PMO progress?	❑	❑	Is there a PMO repository?	❑	❑
Is the PMO management plan approved?	❑	❑	Is there a PIMS?	❑	❑
Does the PMO management plan include processes, roles and responsibilities, and schedule?	❑	❑	Are project deliverables documented in the project repository?	❑	❑
Is the PMO management plan part of the integrated project plan?	❑	❑	Are the project repository and PIMS communicated to the team?	❑	❑
Is the PMO management plan included in the PMO repository?	❑	❑	Is the project team trained in the use of PIMS?	❑	❑
Is the PMIS developed by the PMO?	❑	❑	Is there alignment with PMO performance reporting?	❑	❑
PMO performance report			*Appendixes*		
Are PMO metrics communicated?	❑	❑	Do the appendixes include details on PMO processes, tools, and techniques?	❑	❑
Are PMO performance reporting guidelines available?	❑	❑			
Do variations from the PMO process exist?	❑	❑			
Are stakeholders informed about changes?	❑	❑			
Are PMO processes updated to reflect changes?	❑	❑			
Are lessons learned documented in a project file?	❑	❑			

4.5 Summary

The project management model presented in this chapter describes the component processes of project management. The purpose provides a statement of the derived result, and the policy provides various rules stating the minimum requirements for achieving the purpose statement for each of the project management component processes. The roles and responsibilities of executive management, program management, and project delivery (project team) are presented to demonstrate the integrated nature of these processes in managing and delivering multiple IT projects. The key project management deliverables, process flow, and checklist templates provide some excellent real-world implementation guidelines, which can be referenced and applied to project management processes.

Business organizations that are in the process of managing and delivering multiple IT projects with the goal of integrated or enterprise project management should consider the following project management recommendations as a framework to guide them towards the successful management and delivery of multiple IT projects:

- Establish an IT PDLC that supports an iterative and incremental approach and consists of well-defined phases to ensure orderly management, approval, control, and delivery of IT projects.
- Conform to the PMBOK project management processes, and modify the guidelines based on your existing IT project environment and the framework presented in this chapter.
- Obtain commitments and approvals from executive and senior management by demonstrating the value and benefits of deploying the IT project management processes in your company on one or many projects/program initiatives.
- Establish and deploy a PMO to support the project team with methodology, templates, process deployment, project management training, project metrics, project measurement criteria, project management tools, and project management reporting to increase the project team productivity and reusability of project information. The PMO staffs must advocate the need for consistency, standardization, and integration of processes, deliverables, tools, and techniques to ensure that the IPM-IT environment is effectively implemented.

The PMO methodology and processes should be deliverables based rather than procedure oriented. The processes or methodologies should be SMART—specific, measurable, achievable, realistic, and timely—rather than DUMB—doubtful, unrealistic, massive, and boring. It must be flexible enough to be easily adapted to different project development approaches.

The main focus of this chapter has been to provide the reader with further details on the project management component processes of the IPM framework model. By now, the reader should have gained a fairly good understanding of the integrated nature of these processes. Chapter 6 includes further discussions to show the horizontal integration of these project management processes based on an iterative and incremental IT PDLC model. Project managers and architects, especially those with keen interests in IPM processes, will appreciate the value of these horizontal and vertical integration processes to better prepare them for managing and delivering successful IT projects in this dynamically changing IT industry.

4.6 Questions to think about: management perspectives

1. Think about how your organization identifies, prioritizes, and manages multiple IT projects. How does your organization manage, approve, and control the delivery of IT projects? What are the key rationales in determining the value for an integrated and incremental approach to software development? What are the major components of projects? How do these components relate to your project environment? What is the perception of senior management of the need for consistent, standard, and integrated processes to guide the delivery of projects?
2. Think about how your organization applies project management processes. What are the 10 process components of project management? What are the PMI-PMBOK knowledge areas? How do these processes or knowledge areas relate to your project management processes?
3. Think about how your organization manages projects within scope, schedule, cost, and quality constraints. How are scope and quality defined? What are the components of the WBS? How are schedule and cost estimated? What are the core project management processes? How do the core IT project management processes discussed in this chapter relate to your organization's approach?
4. Think about how your organization manages project risks. What are the four major risk response strategies? How does your organization communicate project status? How does your organization manage staff utilization in and out of the project? How does your organization manage and control issues and change requests? How does your organization administer contracts? What are the facilitating or supporting project management processes? How do the support project management processes discussed in this book relate to your organization's approaches?
5. Think about how your organization deploys project management methodologies, templates, training, and the like to project teams? What are your management perspectives of the value of the PMO? What are consultants' views of the PMO? How do personal greed for power, politics, and personal financial interests affect the PMO office? What are the major responsibilities or process components of the PMO? How do these responsibilities or process components relate to those of your PMO staff?
6. Think about how your organization manages deliverables. What are your management perspectives on deliverables-based versus procedural-based methodologies? What are the components of a SMART methodology versus a DUMB methodology? How does your methodology support different development approaches? How does

the methodology or process discussed in this book relate to your approach?

Selected bibliography

Bennatan, E. M., *On Time, Within Budget, Software Project Management Practices and Techniques,* New York: McGraw-Hill, 1992.

Boehm, B., "Software Risk Management Principles and Practices," *IEEE Software,* Vol. 8, No. 1, January 1991.

ESI International, *Project Framework—A Project Management Maturity Model,* Vol. 1, Arlington, VA: ESI International, 1999.

Fleming, Q. W., and J. M. Koppelman, *Earned Value Project Management,* 2nd ed., Newton Square, PA: PMI, 2000.

Kerzner, H., *Project Management: A Systems Approach to Planning, Scheduling, and Controlling,* 7th ed., New York: John Wiley & Sons, 2001.

Lewis, J. P., *Project Planning Scheduling and Control,* 3rd ed., New York: McGraw-Hill, 2001.

Muller, R. J., *Productive Objects: An Applied Software Project Management Framework,* San Francisco, CA: Morgan Kaufmann, 1998.

Paulk, M. C., et al., *Key Practices of the Capability Maturity Model®,* Version 1.1, Pittsburgh, PA: Software Engineering Institute, Carnegie Mellon University, 1993.

PMI, *Project Management Institute: A Guide to Project Management Body of Knowledge,* 2000 Edition, Newton Square, PA: PMI, 2000.

Royce, W., *Software Project Management—A Unified Framework,* Reading, MA: Addison-Wesley, 1998.

Thonsett, R., *Third Wave Project Management: A Handbook for Managing the Complex Information Systems for the 1990s,* Englewood Cliffs, NJ: Prentice Hall, 1989.

CHAPTER

5

Contents

IT Management Model

5.1 IT management

Technology makes it possible for people to gain control over everything, except over technology.
—John Tudor

The function of IT management is to ensure that resources are optimally allocated, project costs are accurately estimated, and project deliverables are logically developed to support data, applications, and technology architecture components. Projects produce deliverables, and a logical and integrated understanding of how these components fit together allows the project manager to better integrate and optimize the resources to effectively meet stakeholders' expectations. IT management consists of the following processes:

- Resource allocations;
- Cost estimating;
- DA support;
- AA support;
- TA support;
- Applications support services.

Figure 5.1 is a graphical representation of the modeling concept that is used to demonstrate the structure and relationships of the processing components for IT management from a project management perspective.

The intent of this chapter is not to provide detailed processes or procedures on the implementation of IT management processes. There are many excellent books on IT management processes, concepts, theory, and applicability, some of which are mentioned in the selected bibliography at the end of this chapter. The main objective of this chapter is to provide an overview of the contents (what), purpose (why), roles and responsibilities (who, what), and procedures (how) of the

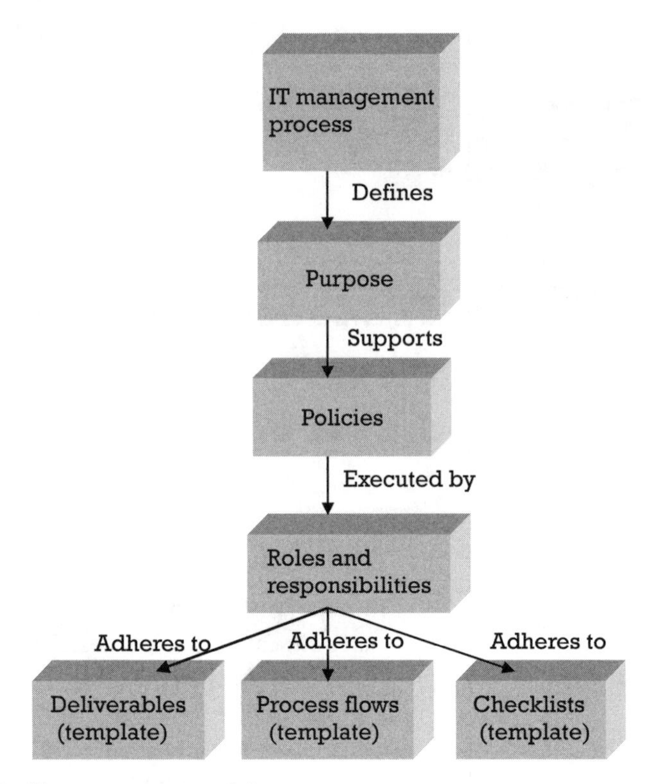

Figure 5.1 IT management model.

components of a typical IT management process and to demonstrate how IT management fits within the context of the overall IPM-IT framework. The procedure sections for each of the IT management component processes provide baseline templates for the deliverables (what), process flows (how), and checklists (measurement criteria) for the practicing project manager during the management, delivery, execution, and integration of the IT management component processes.

The materials presented in the following sections is organized to support the structure represented in Figure 5.1. This content structure will provide readers with a logical and consistent flow of information to enable them to gain a better understanding of and appreciation for the integrated nature of IT project management as presented in this book.

5.2 Resource allocations

You have to learn to treat people as a resource—you have to ask not what do they cost, but what is the yield, what can they produce?
—Peter F. Drucker

Executive management shall use the resource allocations report to approve the allocation of people, technology, and facility resources at each phase of

- Act as the program sponsor with integrated business and IT responsibility and PM accountability; champion PM processes, tools, and techniques.
- Provide resource allocations guidance, direction, and advice to the program manager to ensure that all mission-critical resources are productively aligned with high-priority projects.

Program business manager responsibilities include the following:

- Submit details on business guidance, direction, and advice to the program delivery manager to ensure the required business staffing plan and availability are productively aligned to deliver maximum business benefits to the company within project scope, time, and quality constraints.
- Report on business commitment to the approved staffing plan.
- Report on all business changes to the approved staffing plan.
- Assign business resources for supporting the delivery of the staffing plan.
- Report to the program steering committee with direct accountability to the business executive, and keep the program working committee informed of the forecast and current staffing plan.

Program IT manager responsibilities include the following:

- Submit details on IT guidance, direction, and advice to the program delivery manager to ensure the required IT staffing plan delivers maximum overall business and IT benefits to the company in support of project scope, time, and quality constraints.
- Report on IT commitment to allocated resources–staffing plan.
- Report on all IT support deviations to allocated resources–staffing plan.
- Assign IT resources for supporting the delivery of the staffing plan.
- Report to the program steering committee with direct accountability to the business executive, and keep the program working committee informed of availability of allocated resources–staffing plan.

Program delivery manager responsibilities include the following:

- Define what the project managers can and cannot do when allocating resources.
- Allocate labor resources to business and IT projects based on the value to the business.
- Establish an agreed-upon process for allocating resources and evaluating the impact on the current and forecast staffing plan.
- Verify that the project team adheres to PMO resource allocation policies, guidelines, and procedures by utilizing checklists that are included in the staffing plan.
- Manage the resource allocations by reporting on the contents and recommended solutions to the appropriate management team.

- Update the resource allocations processes as needed to keep them effective.
- Report to the program steering committee with direct accountability to the PM executive, and keep project managers and stakeholders informed of the availability of allocated resources–staffing plan.

Project team (project manager and team) responsibilities include the following:

- Project managers to produce resource allocations in accordance with PMO staffing guidelines and company's business and technology (hardware, software and network) standards.
- Project managers to use resource allocations as the basis to determine cost estimates for the project charter and integrated project plan.
- Project managers to integrate, assemble, and report to the program working committee on the progress and accuracy level of the allocated resources, issues, corrective actions, risks to the project, and impact to cost estimates with direct accountability to the program delivery manager.
- Project team to develop phased deliverables based on the allocated resources and supporting reserved and approved cost estimates.

5.2.4 Procedures: resource allocations

Roles and responsibilities discussed in Section 5.2.3 are brief because they are intended to serve as a reference as to who does what during the execution of the resource allocations process. Business procedures are now introduced to demonstrate how this resource allocations process is executed using the major standard deliverables (what), process flow (how), and checklist or measurement criteria (why) templates. The process is intended to serve as a reference based on the English-language interrogatives: who does what, how, and why. The deliverables, process flow, and checklist templates can serve as guidelines for practicing project managers during the execution of the resource allocations process.

The deliverables template in Table 5.1 highlights the key deliverables that the practicing project manager can execute to enable optimum allocation of labor resources. The guidelines provided in Table 5.2 are based on real-world practical implementations, which should provide some insights to project managers. The effort percentages represent foundation guiding principles, or rules of thumb, that can be readily applied to IT projects that adopt iterative custom-based development, iterative packaged-based development, traditional waterfall-based development, and traditional packaged-based development approaches to software developments.

The process flow template in Figure 5.2 highlights the major processes to support the delivery of the key deliverables for each of the defined phases. These processes reference the cost estimating deliverables to show the integrated nature and dependencies of resource allocations and cost estimating processes for reserving and approving funds.

Table 5.1 Deliverables Template: Resource Allocation Report

Section 1—Introduction
This section includes an executive summary, resource allocation—labor, technology, and facilities—policies, and approach, and introduces the structure and contents of the report.
Section 2—Project Business Definition Phase
This section highlights the major deliverables of the resource allocations process by providing the major components of the staffing plan, technology plan, and facilities plan that will be utilized during this phase. The staffing plan includes the business, project management, and IT management skills, roles and responsibilities, names of resources, and efforts allocated for contract and in-house resources. The technology plan includes the hardware, software and network configuration, and the facilities plan includes space, telephone, and system management resource allocations. These plans are an integral component of the integrated project plan, and are updated during subsequent project phases to support reserved and allocated cost estimates.
Section 3—Project Requirements Analysis Phase
This section highlights the major deliverables of the resource allocations process by providing the major components of the staffing plan, technology plan, and facilities plan that will be utilized during this phase. These plans reflect the current integrated project plan updates and forecasts for the duration of this phase for further refinements during subsequent project phases in support of reserved and allocated cost estimates.
Section 4—Project Architecture Design Phase
This section highlights the major deliverables of the resource allocations process by providing the major components of the staffing plan, technology plan, and facilities plan that will be utilized during this phase. These plans reflect the current integrated project plan updates and forecasts for the duration of this phase for further refinements during subsequent project phases to support reserved and allocated cost estimates.
Section 5—Project IDPs
This section highlights the major deliverables of the resource allocations process by providing the major components of the staffing plan, technology plan, and facilities plan that will be utilized during this phase. These plans reflect the current integrated project plan updates and forecasts for the duration of this phase for further refinements during subsequent project subphases to support reserved and allocated cost estimates.
Appendixes—Project Resource Allocation Model
Appendixes provide further details on resource allocation guidelines.

The checklist template in Table 5.3 highlights the major measurement criteria to ensure delivery of quality deliverables identified in the deliverables template and adherence to the process flows identified in the process flow template. The checklist is used to objectively determine success criteria for allocating resources and for reserving and approving funds.

IT resource estimating model—guiding principles

Roles defined in companies' IT SDLC guidelines are as follows:

- AA—applications architecture; TA—technical architecture; DA–data analyst; DBA—database administrator (IT infrastructure); BA—business analyst; SA—systems analyst;
- DV—developer; OP—operations (core delivery team); PM—project manager; SC—steering committee (project management/senior management); UC—user community;
- TR—trainer/user tester (business users);

Table 5.2 Deliverables Template: Labor Resource Allocation Model

Phases	*IT Infrastructure*	*Core Delivery Team*	*Project Management*	*User Community*	*SDLC Guidelines*
Total resources %	AA 5%; DA 5%; DBA 10%; TA 5%	DV 20%; SA 5%; BA 5%; OP 5%	PM 15%; SC 5%	UC 15%; TR 5%	
Definition	1%	1%	2%	1%	5%
	AA 1%;	BA 1%;	PM 1%;	UC 1%	
			SC 1%		
Analysis	4%	4%	4%	3%	15%
	AA 1%; TA 1%;	BA 2%; SA 1%;	PM 3%; SC 1%	UC 3%	
	DA 1%; DBA 1%	OP 1%			
Architecture	6%	9%	5%	5%	25%
	AA 1%; TA 1%;	BA 2%; SA 2%;	PM 4%; SC 1%	UC 5%	
	DA 2%; DBA 2%	DV 3%; OP 2%			
IDP–construction	7%	12%	3%	3%	25%
	AA 1%; TA 1%;	SA 2%; DV 10%	PM 2%; SC 1%	UC 3%	
	DA 1%; DBA 4%				
IDP–integration	5%	7%	3%	4%	20%
	AA 1%; DBA 2%;	DV 6%; OP 1%	PM 3%;	UC 2%; TR 2%	
	DA 1%; TA 1%				
IDP–deployment	2%	2%	2%	4%	10%
	DBA 1%; TA 1%	OP 1%; DV 1%	PM 1%; SC 1%	UC 1%; TR 3%	
PM % Guidelines	25%	35%	20%	20%	100%

According to standard IT project management guidelines:

- 25% of total effort is consumed by IT infrastructure team (AA; TA; DA; DBA).
- 35% of total effort is consumed by core delivery team (BA; SA; DV; OP).
- 20% of total effort is consumed by project management/senior management team (PM; SC).
- 20% of total effort is consumed by user community team (UC; TR)

According to companies' IT systems delivery life-cycle (SDLC) guidelines:

- 5% of total effort is consumed in the definition phase.
- 15% of total effort is consumed in the analysis phase.
- 25% of total effort is consumed in the architecture phase.

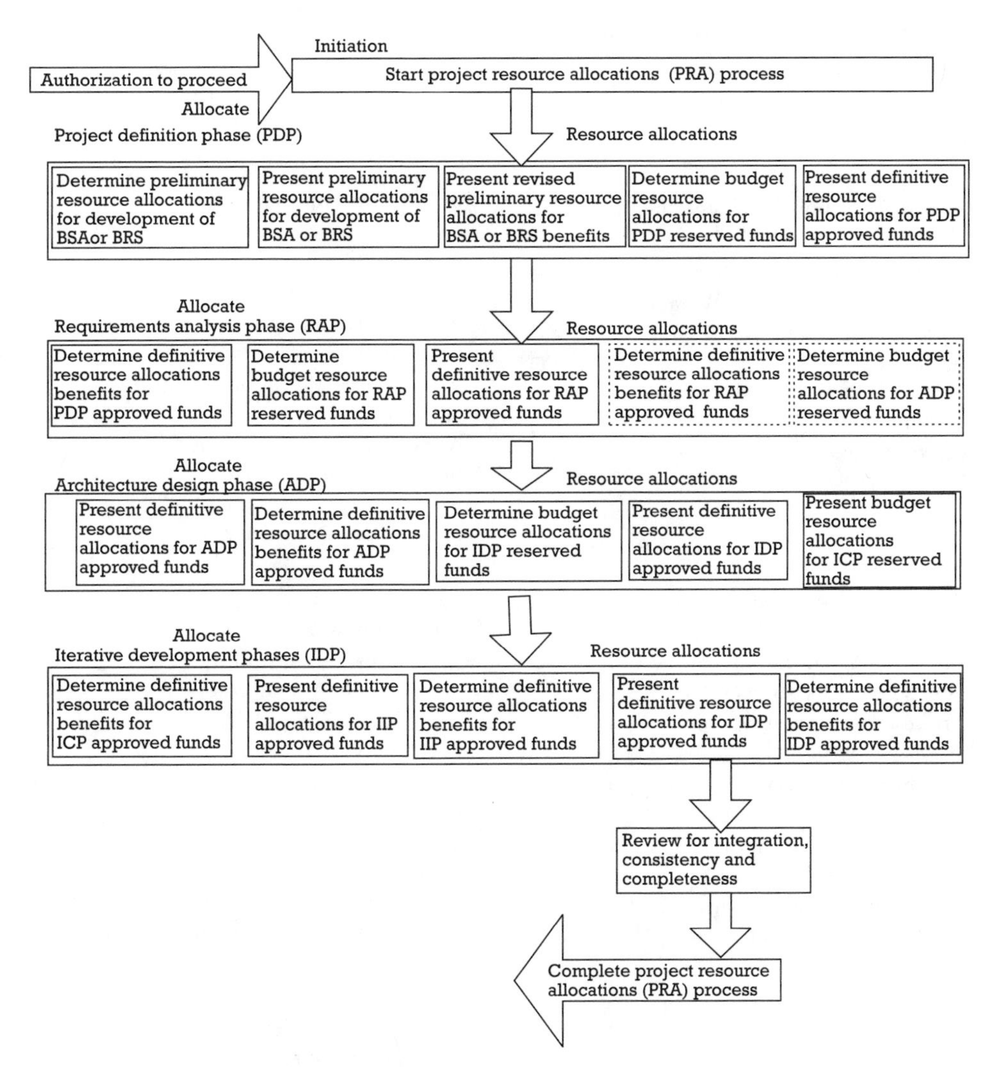

Figure 5.2 Resource allocation process flow.

- 25% of total effort is consumed in the iterative development–construction phase.
- 20% of total effort is consumed in the iterative development–integration phase.
- 10% of total effort is consumed in the iterative development–deployment phase.

Roles and responsibility matrixes in the companies' SDLC guidelines formed the basis for resource distribution/allocation during each of the phases.

Resource estimating method/approach

The guidelines and approach discussed in this section can be applied to any of the software development approaches:

Table 5.3 Resource Allocations Checklist

Project Resource Allocation (PRA) Criteria	*Yes*	*No*	*Project Resource Allocation (PRA) Criteria*	*Yes*	*No*
Initiation			*PDP*		
Is management committed?	❑	❑	Are resource allocations for BSA or business requirements statement (BRS) approved?	❑	❑
Have a PM sponsor and steering committee been established?	❑	❑	Are benefits revisited after completion of BSA or BRS?	❑	❑
Is there a PM methodology?	❑	❑	Are resource allocations for development of PDP reserved?	❑	❑
Does the PM methodology have a PRA component?	❑	❑	Are resource allocations for PDP approved?	❑	❑
Has a program delivery manager been assigned?	❑	❑	Are PDP resource allocations included in project plans?	❑	❑
Does the assigned program delivery manager understand the PRA process?	❑	❑	Are benefits revisited with business involvements?	❑	❑
RAP			*ADP*		
Are benefits revisited after completion of PDP?	❑	❑	Are benefits revisited after completion of RAP?	❑	❑
Are resource allocations for development of RAP reserved?	❑	❑	Are resource allocations for development of ADP reserved?	❑	❑
Are resource allocations for development of RAP approved?	❑	❑	Are resource allocations for development of ADP approved?	❑	❑
Are there any deviations from reserved resource allocations?	❑	❑	Are there deviations from reserved resource allocations?	❑	❑
Are RAP resource allocations included in project plans?	❑	❑	Are resource allocations for ADP included in the integrated project plan?	❑	❑
Are benefits revisited with business involvement?	❑	❑	Are benefits revisited with business involvement?	❑	❑
IDPs			*Appendixes*		
Are benefits revisited after completion of ADP?	❑	❑	Do the appendixes include details on resource allocations guidelines?	❑	❑
Are resource allocations for development of IDP reserved?	❑	❑			
Are there deviations from reserved resource allocations?	❑	❑			
Are resource allocations for IDP included in project plans?	❑	❑			
Are benefits revisited with business involvements?	❑	❑			

- Custom iterative developments[2];
- Purchased-packaged iterative development[3];
- Custom waterfall developments[4];
- Purchased-packaged waterfall developments[5].

2. Custom–iterative development: An approach used to develop software applications by constructing applications code in house and using iterative development processes during development and deployment of the solution.

The approach is as follows:

1. List roles outlined in company's systems delivery life cycle (SDLC)[6] guidelines.
2. Classify roles into infrastructure, CORE project delivery, project management, end users, as outlined above.
3. Use standard PM guidelines as outlined above: PM percentage guidelines: Infrastructure—25%; Core—35%; PM—20%; Users—20%.
4. Use company's SDLC guidelines for effort according to phases: definition—5%; analysis—15%; architecture—25%; construction—25%; integration—20%; deployment—10%.
5. Assign overall resource percentages for each phase by cross-referencing PM% guidelines with the company's SDLC guidelines. For example, a 5% resource allocation for infrastructure team efforts during the integration phase represents 20% (SDLC guidelines) of 25% (PM guidelines).
6. Assign individual resource percentages during each phase for each role based on the company's SDLC guidelines and the roles and responsibility matrix. For example, in the definition phase, the roles (responsible/contributors) are BA, PM, UC, SC, and AA. Distribute resource percentages based on SDLC guidelines roles and responsibility matrix.
7. Adjust resource percentage allocations to reflect optimum resource distribution (if necessary).
8. Aggregate to obtain resource role percentages for all phases.
9. Assign effort for each project in project management tool resources according to overall percentage:
   - IT infrastructure: AA 5%; DA 5%; DBA 10%; TA 5%;
   - Core delivery team: DV 20%; SA 5%; BA 5%; OP 5%;
   - Project management: PM 15%; SC 5%;
   - SDLC guidelines: UC 15%; TR 5%.

3. Purchased package–iterative development: An approach used to develop software applications by purchasing software packages–applications code and using iterative development processes during development and deployment of the solution.
4. Custom–waterfall development: An approach used to develop software applications by constructing applications code in house and using traditional sequential development processes during development and deployment of the solution.
5. Purchased package–waterfall development: An approach used to develop software applications by purchasing software packages–applications code and using traditional sequential processes development and deployment of the solution.
6. SDLC: The systems delivery life cycle, which supports a set of established phases to manage and control software development; synonymous with PDLC.

10. For all projects in the program, use the estimated effort allocations from the BSA project as the basis and assign the above role resource percentage.
11. For new projects, categorize projects as follows:
    - Small project < 1,000 effort hours or < 65 days in duration;
    - Medium project > 1,000 and < 2,000 effort hours or > 65 and < 120 days in duration;
    - Large project > 2,000 effort hours or > 120 days in duration.
12. Determine efforts allocations by using function point analysis (FPA) or another effort estimating technique and assign the above resource percentage.

Note: FPA is another technique that is normally used to estimate effort.

5.3 Project cost estimating

Predictions are hard, especially about the future. There are two types: lucky or lousy.
—Yogi Berra

Executive management shall use the cost estimates to reserve and approve funds at each phase of the IT PDLC. Cost estimates are reserved and approved at the beginning of each phase based on the allocated resources. The components of costs estimating for the phases of the IT PDLC are as follows:

- *Preliminary cost estimates or smart wild-ass guess (SWAG):* –25% to +75% level of accuracy. These estimates are made during the development of the BSA or BRS, using project benefits and project justification estimates derived from the business management function to determine the initial justification for the project. Management reserves are used to cover these project initiation and identification efforts. These estimates provide preliminary project funding estimates for further refinements during the business definition phase.
- *Budget cost estimates:* –10% to +25% level of accuracy. These estimates are developed during the PDP and are used to reserve funds for the entire project, with a budget estimate reserved for each phase. Management reserves to allow for future situations that are impossible to predict are incorporated into these estimates. Management reserves are usually used to cover cost or schedule overruns and are intended to reduce the risk of missing unpredictable cost or schedule objectives.
- *Definitive cost estimates:* –5% to +10% level of accuracy. These detailed phased estimates are developed at the end of each phase of the project and are used to approve funds for the beginning of each subsequent phase within the constraints of the reserved funds from the budget cost estimates.

5.3.1 Purpose

The purpose of this process is to establish reasonably accurate cost estimates for the entire project and for each of the defined phases to better manage and control the allocation and approval of funds. This cost estimating model will form the baseline to manage cost and effort objectives of development projects. The program steering committee will use these cost estimates and potential benefits as deciding factors to approve or disapprove funds.

5.3.2 Policy

The policy of project cost estimating is the following:

- Project preliminary cost estimates shall be within a range of –25% to +75% of the project costs and shall include the entire project costs. Project funds shall be allocated or reserved in principle for these cost estimates, but not approved.
- Project budget cost estimates shall be within a range of –10% to +25% for the phases of the project being estimated and shall include the entire project costs. Project funds shall be allocated or reserved for these cost estimates, but not approved.
- Definitive cost estimates shall be within a range of –5% to + 10% for the phases of the project being estimated. Project funds shall be approved for these cost estimates.

5.3.3 Roles and responsibilities

Executive business manager responsibilities include the following:

- Provide executive-level business approvals to the program business manager to ensure that the business cost estimates are reasonably accurate and to review the estimating process based on the assumptions within budgetary and schedule constraints.
- Provide executive-level business commitment for the business investments.
- Approve all business changes to the approved business investments.
- Approve business resources to support development of the business investment estimates.
- Act as the overall program sponsor and chairman of the program steering committee.

Executive IT manager responsibilities include the following:

- Provide executive-level IT approvals to the program IT manager to ensure that the IT investment cost estimates are reasonably accurate and to review the estimating process based on the assumptions within budgetary and schedule constraints.
- Provide executive-level IT commitment for IT investments–cost estimates.

- Approve all IT changes to IT investments–cost estimates.
- Provide IT resources to support development of IT investments–cost estimates.
- Act as the program sponsor from the IT perspective.

Executive project management manager responsibilities include the following:

- Provide executive-level PM guidance, direction, and advice to the program delivery manager to ensure that the required levels of IT investments–cost estimates and the costs estimating processes are effectively communicated, managed, and controlled.
- Integrate approved business and IT investments–cost estimates.
- Integrate changes to business and IT investments–cost estimates.
- Provide PM resources for delivering cost estimates.
- Act as the program sponsor with integrated business and IT responsibility and PM accountability; champion PM processes, tools, and techniques.
- Provide cost estimating guidance, direction, and advice to the program manager to ensure that the projects stay within the cost estimating range and provide overall business benefits to the company.

Program business manager responsibilities include the following:

- Submit details on business guidance, direction, and advice to the program delivery manager to ensure that the required business cost estimates and assumptions deliver maximum business benefits to the company in support of project scope, time, and quality constraints.
- Report on business commitment to reserved and approved cost estimates.
- Report on all business changes to reserved and approved cost estimates.
- Assign business resources to support delivery of the cost estimates.
- Report to the program steering committee with direct accountability to the business executive, and keep the program working committee informed of the accuracy of cost estimates.

Program IT manager responsibilities include the following:

- Submit details on IT guidance, direction, and advice to the program delivery manager to ensure the required IT cost estimates and processes deliver maximum overall business and IT benefits to the company in support of project scope, time, and quality constraints.
- Report on IT commitment to reserved and approved cost estimates
- Report on IT support deviations to the cost estimates.
- Assign IT resources to support delivery of the cost estimates.
- Report to the program steering committee with direct accountability to the business executive, and keep the program working committee informed of the accuracy of cost estimates.

Program delivery manager responsibilities include the following:

- Define what the project managers can and cannot do when producing cost estimates.
- Determine cost performance measurement to business and IT projects based on the value to the business and budgeting constraints.
- Establish an agreed-upon process for submitting cost estimates and evaluating their impact on the current cost baseline.
- Verify that the project team adheres to PMO cost estimating policies, guidelines, and procedures by utilizing checklists that are included in the cost management plan.
- Manage the cost estimates by reporting on the contents, assumptions, and recommended estimates to the appropriate management team.
- Update cost estimating processes as needed to keep them effective.
- Report to the program steering committee with direct accountability to the PM executive, and keep project managers and stakeholders informed of the accuracy of cost estimates.

Project team (project manager and team) responsibilities include the following:

- Project managers produce cost estimates in accordance with PMO cost estimating guidelines.
- Project managers incorporate cost estimates into the project charter and integrated project plan.
- Project managers integrate, assemble, and report to the program working committee on the progress and accuracy level of the cost estimates, issues, corrective actions, and risks to the project with direct accountability to the program delivery manager.
- Project team develops phased deliverables based on the reserved and approved cost estimates.

5.3.4 Procedures: cost estimating

The roles and responsibilities discussed in Section 5.3.3 are brief because they are intended to serve as a reference as to who does what during the execution of the cost estimating process. Business procedures are now introduced to demonstrate how this cost estimating process is executed using the major standard deliverables (what), process flow (how), and checklist or measurement criteria (why) templates. The process is intended to serve as a reference based on the English-language interrogatives: who does what, how, and why. The deliverables, process flow, and checklist templates can serve as a guideline for the practicing project managers during the execution of the cost estimating process.

The deliverables template in Table 5.4 highlights the key deliverables that the practicing project manager can execute to enable optimum allocation of labor resources. The guidelines provided in Figure 5.3 are based on

Table 5.4 Deliverables Template: Cost Estimating Report

Section 1—Introduction
This section includes an executive summary, cost estimating policies, and approach and introduces the report.
Section 2—Project Definition Phase
This section highlights the major deliverables of the cost estimating process by providing preliminary cost estimates, budget cost estimates, and definitive cost estimates for this phase. Preliminary cost estimates are determined during the preceding BSA or BRS, budget cost estimates are determined at the beginning of this phase, and definitive cost estimates are determined at the end of this phase. These cost estimates establish the baseline for allocation—reserved—and release—approved—of funds for subsequent phases and integrate with the project plan and cost estimates reports.
Section 3—Project Requirements Analysis Phase
This section highlights the major deliverables of the cost estimating process by providing budget cost estimates and definitive cost estimates for this phase. Budget cost estimates are determined during the definition phase, and definitive cost estimates are determined at the end of this phase. These cost estimates establish the baseline for allocation—reserved—and release—approved—of funds for subsequent phases and integrate with the project plan and cost estimates reports.
Section 4—Project Architecture Design Phase
This section highlights the major deliverables of the cost estimating process by providing budget cost estimates and definitive cost estimates for this phase. Budget cost estimates are determined at the beginning of this phase, and definitive cost estimates are determined at the end of this phase. These cost estimates establish the baseline for allocation—reserved—and release—approved—of funds for subsequent phases and integrate with the project plan and cost estimates reports.
Section 5—Project IDPs
This section highlights the major deliverables of the cost estimating process by providing budget cost estimates and definitive cost estimates for these subphases or releases. Budget cost estimates are determined at the beginning of each release, and definitive cost estimates are determined at the end of each release. These cost estimates establish the baseline for allocation—reserved—and release—approved—of funds for subsequent releases and integrate with the project plan and cost estimates reports.
Appendixes—Project Cost Estimating Model
Appendixes provide further details on the cost estimating.

real-world practical implementations, which should provide some insights to project managers. The effort percentages represent foundation guiding principles, or rules of thumb, that can be readily applied to IT projects that adopt iterative custom-based development, iterative packaged-based development, traditional waterfall-based development, and traditional packaged-based development approaches to software developments.

The process flow template in Figure 5.4 highlights the major processes to support the delivery of the key deliverables for each of the defined phases. These processes reference the cost estimating deliverables to show the integrated nature and dependencies of resource allocations and cost estimating processes for reserving and approving funds.

The checklist template in Table 5.5 highlights the major measurement criteria to ensure delivery of quality deliverables identified in the deliverables template and adherence to the process flows identified in the process flow template. The checklist is used to objectively determine success criteria for allocating resources and for reserving and approving funds.

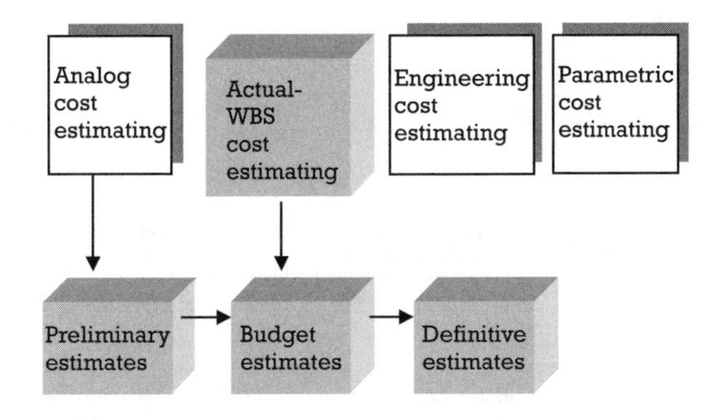

Preliminary-SWAG (one-year project);

Budget component	SWAG estimate
Technology	9%
· Hardware	3%
· Software	4%
· Network	2%
Facilities	1%
· Office space, telephone, administrative	1%
Labor	90%
· Definition, analysis and design	40%
· Construction, integration and deploymen	50%
· Maintenancel year after deployment	10%-35%

Figure 5.3 Cost-estimating model.

The costs of maintaining and enhancing IT systems are buried within the company's operations and support budget costs; as a result, it is easy from a management perspective to regard these costs as annual expenses, or overhead, rather than investments that need to be quantified and measured.

5.4 Data architectures

We have for the first time an economy based on a key resource [information] that is not only renewable, but self-generating. Running out of it is not a problem, but drowning in it is.
—John Naisbitt

Executive management shall use the DA deliverables and the related checklists to approve the Data deliverables at each phase of the IT PDLC. DA support services are performed during each of the phases based on the data requirements of the project. These deliverables and supporting activities are

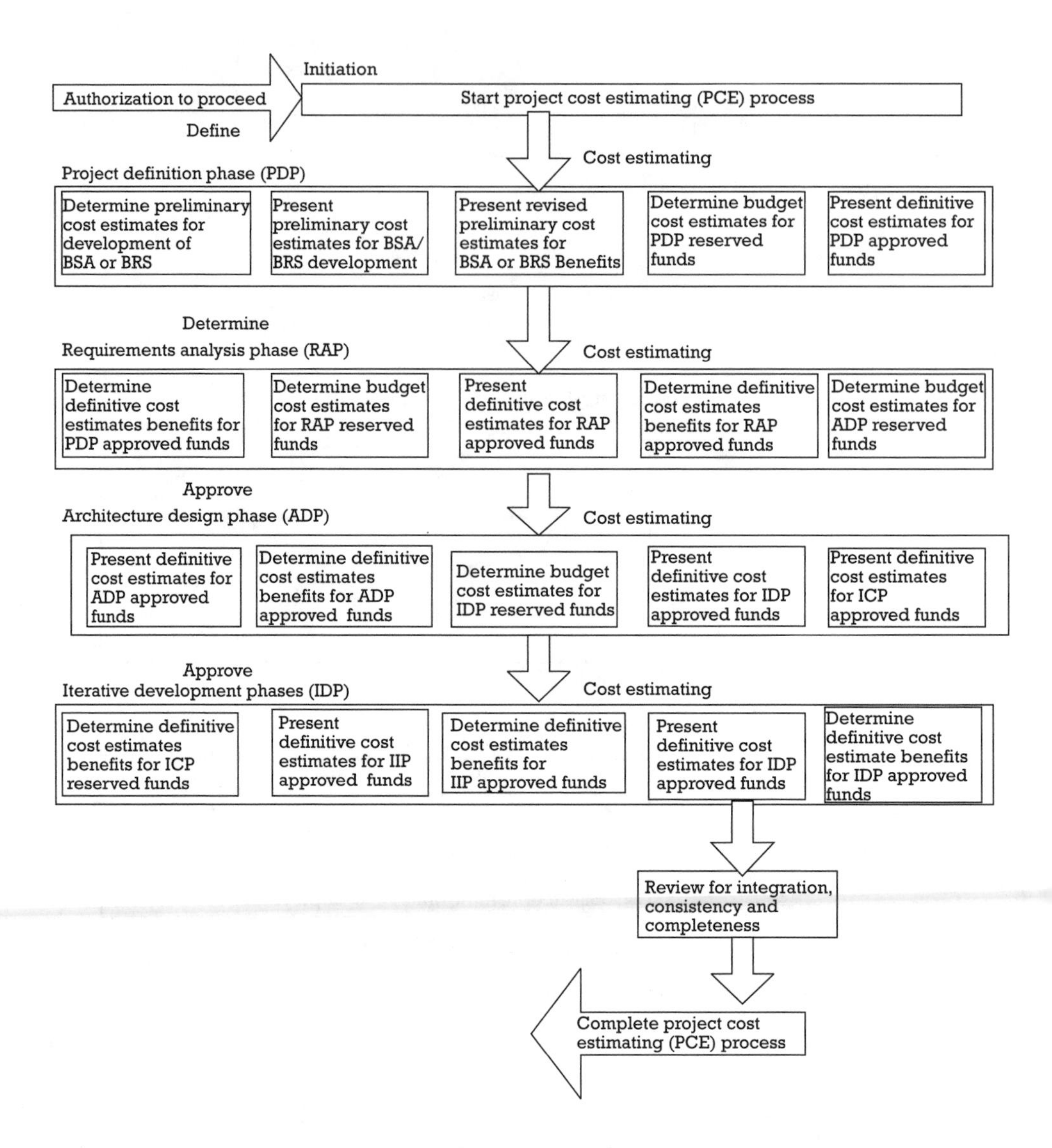

Figure 5.4 Cost-estimating process flow.

identified, managed, and integrated within the core components of the master WBS. The major DA deliverables are as follows:

- *Project data management infrastructure support:* These services provide directory information on what, who, why, where, and how to access and use existing data management strategies, standards, and procedures, data repository,[7] database management software tools, data modeling software tools, and data communications middleware tools.

7. Data repository: A database containing data deliverables that is integrated with the PMO project repository to ensure completeness, consistency, integration, and reusability of the project deliverables. It also integrates with the applications and technology repositories.

Table 5.5 Deliverables Template: Cost Estimating Checklist

Project Cost Estimating (PCE) Criteria	*Yes*	*No*	*Project Cost Estimating (PCE) Criteria*	*Yes*	*No*
Initiation			*PDP*		
Is management committed?	❑	❑	Are cost estimates for BSA or BRS approved?	❑	❑
Have a PM sponsor and steering committee been established?	❑	❑	Are benefits revisited after completion of BSA or BRS?	❑	❑
Is there a PM methodology?	❑	❑	Are cost estimates for development of PDP reserved?	❑	❑
Does the PM methodology have PCE component?	❑	❑	Are cost estimates for PDP approved?	❑	❑
Has a program delivery manager been assigned?	❑	❑	Are PDP cost estimates included in project plans?	❑	❑
Does the assigned program delivery manager understand the PCE process?	❑	❑	Are benefits revisited with business involvements?	❑	❑
RAP			*ADP*		
Are benefits revisited after completion of PDP?	❑	❑	Are benefits revisited after completion of RAP?	❑	❑
Are cost estimates for development of RAP reserved?	❑	❑	Are cost estimates for development of ADP reserved?	❑	❑
Are cost estimates for development of RAP approved?	❑	❑	Are cost estimates for development of ADP approved?	❑	❑
Are there deviations from reserved cost estimates?	❑	❑	Are there deviations from reserved cost estimates?	❑	❑
Are RAP cost estimates included in project plans?	❑	❑	Are cost estimates for ADP included in the project plan?	❑	❑
Are benefits revisited with business involvements?	❑	❑	Are benefits revisited with business involvements?	❑	❑
IDPs			*Appendixes*		
Are benefits revisited after completion of the ADP?	❑	❑	Do the appendixes include details on cost estimating techniques?	❑	❑
Are cost estimates for development of IDP reserved?	❑	❑			
Are cost estimates for development of IDP approved?	❑	❑			
Are there deviations from reserved cost estimates?	❑	❑			
Are cost estimates for IDP included in project plans?	❑	❑			
Are benefits revisited with business involvements?	❑	❑			

- *Project DA development support:* These services produce data component deliverables that include business subject area and entity models and inventory of existing logical and physical databases, used during the development of the BSA.
- *Project data modeling support:* These services produce detailed data component deliverables that include data entity models for implementation using the DBMS software.

- *Project DBMS support:* These services produce the database structures for storing, retrieving, accessing, reporting, and deleting data using the standard DBMS software.
- *Project data conversion, access, and security support:* These services produce the data component deliverables to convert, access, and secure the databases.

The major deliverables of DA support services as presented in this chapter align with business management and project management and consist of the following components:

- Project DA requirements;
- Project DA management plan;
- Project DA deliverables;
- Project integrated DA and PMO repository services;
- Project DA metrics and performance reports.

5.4.1 Purpose

The purpose of this process is to deliver all the data component deliverables identified in the integrated project plan and WBS to meet the data requirements of the project. These product scope–data deliverables will form the baseline to manage the quality objectives of the development projects. The program steering committee will use these data deliverables and potential benefits as deciding factors during approval or disapproval to proceed with each subsequent phases of the project based on the PDLC.

5.4.2 Policy

The policy of data architectures is the following:

- The DA deliverables shall be valued and protected as one of the most critical assets of the project and the company by documenting and maintaining the components in the project repository, according to company's DA standards and PMO guidelines.
- The DA shall be business driven rather than technology driven and shall be aligned with the AA and developed according to company's DA standards.
- Data shall be effectively managed by establishing data management roles and responsibilities that require good working relationships between business management, project management, and IT DA services.
- Logical and physical data models shall be defined using data modeling and data repository tools.
- Logical data models shall be developed prior to development of the physical models to ensure proper integration with business data requirements.

- The IT program manager, prior to development and implementation, shall approve data conversion, access, and security processes and deliverables.

5.4.3 Roles and responsibilities

Executive business manager responsibilities include the following:

- Provide executive-level business approval to the program business manager to ensure that the business architecture is aligned with the DA and to review the BSA document based on the business direction within budgetary and schedule constraints.
- Provide executive-level business commitment for the business DA components within the overall BSA document.
- Approve all business changes to the business DA components within the overall BSA document.
- Approve business resources to support development of the business DA components within the overall BSA document.
- Act as the overall program sponsor and chairman of the program steering committee.

Executive IT manager responsibilities include the following:

- Provide executive-level IT approval to the program business manager to ensure that the business architecture is aligned with the DA and to review the BSA document based on the business direction–business plan within budgetary and schedule constraints.
- Provide executive-level IT commitment for the DA components within the overall BSA document.
- Approve all IT changes to the DA components within the overall BSA document.
- Approve IT resources to support development of the DA components within the overall BSA document.
- Act as the program sponsor from the IT perspective.

Executive project management manager responsibilities include the following:

- Provide executive-level PM guidance, direction, and advice to the program manager to ensure that the business architecture is aligned with the DA and to review the BSA document based on the business direction–business plan within budgetary and schedule constraints.
- Integrate business, data, applications, and technology architectures into the overall BSA document.
- Integrate changes to business, data, applications, and technology architectures into the overall BSA document.
- Approve PM resources for DA components of the overall BSA document.

- Act as the program sponsor with integrated business and IT responsibility and PM accountability; champion PM processes, tools, and techniques.
- Provide DA development guidance, direction, and advice to the program delivery manager to ensure that business and technology architectures are effectively aligned with high-priority projects.

Program business manager responsibilities include the following:

- Submit details on business guidance, direction, and advice to the program delivery manager to ensure that the business architecture is aligned with the DA to deliver maximum business benefits to the company in support of project scope, time, and quality constraints.
- Report on business commitment to the DA within the overall BSA.
- Report on all business changes to the DA within the overall BSA.
- Assign business resources for supporting the delivery of the DA within the overall BSA.
- Report to the program steering committee with direct accountability to the business executive, and keep the program working committee informed of the implementation of the DA within the overall BSA.

Program IT manager responsibilities include the following:

- Submit details on IT guidance, direction, and advice to the program delivery manager to ensure that the business architecture is aligned with the DA to deliver maximum business benefits to the company within the project scope, time, and quality constraints.
- Report on IT commitments to the DA within the overall BSA.
- Report on all IT changes to the DA within the overall BSA.
- Assign IT resources for supporting the delivery of the DA within the overall BSA.
- Report to the program steering committee with direct accountability to the business executive, and keep the program working committee informed of the implementation of the DA within the overall BSA.

Program delivery manager responsibilities include the following:

- Review and communicate business and IT (physical) DAs/models and ensure linkage with business DAs.
- Communicate data models to project teams.
- Verify that the project team adheres to DA policies, guidelines, and procedures by utilizing checklists that are included in the DA deliverables.
- Manage the integrated DA by reporting on the contents and recommended solutions to the appropriate management team.
- Update DA processes as needed to keep them effective.
- Report to the program steering committee with direct accountability to the PM executive, and keep project managers and stakeholders

informed of the effectiveness and usefulness of the DA during project implementation.

Project team (project manager and team) responsibilities include the following:

- Project managers to manage DA deliverables in accordance with PMO quality guidelines.
- Project managers to incorporate DA deliverables and activities into the project charter and integrated project plan.
- Project managers to integrate, assemble, and report to the program working committee on the progress and quality of the DA deliverables, issues, corrective actions, and risks to the project with direct accountability to the program delivery manager.
- Project team to develop DA deliverables based on PMO and DA standards.

5.4.4 Procedures: DA

The roles and responsibilities discussed in Section 5.4.3 are brief because they are intended to serve as a reference as to who does what during the execution of the DA process. Business procedures are now introduced to demonstrate how this DA process is executed using the major standard deliverables (what), process flow (how), and checklist or measurement criteria (why) templates. The process is intended to serve as a reference based on the English-language interrogatives: who does what, how, and why. The deliverables, process flow, and checklist templates can serve as guidelines for practicing project managers during the execution of the DA process.

The deliverables template in Table 5.6 highlights the key deliverables that the practicing project manager must manage to enable the delivery of quality deliverables. The guidelines provided in Figure 5.5 are based on real-world practical implementations, which should provide some insights to project managers. The *N*-tier software development environment shows how the key data, applications, and technology architecture components integrate.

The process flow template in Figure 5.6 highlights the major processes to support the delivery of the key deliverables. These processes reference the data repository and project management repository to show the integrated nature and dependencies of these repositories and to ensure the high quality, reusability, integrity, consistency, and completeness of these deliverables.

The checklist template in Table 5.7 highlights the major measurement criteria to ensure delivery of quality deliverables identified in the deliverables template and adherence to the process flows identified in the process flow template. The checklist is used to objectively determine success criteria for commitment and approval of DA deliverables and process to proceed with further project developments.

Table 5.6 Deliverables Template: DA Report

Section 1—Introduction
This section includes an executive summary, data management policies, and approach and introduces the report.
Section 2—Project Data Architecture Requirements
This section highlights the project DA requirements, deliverables, and DA processes to support the project team. These DA requirements establish the baseline for development of the DA management plan in Section 3 and the development of DA deliverables in Section 4.
Section 3—Project Data Architecture Management Plan
This section highlights the plan for managing and developing the DA deliverables by identifying the deliverables—what—the activities and management processes—how—schedule—when—and DA staff—who—to manage and deliver the DA deliverables to support the project team requirements. This DA management plan integrates with the WBS and project plan produced during the definition phase. It describes the baseline plan to support development and management of the project data DA requirements in Section 4.
Section 4—Project Data Architecture Deliverables
This section highlights the major deliverables of DA support by providing an overview on the contents of these deliverables and the location where they are stored. These deliverables are documented in the project repository and integrate with the PMO and DA repositories detailed in Section 5.
Section 5—Integrated Data Architecture and PMO Repository Services
This section highlights the major deliverables of the DA architecture support services by providing the repository contents and information retrieval system for storing, retrieving, accessing, disseminating, and deleting DA deliverables. The DA deliverables are stored in the repository, which uses the PMIS for maintaining the integrity and consistency of the contents. This repository forms the baseline for producing DA metrics and performance reports discussed in Section 6.
Section 6—Project Data Architecture Metrics and Performance Reports
This section highlights the overall performance of the DA deliverables, the impact on the other project deliverables, and suggested corrective actions for deviations in scope, schedule, cost, and quality performance, based on the integration features of the repository services and PMIS outlined in Section 5.

5.5 Applications architecture

When a system becomes completely defined, some damn fool discovers something which either abolishes the system or expands it beyond recognition.
—Murphy's Law

Executive management shall use the AA deliverables and the related checklists to approve the AA deliverables at each phase of the IT PDLC. AA support services are performed during each of the phases based on the business application requirements of the project. These deliverables and supporting activities are identified, managed, and integrated within the core components of the integrated WBS. The major components of AA services are as follows:

- *Project application management infrastructure support:* These services provide directory information on what, who, why, where, and how to access and use existing application management strategies, standards

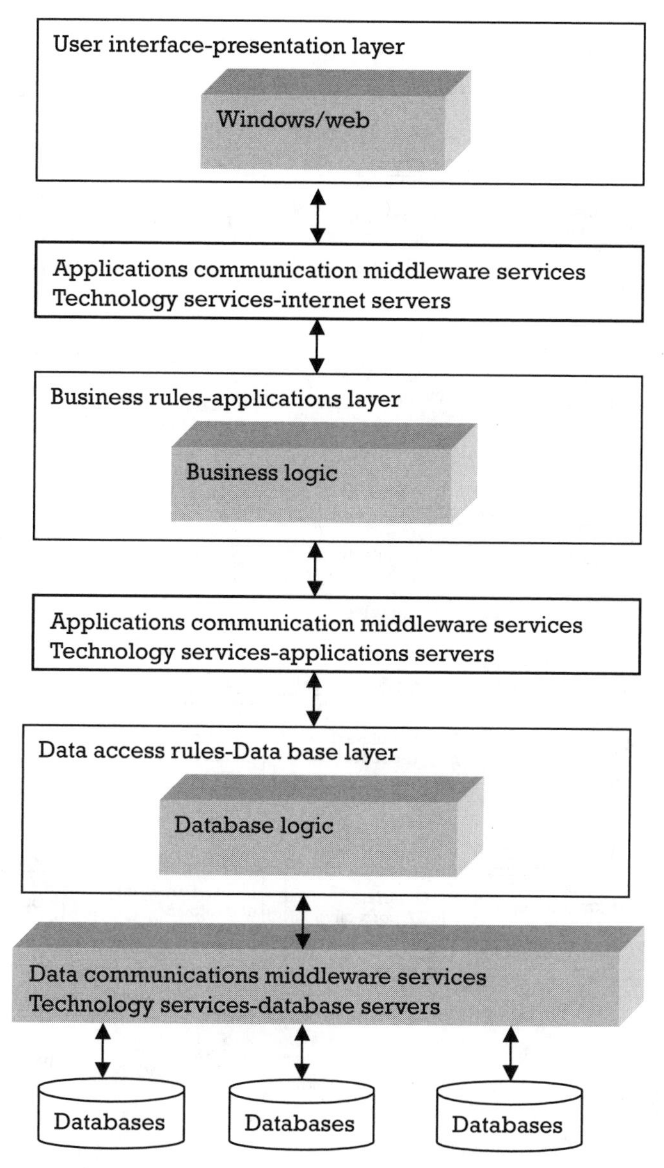

Figure 5.5 DA communications middleware model.

and procedures, application repository, systems management software tools, applications development software tools, and applications communications middleware tools.

- *Project AA development support:* These services produce application component deliverables that include business strategies and requirements, business process models, and inventory of existing logical and physical applications used during the development of the BSA.
- *Project application user interface model support:* These services produce detailed user interface component deliverables that include Web-based

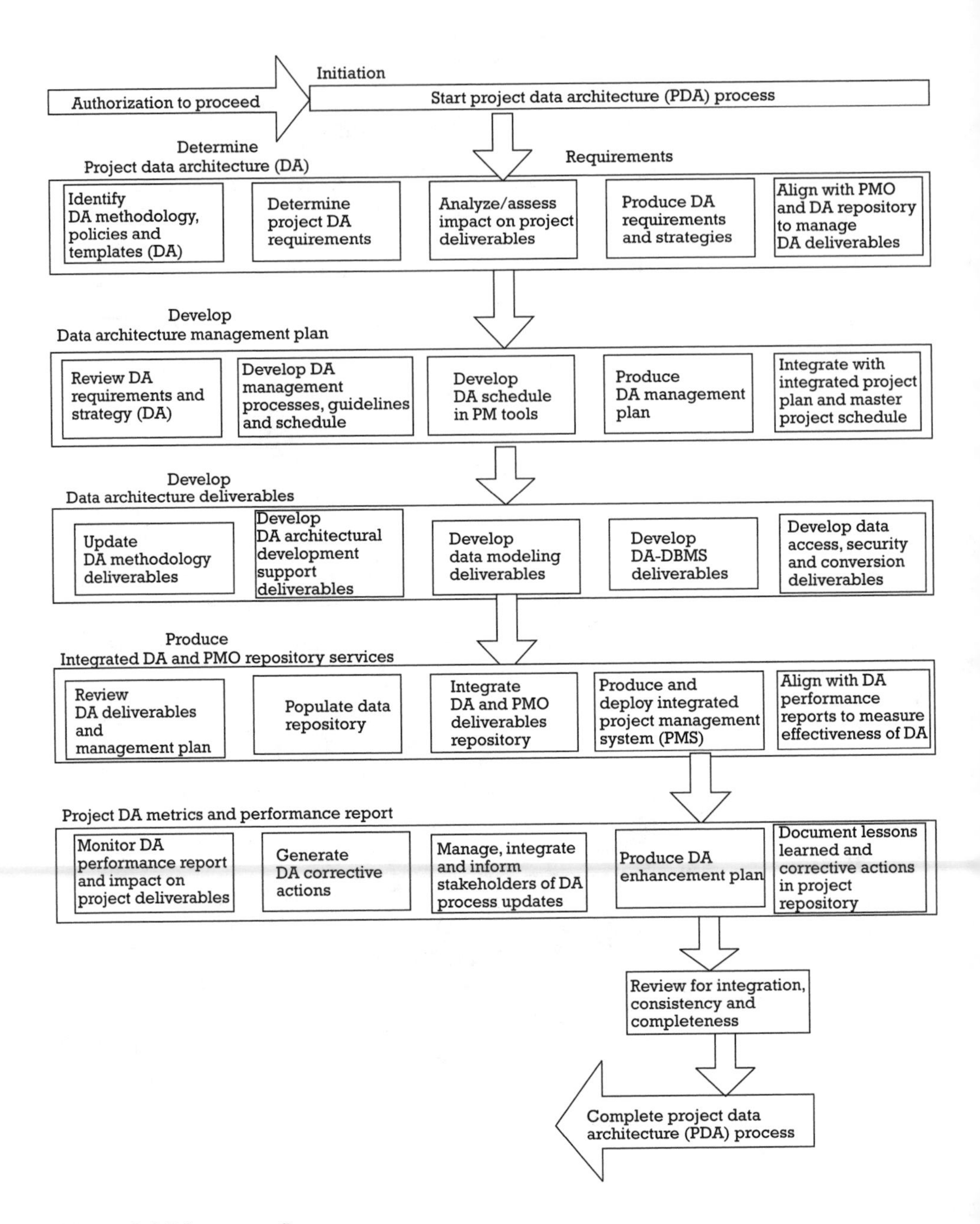

Figure 5.6 DA process flow.

and Windows prototypes for implementation using the presentation software and the applications communications middleware.

- *Project application development support:* These services produce the application code based on the business rules for executing the application, using the application development software—custom development or purchased package—applications communications middleware, and systems management software.

Table 5.7 DA Checklist

Project DA Criteria	*Yes*	*No*	*Project DA Criteria*	*Yes*	*No*
Initiation			*DA requirements*		
Is management committed?	❑	❑	Do DA policies, processes, and templates exist?	❑	❑
Have a PM sponsor and steering committee been established?	❑	❑	Are DA requirements determined?	❑	❑
Is there a PM methodology?	❑	❑	Is the impact on project deliverables determined?	❑	❑
Does the PM methodology have a DA component?	❑	❑	Are DA strategies accepted?	❑	❑
Has a program delivery manager been assigned?	❑	❑	Are DA and PMO deliverables integrated?	❑	❑
Does the assigned program delivery manager understand the DA process?	❑	❑	Are PMO and DA involved in DA strategy development?	❑	❑
DA management plan			*DA deliverables*		
Is there a checklist for measuring DA progress?	❑	❑	Do deliverables exist?	❑	❑
Is the DA management plan approved?	❑	❑	Does they contain a logical data model?	❑	❑
Does the DA management plan include processes, roles and responsibilities, and schedule?	❑	❑	Do they contain physical DBMS components?	❑	❑
Is the DA plan included within the integrated project plan?	❑	❑	Do they contain data security, integrity, and conversion utilities?	❑	❑
Is the DA schedule integrated with project schedule?	❑	❑	Is there a checklist for determining quality?	❑	❑
Is the DA plan developed with DA and PMO involvement?	❑	❑	Are deliverables documented in the PMO and DA repository?	❑	❑
DA and PMO repository			*DA metrics and performance*		
Are the DA and PMO repository integrated?	❑	❑	Does a DA performance process exist?	❑	❑
Does the PMIS maintain project repository data?	❑	❑	Are DA performance reporting guidelines available?	❑	❑
Does duplication exists between these repositories?	❑	❑	Do variations from the DA baseline exist?	❑	❑
Is there a master DA repository model?	❑	❑	Are stakeholders informed about changes?	❑	❑
Is the model communicated to the project team?	❑	❑	Is there a DA enhancement plan to implement changes?	❑	❑
Are the PMIS and DA repository deployed to project team?	❑	❑	Are lessons learned documented in a project file?	❑	❑

- *Project application common services directory support:* These services produce the applications repository component deliverables to support code reusability, integrity, consistency, and completeness. The common services directory is one of the key features of the applications communications middleware software.

The major deliverables of the AA support services as presented in this chapter align with business management and project management and consist of the following components:

- Project AA requirements;
- Project AA management plan;
- Project AA deliverables;
- Project integrated DA and PMO repository services;
- Project AA metrics and performance reports.

5.5.1 Purpose

The purpose of this process is to deliver all the application component deliverables identified in the integrated project plan and WBS to meet the application requirements of the project. These product scope deliverables will form the baseline to manage the quality objectives of the development projects. The program steering committee will use these application deliverables and potential benefits as deciding factors during approval or disapproval to proceed with each subsequent phase of the project based on the PDLC.

5.5.2 Policy

The policy of applications architecture is the following:

- The AA deliverables shall be valued and protected as one of the key critical assets of the project and the company by documenting and maintaining the components in the project applications repository[8] in accordance with company's AA standards and PMO guidelines.
- The AA shall be business driven rather than technology driven, aligned with the DA, and developed according to company's AA standards.
- Applications shall be effectively managed by establishing applications development roles and responsibilities that require good working relationships among business, project management, and IT AA services.
- Logical and physical application models and code shall be defined using applications modeling tools, applications communications middleware, and the application common services repository. Applications interfaces shall be message based, using the applications communications middleware software.
- Logical business models shall be developed prior to development or purchase of the code to ensure proper integration with business and data requirements.

8. Applications repository: A database containing applications deliverables that is integrated with the PMO project repository to ensure completeness, consistency, integration, and reusability of the project deliverables. It also integrates with the data and technology repositories.

- The IT program manager, prior to development and implementation, shall approve applications migration, applications interface communications, common services code components, access and security processes, and deliverables.

5.5.3 Roles and responsibilities

Executive business manager responsibilities include the following:

- Provide executive-level business approval to the program business manager to ensure that the business architecture is aligned with the AA and to review the BSA document based on the business direction within budgetary and schedule constraints.
- Provide executive-level business commitment for the business AA components within the overall BSA document.
- Approve all business changes to the business AA components within the overall BSA document.
- Approve business resources to support development of the business AA components within the overall BSA document.
- Act as the overall program sponsor and chairman of the program steering committee.

Executive IT manager responsibilities include the following:

- Provide executive-level IT approval to the program IT manager to ensure that the business architecture is aligned with the AA and to review the BSA document based on the business direction–business plan within budgetary and schedule constraints.
- Provide executive-level IT commitment for the AA components within the overall BSA document.
- Approve all IT changes to the AA components within the overall BSA document.
- Approve IT resources to support development of the AA components within the overall BSA document.
- Act as the program sponsor from the IT perspective.

Executive project management manager responsibilities include the following:

- Provide executive-level PM guidance, direction, and advice to the program delivery manager to ensure that the business architecture is aligned with the AA and to review the BSA document based on the business direction–business plan within budgetary and schedule constraints.
- Integrate business, data, applications, and technology architectures into the overall BSA document.
- Integrate changes to business, data, applications, and technology architectures into the overall BSA document.

- Provide PM resources for AA components of the overall BSA document.
- Act as the program sponsor with integrated business and IT responsibility and PM accountability; champion PM processes, tools, and techniques.
- Provide AA development guidance, direction, and advice to the program delivery manager to ensure that business and technology architectures are effectively aligned with high-priority projects.

Program business manager responsibilities include the following:

- Submit details on business guidance, direction, and advice to the program delivery manager to ensure that the business architecture is aligned with the AA to deliver maximum business benefits to the company in support of project scope, time, and quality constraints.
- Report on business commitment to the AA within the overall BSA.
- Report on all business changes to the AA within the overall BSA.
- Assign business resources for supporting the delivery of the AA within the overall BSA.
- Report to the program steering committee with direct accountability to the business executive, and keep the program working committee informed of the implementation of the AA within the overall BSA.

Program IT manager responsibilities include the following:

- Submit details on IT guidance, direction, and advice to the program delivery manager to ensure that the business architecture is aligned with the AA to deliver maximum business benefits to the company in support of project scope, time, and quality constraints.
- Report on IT commitment to the AA within the overall BSA.
- Report on all IT changes to the AA within the overall BSA.
- Assign IT resources for supporting the delivery of the AA within the overall BSA.
- Report to the program steering committee with direct accountability to the business executive, and keep the program working committee informed of the implementation of the AA within the overall BSA.

Program delivery manager responsibilities include the following:

- Review and communicate business and IT (physical) AAs/models and ensure linkage with business DAs.
- Communicate AA policies and ensure the project team adheres to AA methodology, processes, tools, and techniques.
- Verify that the project team adheres to AA policies, guidelines, and procedures by utilizing checklists that are included in the AA deliverables.
- Manage the integrated AA by reporting on the contents and recommended solutions to the appropriate management team.
- Update AA processes as needed to keep them effective.

- Report to the program steering committee with direct accountability to the PM executive, and keep project managers and stakeholders informed of the effectiveness and usefulness of the AA during the project's implementation.

Project team (project manager and team) responsibilities include the following:

- Project managers to manage applications deliverables in accordance with PMO quality guidelines;
- Project managers to incorporate applications deliverables and activities into the project charter and integrated project plan;
- Project managers to integrate, assemble, and report to the program working committee on the progress and quality of the applications deliverables, issues, corrective actions, and risks to the project with direct accountability to the program delivery manager;
- Project team to develop applications deliverables based on PMO and AA standards.

5.5.4 Procedures: AA

The roles and responsibilities discussed in Section 5.5.3 are brief because they are intended to serve as a reference as to who does what during the execution and delivery of the AA processes and deliverables. Business procedures are now introduced to demonstrate how this AA process is executed using the major standard deliverables (what), process flow (how), and checklist or measurement criteria (why) templates. The process is intended to serve as a reference based on the English-language interrogatives: who does what, how, and why. The deliverables, process flow, and checklist templates can serve as guidelines for practicing project managers during the management of the AA processes and deliverables.

The deliverables template in Table 5.8 highlights the key deliverables that the practicing project manager must manage to enable the delivery of quality deliverables. The guidelines provided in Figure 5.7, based on real-world practical implementations, should provide some insights to project managers. The *N*-tier software development environment shows how the key DA components integrated with applications and technology architectures.

The process flow template in Figure 5.8 highlights the major processes to support the delivery of the key deliverables. These processes reference the common services applications repository and project management repository to show the integrated nature and dependencies of these repositories and to ensuring the high quality, reusability, integrity, consistency, and completeness of these deliverables.

The checklist template in Table 5.9 highlights the major measurement criteria to ensure delivery of quality deliverables identified in the deliverables template and adherence to the process flows identified in the process

Table 5.8 Deliverables Template: AA Report

Section 1—Introduction
This section includes an executive summary, applications management policies, and approach and introduces the report.
Section 2—Project Applications Architecture Requirements
This section highlights the project AA requirements, deliverables, and AA processes to support the project team. These AA requirements establish the baseline for the development of the AA management plan in Section 3 and the development of AA deliverables in Section 4.
Section 3—Project Applications Architecture Management Plan
This section highlights the plan for managing and developing the AA deliverables by identifying the deliverables—what—the activities and management processes—how—the schedule—when—and the AA staff—who—that will manage and deliver the AA deliverables to support the project team requirements. This AA management plan integrates with the WBS and project plan produced during the definition phase. It describes the baseline plan to support development and management of the project AA requirements in Section 4.
Section 4—Project Applications Architecture Deliverables
This section highlights the major deliverables of AA support by providing an overview on the contents of these deliverables and the location where they are stored. These deliverables are documented in the project repository and integrate with PMO and AA repositories in Section 5.
Section 5—Integrated AA and PMO Repository Services
This section highlights the major deliverables of the AA support services by providing the repository contents and information retrieval system for storing, retrieving, accessing, disseminating, and deleting AA deliverables. The AA deliverables are stored in the repository, which uses the PMIS for maintaining the integrity and consistency of the contents. This repository forms the baseline for producing AA metrics and performance reports in Section 6.
Section 6—Project AA Metrics and Performance Reports
This section highlights the overall performance of the AA deliverables, the impact on the other project deliverables, and suggested corrective actions for deviations in scope, schedule, cost, and quality performance, based on the integration features of the AA repository services and PMIS outlined in Section 5.

flow template. The checklist is used to objectively determine success criteria for commitment and approval of AA deliverables and process to proceed with further project developments.

5.6 Technology architecture

The real problem is not whether machines think, but whether men do.
—B. F. Skinner

Executive management shall use the TA deliverables and the related checklists to approve the technology deliverables at each phase of the IT PDLC. TA support services are performed during each of the phases based on the DA and AA requirements of the project. These deliverables and supporting activities are identified, managed, and integrated within the core components of the integrated WBS. The major components of TA support services are as follows:

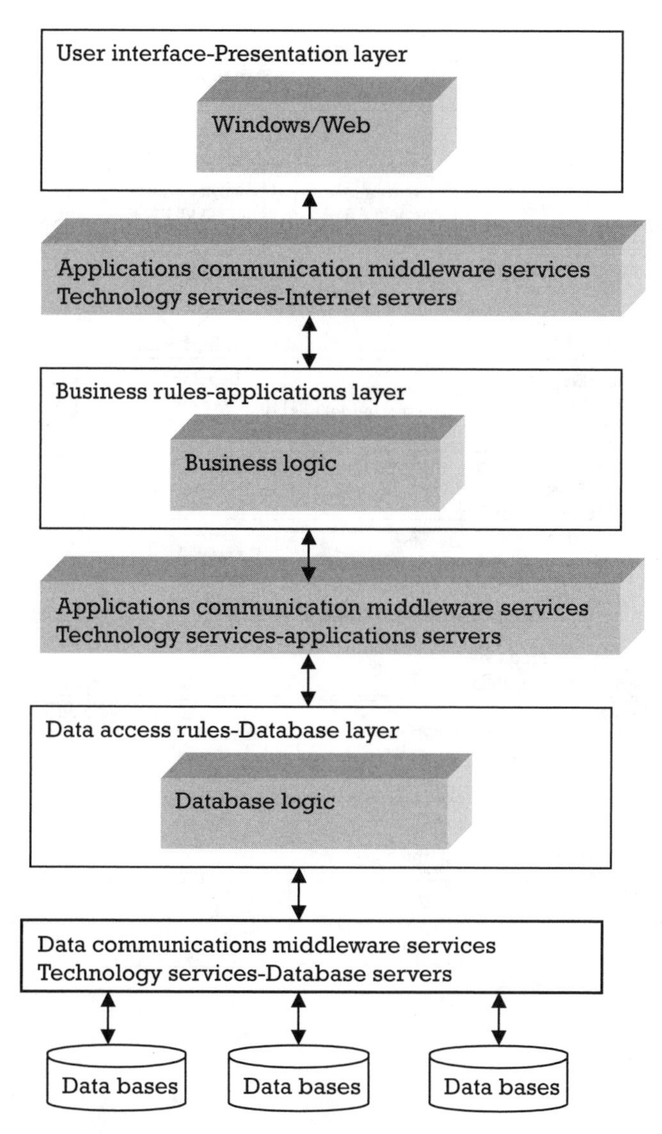

Figure 5.7 AA middleware model.

- *Project technology management infrastructure support:* These services provide directory information on what, who, why, where, and how to access and use existing technology management strategies, standards and procedures, technology repository, operating systems and network software tools, technology performance software tools, and technology hardware and network components.
- *Project TA development support:* These services produce technology component deliverables that include technology (hardware, systems software, and network) strategies and requirements, technology configuration map, and inventory of existing technology environment used during the development of the BSA.

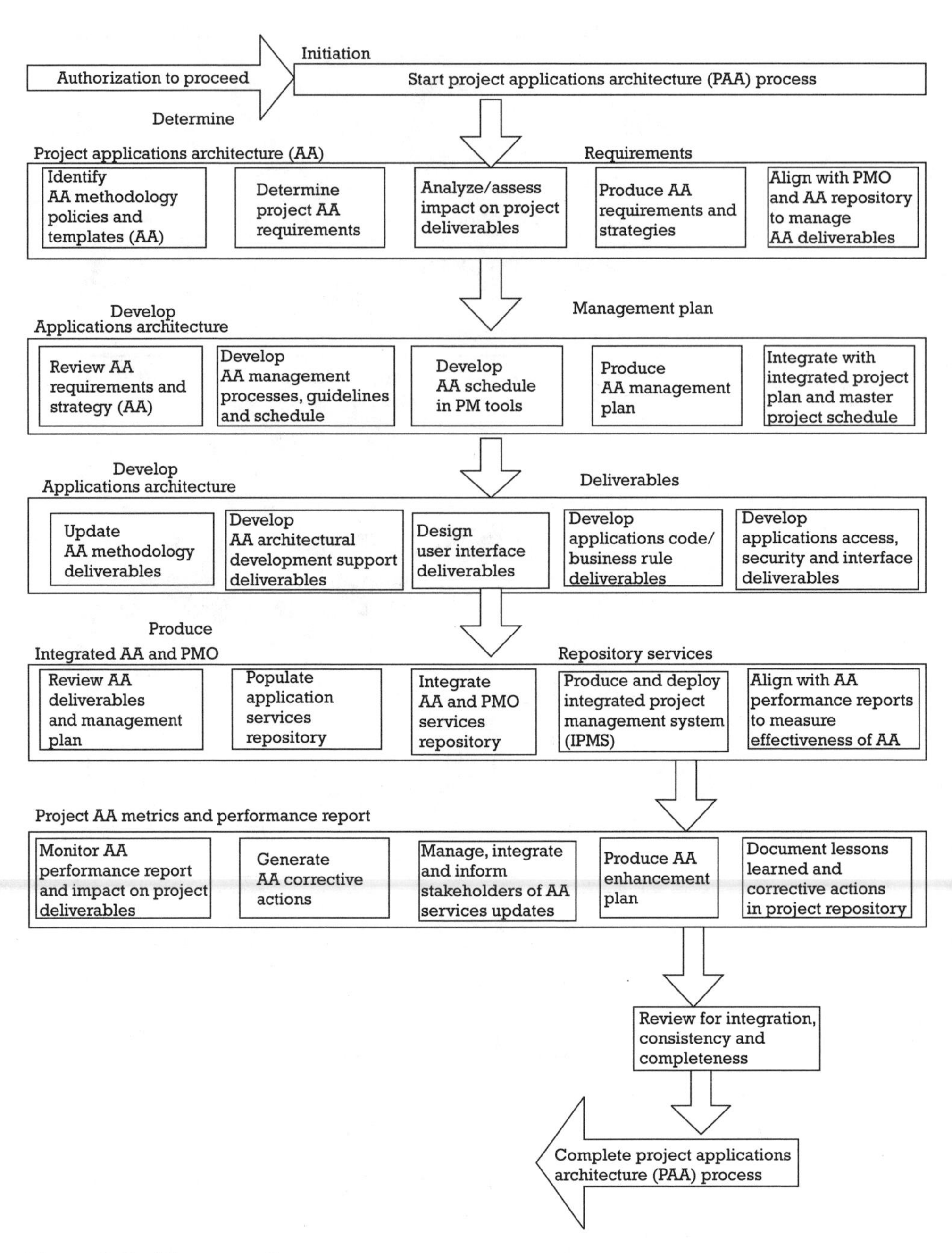

Figure 5.8 AA process flow.

- *Project technology implementation model support:* These services design and install the detailed technology configuration component deliverables that include hardware, systems software, and network configuration for implementation, using the DA and AA components.
- *Project technology integration support:* These services provide the technology repository services to permit integration of data, applications, and

Table 5.9 AA Checklist

Project AA Criteria	*Yes*	*No*	*AA Criteria*	*Yes*	*No*
Initiation			*AA requirements*		
Is management committed?	❑	❑	Do AA policies, processes, and templates exist?	❑	❑
Have a PM sponsor and steering committee been established?	❑	❑	Are AA requirements determined?	❑	❑
Is there a PM methodology?	❑	❑	Is the impact on project deliverables determined?	❑	❑
Does the PM methodology have an AA component?	❑	❑	Are AA strategies accepted?	❑	❑
Has a program delivery manager been assigned?	❑	❑	Are AA and PMO deliverables integrated?	❑	❑
Does the assigned program delivery manager understand the AA process?	❑	❑	Are PMO and AA involved in AA strategy development?	❑	❑
AA management plan			*AA deliverables*		
Is there a checklist for measuring AA progress?	❑	❑	Do AA deliverables exist?	❑	❑
Is the AA management plan approved?	❑	❑	Do they contain user interface design?	❑	❑
Does the AA management plan include processes, roles and responsibilities, and schedule?	❑	❑	Do they contain code/business rules implementation?	❑	❑
Is the AA plan included within the integrated project plan?	❑	❑	Do they contain application security, integrity, and conversion utilities?	❑	❑
Is the AA schedule integrated with project schedule?	❑	❑	Is there a checklist for determining quality?	❑	❑
Is the AA plan developed with AA and PMO involvement?	❑	❑	Are deliverables documented in the PMO and AA repository?	❑	❑
AA and PMO repository			*AA metrics and performance*		
Is the AA and PMO repository integrated?	❑	❑	Does an AA performance process exist?	❑	❑
Does the PMIS maintain project repository data?	❑	❑	Are AA performance reporting guidelines available?	❑	❑
Does duplication exists between these repositories?	❑	❑	Do variations from the AA baseline exist?	❑	❑
Is there a master AA repository model?	❑	❑	Are stakeholders informed about changes?	❑	❑
Is the model communicated to the project team?	❑	❑	Is there AA enhancement plans to implement changes?	❑	❑
Are PMIS and the AA repository deployed to the project team?	❑	❑	Are lessons learned documented in a project file?	❑	❑

technology architectures, as shown in Figure 5.9. These technology deliverables support technology reusability, integration, consistency, and completeness.

- *Project technology performance support:* These services monitor and optimize the performance of hardware, network, and systems software to support data and applications architecture requirements. The technology performance services ensure that the installed components are

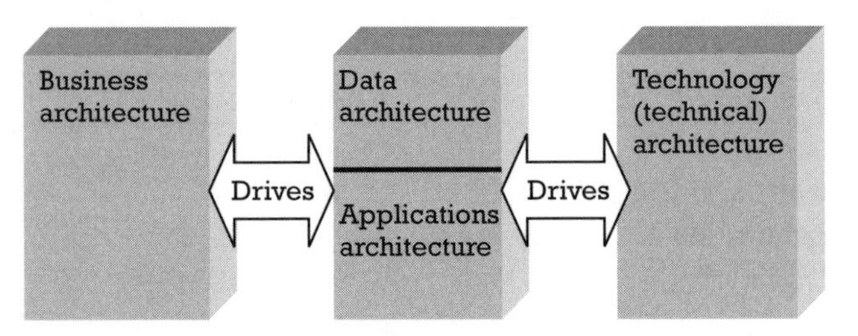

Figure 5.9 TA integration.

performing according to data, applications, and technology specifications.

The major deliverables of the TA support services as presented in this chapter align with business management and project management and consist of the following components:

- Project TA requirements;
- Project TA management plan;
- Project TA deliverables;
- Project integrated TA and PMO repository services;
- Project TA metrics and performance reports.

5.6.1 Purpose

The purpose of this process is to deliver all the technology component deliverables identified in the integrated project plan and WBS to meet the technical requirements of the project. These product scope deliverables will form the baseline to manage the technical quality objectives of the development projects. The program steering committee will use these technology deliverables and potential benefits as deciding factors during approval or disapproval to proceed with each subsequent phases of the project based on the PDLC.

5.6.2 Policy

The policy of the technology architecture is the following:

- The TA deliverables shall be valued and protected as one of the key critical assets of the project and the company by documenting and maintaining the components in the technology repository,[9] according to the company's TA standards and PMO guidelines.
- The TA shall be business driven, rather than technology driven, and shall be aligned with the data and applications architecture, and developed according to the company's TA standards.

9. Technology repository: A database containing technology deliverables that is integrated with the PMO project repository to ensure completeness, consistency, integration, and reusability of the project deliverables. It also integrated with the data and applications repositories.

- Technology shall be effectively managed by establishing technology architects roles and responsibilities that require good working relationships among business management, project management and IT technical architecture services.
- Logical and physical technology models and code shall be defined using existing technology inventory, technology performance monitoring tools, applications, and data communications middleware, and applications and data common services repository.
- Logical technology configuration models shall be developed prior to installation of the technical infrastructure environment to ensure proper integration with business, applications, and data requirements.

The IT program manager, prior to installation and implementation, shall approve the proposed technical environment, interface communications, common services code components, access and security processes and deliverables.

5.6.3 Roles and responsibilities

Executive business manager responsibilities include the following:

- Provide executive-level business approval to the program business manager to ensure that the business architecture is aligned with the TA and to review the BSA document based on the business direction within budgetary and schedule constraints.
- Provide executive-level business commitment for the business TA components within the overall BSA document.
- Approve all business changes to the business TA components within the overall BSA document.
- Approve business resources to support development of the business AA components within the overall BSA document.
- Act as the overall program sponsor and chairman of the program steering committee.

Executive IT manager responsibilities include the following:

- Provide executive-level IT approval to the program IT manager to ensure that the business architecture is aligned with the TA and to review the BSA document based on the business direction–business plan within budgetary and schedule constraints.
- Provide executive-level IT commitment for the TA components within the overall BSA document.
- Approve all IT changes to the TA components within the overall BSA document.
- Approve IT resources to support development of the TA components within the overall BSA document.
- Act as the program sponsor from the IT perspective.

Executive project management manager responsibilities include the following:

- Provide executive-level PM guidance, direction, and advice to the program delivery manager to ensure that the business architecture is aligned with the TA and to review the BSA document based on the business direction–business plan within budgetary and schedule constraints.
- Integrate business, data, applications, and technology architectures into the overall BSA document.
- Integrate changes to business, data, applications, and technology architectures into the overall BSA document.
- Approve PM resources for TA components of the overall BSA document.
- Act as the program sponsor with integrated business and IT responsibility and PM accountability; champion PM processes, tools, and techniques.
- Provide TA development guidance, direction, and advice to the program delivery manager to ensure that business and technology architectures are effectively aligned with high-priority projects.

Program business manager responsibilities include the following:

- Submit details on business guidance, direction, and advice to the program delivery manager to ensure that the business architecture is aligned with the AA to deliver maximum business benefits to the company in support of project scope, time, and quality constraints.
- Report on business commitment to the TA within the overall BSA.
- Report on all business changes to the TA within the overall BSA.
- Assign business resources for supporting the delivery of the TA within the overall BSA.
- Report to the program steering committee with direct accountability to the business executive, and keep the program working committee informed of the implementation of the TA within the overall BSA.

Program IT manager responsibilities include the following:

- Submit details on IT guidance, direction, and advice to the program delivery manager to ensure that the business architecture is aligned with the TA to deliver maximum business benefits to the company in support of project scope, time, and quality constraints.
- Report on IT commitment to the TA within the overall BSA.
- Report on all IT changes to the TA within the overall BSA.
- Assign IT resources for supporting the delivery of the TA within the overall BSA.
- Report to the program steering committee with direct accountability to the business executive, and keep the program working committee informed of the implementation of the TA within the overall BSA.

Program delivery manager responsibilities include the following:

- Review and communicate business and IT (physical) TAs/models and ensure linkage with business, data, and application architectures.
- Communicate TA policies and ensure project teams adhere to TA methodology, processes, tools, and techniques.
- Verify that the project team adheres to TA policies, guidelines, and procedures by utilizing checklists that are included in the TA deliverables.
- Manage the integrated TA by reporting on the contents and recommended solutions to the appropriate management team.
- Update TA processes as needed to keep them effective.
- Report to the program steering committee with direct accountability to the PM executive, and keep project managers and stakeholders informed of the effectiveness and usefulness of the TA during projects implementation.

Project team (project manager and team) responsibilities include the following:

- Project managers to manage technology deliverables in accordance with PMO quality guidelines and IT TA standards.
- Project managers to incorporate technology deliverables and activities into the project charter and integrated project plan.
- Project managers to integrate, assemble, and report to the program working committee on the progress and quality of the technology deliverables, issues, corrective actions, and risks to the project with direct accountability to the program delivery manager.
- Project team to develop technology deliverables, based on PMO guidelines and TA standards.

5.6.4 Procedures: TA

The roles and responsibilities discussed in Section 5.6.3 are brief because they are intended to serve as a reference as to who does what during the management and execution of the TA deliverables and process. Business procedures are now introduced to demonstrate how this TA process is executed using the major standard deliverables (what), process flow (how), and checklist or measurement criteria (why) templates. The process is intended to serve as a reference based on the English-language interrogatives: who does what, how, and why. The deliverables, process flow, and checklist templates can serve as guidelines for the practicing project managers during the execution of the TA process.

The deliverables template in Table 5.10 highlights the key deliverables that the practicing project manager must manage to enable the delivery of quality deliverables. The guidelines provided in Figure 5.10, based on real-

Table 5.10 Deliverables Template: TA Report

Section 1—Introduction
This section includes an executive summary, technology management policies, and approach and introduces the report.
Section 2—Project Technology Architecture Requirements
This section highlights the project TA requirements, deliverables, and TA processes to support the project team. The TA requirements establish the baseline for development of the TA management plan in Section 3 and the development of TA deliverables in Section 4.
Section 3—Project Technology Architecture Management Plan
This section highlights the plan for managing and developing the TA deliverables by identifying the deliverables—what—the activities and management processes—how—schedule—when—and TA staff—who—that will manage and deliver the TA deliverables to support the project team requirements. This TA management plan integrates with the WBS and project plan produced during the definition phase. It describes the baseline plan to support development and management of the project TA requirements in Section 4.
Section 4—Project Technology Architecture Deliverables
This section highlights the major deliverables of TA support by providing an overview on the contents of these deliverables and the location where they are stored. These deliverables are documented in the project repository and integrate with PMO and TA repositories in Section 5.
Section 5—Integrated TA and PMO Repository Services
This section highlights the major deliverables of the TA support services by providing the repository contents and information retrieval system for storing, retrieving, accessing, disseminating, and deleting TA deliverables. The TA deliverables are stored in the repository, which uses the PMIS for maintaining the integrity and consistency of the contents. This repository forms the baseline for producing TA metrics and performance reports in Section 6.
Section 6—Project TA Metrics and Performance Reports
This section highlights the overall performance of the TA deliverables, the impact on the other project deliverables, and suggested corrective actions for deviations from scope, schedule, cost, and quality performance, based on the integration features of the technology repository services and PMIS outlined in Section 5.

world practical implementations should provide some insights to project managers. The N-tier software development environment shows how the key DA, AA, and TA components integrate, using the technology hardware or server components.

The process flow template in Figure 5.11 highlights the major processes to support the delivery of the key deliverables. These processes reference the technology repository and project management repository to show the integrated nature and dependencies of these repositories and to ensure the high quality, reusability, integrity, consistency, and completeness of these deliverables.

The checklist template in Table 5.11 highlights the major measurement criteria to ensure delivery of quality deliverables identified in the deliverables template and adherence to the process flows identified in the process flow template. The checklist is used to objectively determine success criteria for commitment and approval for technology deliverables to proceed with further developments from a technology perspective.

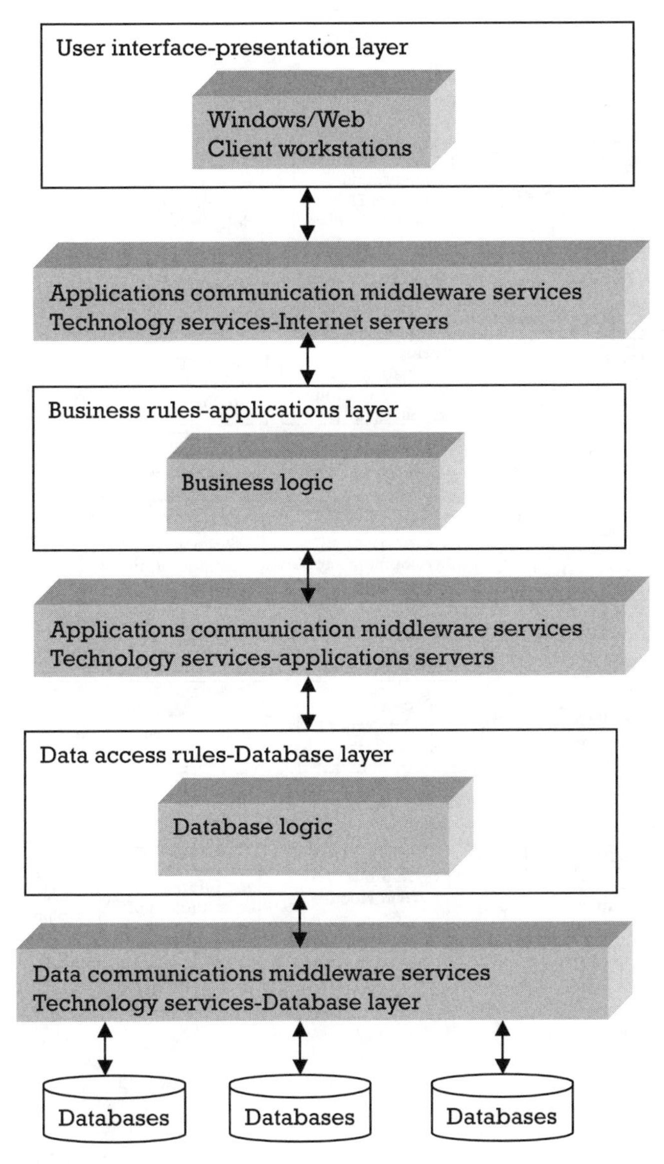

Figure 5.10 TA model.

5.7 Applications support services

If everything seems under control, you are just not going fast enough.
—Mario Andretti

Executive management shall use the applications support services deliverables and the related checklists to approve the deliverables during the deployment phase of the IT PDLC. Applications support services are mainly performed during the iterative development–deployment phases to determine the applications production support requirements of the project. These deliverables and supporting activities are identified, managed, and integrated within the core components of the master WBS.

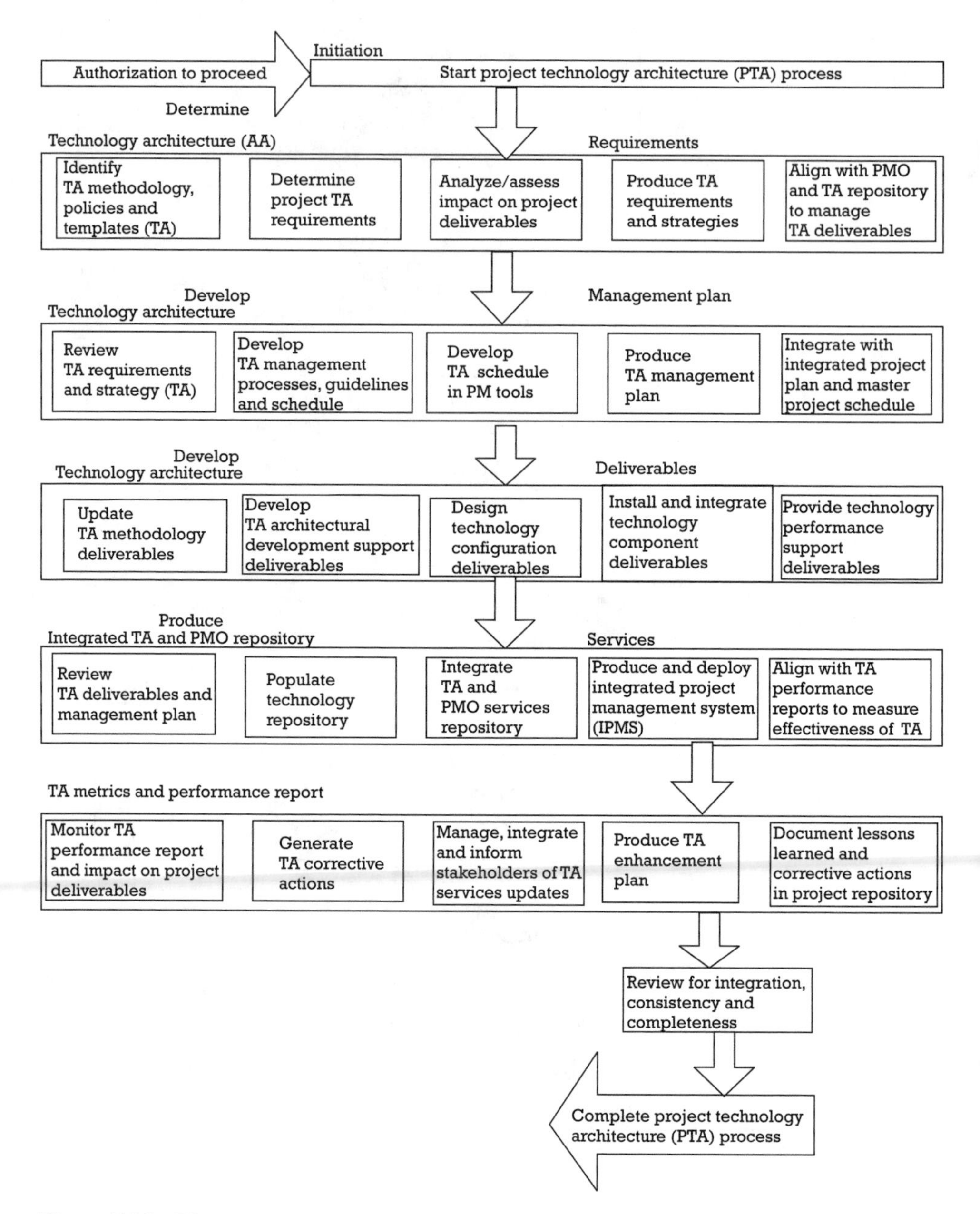

Figure 5.11 TA process flow.

The major deliverables of applications support services are:

- *Project application service-level agreement model (SLAM)*[10]*:* This model provides details on an agreement between the project team-business and IT, and the IT Applications Support group-IT service providers for the

10. SLAM: A service-level agreement model that provides details on an agreement between the project team–business and IT, and the IT applications support group–IT service providers, for the type and level of support. Refer to Figure 5.12.

Table 5.11 TA Checklist

Project TA Criteria	*Yes*	*No*	*Project TA Criteria*	*Yes*	*No*
Initiation			*TA requirements*		
Is management committed?	❑	❑	Do TA policies, processes, and templates exist?	❑	❑
Have a PM sponsor and steering committee been established?	❑	❑	Are TA requirements determined?	❑	❑
Is there a PM methodology?	❑	❑	Is the impact on project deliverables determined?	❑	❑
Does the PM methodology have a TA component?	❑	❑	Are TA strategies accepted?	❑	❑
Has a program delivery manager been assigned?	❑	❑	Are TA and PMO deliverables integrated?	❑	❑
Does the assigned program delivery manager understand the TA process?	❑	❑	Are PMO and TA involved in TA strategy development?	❑	❑
TA management plan			*TA deliverables*		
Is there a checklist for measuring TA progress?	❑	❑	Do TA deliverables exist?	❑	❑
Is the TA management plan approved?	❑	❑	Do they contain a technology configuration map?	❑	❑
Does the TA management plan include processes, roles and responsibilities, and schedule?	❑	❑	Is the technology installed and deployed?	❑	❑
Is the TA plan included within the integrated project plan?	❑	❑	Does the deliverables contain technology performance statistics?	❑	❑
Is the TA schedule integrated with the project schedule?	❑	❑	Is there a checklist for determining TA quality?	❑	❑
Is the TA plan developed with TA and PMO involvement?	❑	❑	Are deliverables documented in the PMO and TA repository?	❑	❑
TA and PMO repository			*TA metrics and performance*		
Is the TA and PMO repository integrated?	❑	❑	Does TA performance process exist?	❑	❑
Does the PMIS maintain project repository data?	❑	❑	Are TA performance reporting guidelines available?	❑	❑
Does duplication exist between these repositories?	❑	❑	Do variations from the TA baseline exist?	❑	❑
Is there a master TA repository model?	❑	❑	Are stakeholders informed about changes?	❑	❑
Is the model communicated to the project team?	❑	❑	Is there a TA enhancement plan to implement changes?	❑	❑
Is PMIS and TA repository deployed to project team?	❑	❑	Are lessons learned documented in a project file?	❑	❑

type and level of support. This agreement is established prior to disassembling the project team, and consists of the processes (how) the IT service providers will provide warranty and production or operations support, systems management support, and enhancement requests for the applications. Figure 5.12 provides an integrated view.

- *Project application warranty support:* This support model contains details on the processes (how) the IT service provider will use to provide training, organizational change, process redevelopments, production operations

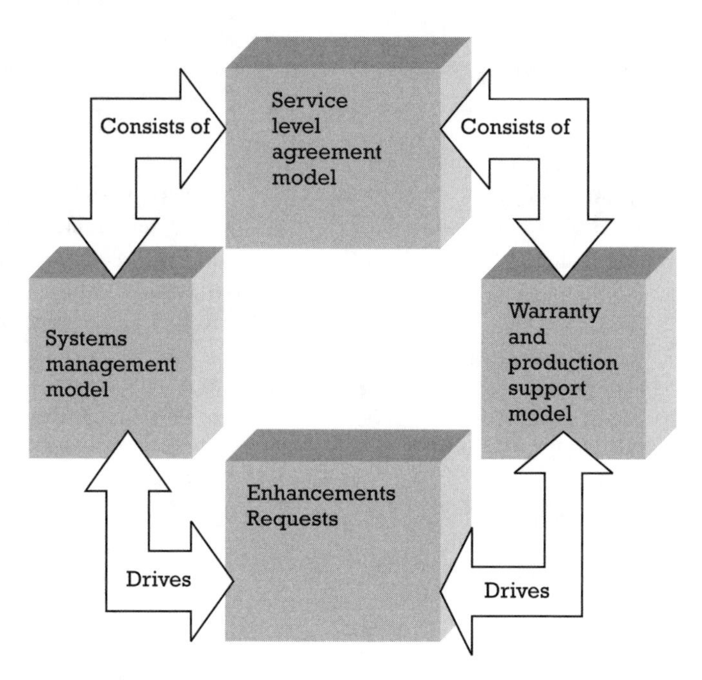

Figure 5.12 SLAM.

management, technology, and facilities infrastructure support services, and applications postimplementation reviews.

- *Project application production support:* This support model contains details on the processes (how) the IT service provider will use to provide systems management support, database support, applications support, and technology support to manage resources after production implementation.
- *Project application systems management support:* This support model contains details on the processes (how) the IT service provider will use to provide help desk, operations management, storage management, performance monitoring and tuning, security services, and disaster recovery support services.
- *Project application enhancement requests:* This support model contains details on the processes (how) the IT service provider will use to provide multiple tiers or levels of client support in order to optimize support resources to provide effective client support.

The major deliverables of the applications support services align with business management, IT management, and project management and consist of the following components:

- Applications support initiatives and alignment with business support initiatives;
- Applications support initiatives alignment with IT TA and business support initiatives alignment with BSA;

- SLAM;
- Service-level delivery plan (SLDP).

5.7.1 Purpose

The purpose of this process is to deliver an SLAM and an SLDP)[11] to deliver the applications support services agreement. These deliverables and activities are included in the integrated project plan and WBS to meet the applications support requirements of the project. This SLAM will form the baseline to manage the quantity and quality objectives and services of the IT service provider. The program steering committee will use this agreement and potential benefits as deciding factors during the approval or disapproval decision to proceed with the IT service provider for production support.

5.7.2 Policy

The policy of application support services is the following:

- The applications service-level agreement and postimplementation reviews shall be delivered and approved prior to production release.
- Systems management decisions shall be based on the priority of the business needs and alignment with applications, data, and technology architectures.
- Production support shall be managed using the applications enhancement processes and the procedures for tier 1 (production support–business focus), tier 2 (new development support–technical focus), and tier 3 (strategic support–strategic focus).
- IT service providers shall limit the tier-1 support efforts to a minimum to reduce long-term tier-1 support costs and optimize capital investments to offset any long-term support costs.
- IT service providers and the business sponsors shall manage the priority of the tier-1, tier-2, and tier-3 requests based on the business needs and integration with the BSA (business, data, applications, and technology).

5.7.3 Roles and responsibilities

Executive business manager responsibilities include the following:

- Provide executive-level business approval to the program business manager to ensure that the applications support model supports the architecture deliverables and to review the applications support model document based on the business direction within budgetary and schedule constraints.

11. SLDP: A Plan used to execute the SLAM.

- Provide executive-level business commitment for the applications support model within the overall applications support service-level agreement.
- Approve all business changes to the applications support model within the overall applications support service-level agreement.
- Approve business resources to support development of the applications support model within the overall applications support service-level agreement.
- Act as the overall program sponsor and chairman of the program steering committee.

Executive IT manager responsibilities include the following:

- Provide executive-level IT approval to the program IT manager to ensure that the applications support model supports the architecture deliverables and to review the applications support model document based on the business direction within budgetary and schedule constraints.
- Provide executive-level IT commitment for the applications support model within the overall applications support service-level agreement.
- Approve all IT changes to the applications support model within the overall applications support service-level agreement.
- Provide IT resources to support development of the applications support model within the overall applications support service-level agreement.
- Act as the program sponsor from the IT perspective.

Executive project management manager responsibilities include the following:

- Provide executive-level PM guidance, direction, and advice to the program delivery manager to ensure that the applications support model supports architecture deliverables and to review the applications support model based on the business direction–business plan within budgetary and schedule constraints.
- Integrate business, data, applications, and technology deliverables into the applications support model within the overall applications support service-level agreement.
- Integrate changes to business, data, applications, and technology deliverables into the applications support model within the overall applications support service-level agreement.
- Provide PM resources for the applications support model within the overall applications support service-level agreement.
- Act as the program sponsor with integrated business and IT responsibility and PM accountability; champion PM processes, tools, and techniques.

- Provide applications support model development guidance, direction, and advice to the program delivery manager to ensure that the applications support service-level agreement is effectively aligned with high-priority projects.

Program business manager responsibilities include the following:

- Submit details on business guidance, direction, and advice to the program delivery manager to ensure that the architecture deliverables are aligned with the applications support model to deliver maximum business benefits to the company in support of project scope, time, and quality constraints.
- Report on business commitment to the applications support model within the overall applications support service-level agreement.
- Report on all business changes to the applications support model within the overall applications support service-level agreement.
- Assign business resources for supporting the delivery applications support model within the overall applications support service-level agreement.
- Report to the program steering committee with direct accountability to the business executive, and keep the program working committee informed of the implementation of the applications support model within the overall applications support service-level agreement.

Program IT manager responsibilities include the following:

- Submit details on IT guidance, direction, and advice to the program delivery manager to ensure that the architecture deliverables are aligned with the applications support model to deliver maximum business benefits to the company in support of project scope, time, and quality constraints.
- Report on IT commitment to the applications support model within the overall applications support service-level agreement.
- Report on all IT changes to the applications support model within the overall applications support service-level agreement.
- Assign IT resources for supporting the delivery of the applications support model within the overall applications support service-level agreement.
- Report to the program steering committee with direct accountability to the business executive, and keep the program working committee informed of the implementation of the applications support model within the overall applications support service-level agreement.

Program delivery manager responsibilities include the following:

- Review and communicate the applications support model within the overall applications support service-level agreement and to ensure linkage with business, data, application, and technology architectures.

- Communicate application support services policies and ensure the project teams adheres to methodology, processes, tools, and techniques.
- Verify that the project team adheres to applications support services policies, guidelines, and procedures by utilizing checklists that are included in the applications support model deliverables.
- Manage the integrated applications support model by reporting on the contents and recommended solutions to the appropriate management team.
- Update applications support model processes as needed to keep them effective.
- Report to the program steering committee with direct accountability to the PM executive, and keep project managers and stakeholders informed of the effectiveness and usefulness of the applications support model during projects maintenance and warranty period.

Project team (project manager and team) responsibilities include the following:

- Project managers to manage applications support services deliverables in accordance with PMO quality guidelines and IT service provider agreement model;
- Project managers to incorporate applications support services deliverables and activities into the project charter and integrated project plan;
- Project managers to integrate, assemble, and report to the program working committee on the progress and quality of the applications support services deliverables, issues, corrective actions, and risks to the project during the warranty period with direct accountability to the program delivery manager;
- Project team to develop and deliver applications support services deliverables based on PMO guidelines and the IT service provider agreement model.

5.7.4 Procedures: applications support services

The roles and responsibilities discussed in Section 5.7.3 are brief because they are intended to serve as a reference as to who does what during the management and execution of the applications support services deliverables and process. Business procedures are now introduced to demonstrate how this applications support services process is executed using the major standard deliverables (what), process flow (how), and checklist or measurement criteria (why) templates. The process is intended to serve as a reference based on the English-language interrogatives: who does what, how, and why. The deliverables, process flow, and checklist templates can serve as guidelines for practicing project managers during the execution of the applications support services process.

The deliverables template in Table 5.12 highlights the key deliverables that the practicing project manager must manage to enable the delivery of quality deliverables. The guidelines provided in Table 5.13, based on real-world practical implementations, should provide some insights to project managers. The applications enhancement requests model in Table 5.13 shows how the IT service provider should manage the requests to control competing priorities based on the business priorities and integration with the data, applications, and technology architecture components.

The process flow template in Figure 5.13 highlights the major processes to support the delivery of the key deliverables. These processes reference the business, applications, data, and technology architectures to show the integrated nature and dependencies of these architectures and to integrate and prioritize business and IT initiatives based on the business needs while ensuring the high quality, reusability, integrity, consistency, and completeness of these deliverables.

The checklist template in Table 5.14 highlights the major measurement criteria to ensure delivery of quality deliverables identified in the deliverables template and adherence to the process flows identified in the process flow template. The checklist is used to objectively determine success criteria for commitment and approval to proceed with the recommended IT service provider and to successfully complete the project.

Table 5.12 Applications Support Services Report

Section 1—Introduction
This section includes an executive summary, applications support services policies, SLAM, and introduces the report.
Section 2—Applications Support and Business Support Initiatives Alignment
This section highlights the major business and applications support initiatives, priority, objectives, and justification business value. It provides directory access to the location of the initiatives. These integrated business and applications support initiatives, documented in the initiatives repository, align with business and IT architectures, and form the basis in determining the priorities of the initiatives. They establish the baseline for projects enhancements (tier 1), new developments (tier 2) or strategic alignment (tier 3) decisions, based on Section 4.
Section 3—IT Architectures and Business Architecture Updates
This section highlights the initiatives alignment to the major deliverables of the IT architectures—data, applications and technology, and business architecture, by providing a matrix showing business value, strategic directives, and priorities of the initiatives described in Section 2.
Section 4—SLAM
This section highlights the major components of the SLAM, which consists of warranty, production, systems management, and enhancements request support. It describes the support services and staffing requirements that establish the basis for the development and execution plan of the SLDP 5.
Section 5—SLDP
This section highlights who, when and how the SLAM will be executed to deliver the support services.
Appendixes—Applications Support Services
Appendixes provide further details on the SLAM, the SLDP, and the selection of the IT service provider(s).

Table 5.13 Deliverables Template: Applications Enhancement Requests Model

Section 1—Enhancement Requests			
Request ID:	Request issue date:		
Requestor name:	Request implementation date:		
Requestor area:	Request approval date:		
Request reason: ❑ IT problem	❑ Business problem	❑ PM problem	
❑ Performance ❑ Maintenance ❑ New ❑ Procedure			
Request severity: ❑ High ❑ Medium ❑ Low			
Request change type: ❑ IT infrastructure	❑ Applications	❑ Data	
❑ Code ❑ Process	❑ Code ❑ Process	❑ Code ❑ Process	
Request change description:			
Business opportunity statement:			
Section 2—Enhancement Assessments			
Estimated business time/effort to implement change:			
Estimated IT time/effort to implement change:			
IT priority ranking: ❑ 1-Urgent ❑ 2-High ❑ 3-Medium ❑ 4-Low			
IT testing level: ❑ Business testers ❑ IT testers			
IT phases affected: ❑ Definition ❑ Analysis ❑ Design ❑ Iterative development			
Support level: ❑ Tier-1 ❑ Tier-2 ❑ Tier-3			
Section 3—Risk Assessments			
Impact on business: ❑ 1-High ❑ 2-Medium ❑ 3-Low			
Impact on technology: ❑ 1-High ❑ 2-Medium ❑ 3-Low			
Impact on people: ❑ 1-High ❑ 2-Medium ❑ 3-Low			
Impact on processes: ❑ 1-High ❑ 2-Medium ❑ 3-Low			
Impact on organization: ❑ 1-High ❑ 2-Medium ❑ 3-Low			

5.8 Summary

The IT management model presented in this chapter describes the component processes of IT management. The purpose provides a statement of the derived result, and the policy provides various rules stating the minimum requirements for achieving the purpose statement for each of the IT management component processes. The roles and responsibilities of executive management, program management, and project delivery (project team) are presented to demonstrate the integrated nature of these processes in managing and delivering multiple IT projects. The key IT management deliverables, process flows, and checklist templates provide excellent real-world implementation guidelines, which can be referenced and applied to IT management processes.

IT organizations that are in the process of managing and delivering multiple IT projects with the goal of integrated or enterprise project management should consider the following IT management recommendations as a framework to guide them towards successful management and delivery of multiple IT projects:

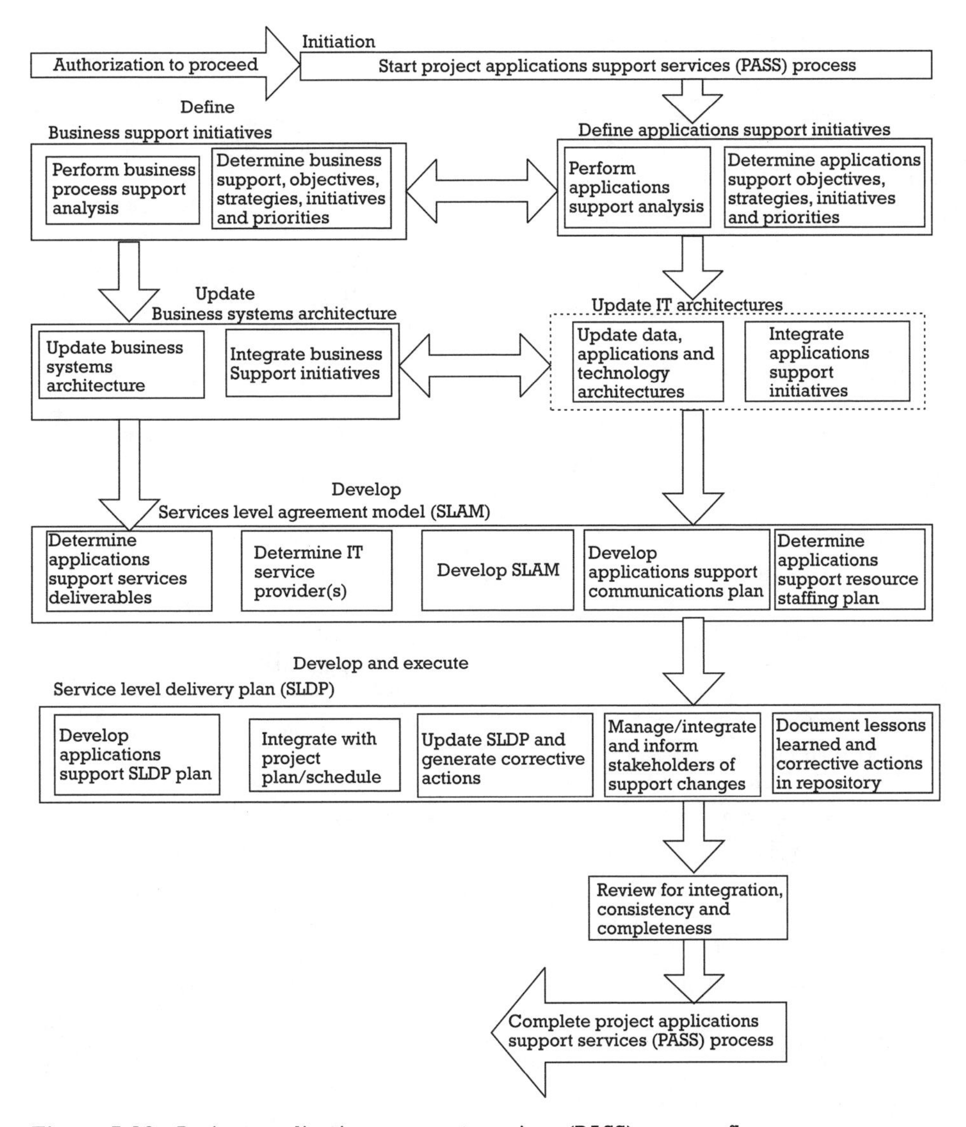

Figure 5.13 Project applications support services (PASS) process flow.

- Establish a resource allocation model, similar to the model presented in Section 5.2. This resource allocation model will form the baseline to better manage the technology and labor costs and effort objectives of the development projects.
- Develop a cost estimating model similar to the model presented in Section 5.3. This cost estimating model will form the baseline to support consistency, standardization, and integration of cost and effort utilizations.
- Implement and deploy a repository to populate data deliverables and integrate with the PMO project repository to ensure completeness, consistency, integrity, and reusability of deliverables.

Table 5.14 Applications Support Services Checklist

Project Applications Support Services (PASS) Criteria	*Yes*	*No*	*PASS Criteria*	*Yes*	*No*
Initiation			*Applications and business support initiatives*		
Is management committed?	❑	❑	Are business initiatives defined and prioritized?	❑	❑
Have a PM sponsor and steering committee been established?	❑	❑	Are applications initiatives defined and prioritized?	❑	❑
Is there a PM methodology?	❑	❑	Are initiatives consolidated? ? Are initiatives generated during PDLC or maintenance?	❑	❑
Does the PM methodology have a PASS component?	❑	❑	Is the enhancement request completed?	❑	❑
Has a program delivery manager been assigned?	❑	❑	Are initiatives generated from business?	❑	❑
Does the assigned program delivery manager understand the PASS process?	❑	❑			
Business and IT architectures			*SLAM*		
Is the business architecture updated?	❑	❑	Is there a SLAM?	❑	❑
Is the IT architecture updated?	❑	❑	Are the IT service providers determined?	❑	❑
Does the business architecture reflect the current and future state of the business?	❑	❑	Are the applications support services deliverables determined?	❑	❑
Does the IT architecture represent the current and future state of IT?	❑	❑	Is there a communications plan?	❑	❑
Is there an AA support group?	❑	❑	Is there a staffing plan?	❑	❑
Does this group maintain the integrity of the AA?	❑	❑	Is the SLAM accepted by client and provider(s)?	❑	❑
SLDP			*Appendixes*		
Is there an SLDP?	❑	❑	Do the appendixes include details on SLDM and SLDP?	❑	❑
Are SDLP performance reporting guidelines available?	❑	❑			
Do variations from the SLDM baseline exist?	❑	❑			
Are stakeholders informed about changes?	❑	❑			
Is the SLDP updated?	❑	❑			
Are lessons learned documented in a project file?	❑	❑			

- Implement and deploy a repository to populate applications deliverables and integrate with the PMO project repository to ensure the completeness, consistency, integrity, and reusability of deliverables.
- Implement and deploy a repository to populate technology deliverables and integrate with the PMO project repository to ensure completeness, consistency, integrity, and reusability of deliverables.
- Establish a SLAM and SLDP, which will form the baseline to better manage the quantity and quality of the objectives and services of the IT service provider.

The main focus of this chapter has been to provide readers with further details on the IT management component processes of the IPM framework model. By now, readers should have gained a fairly good understanding of the integrated nature of these processes. Chapter 6 includes further discussions to show the horizontal integration of the IT management processes based on an iterative and incremental IT PDLC model. IT managers, especially those with keen interests in IT and business process integration, will be able to appreciate the value of these horizontal and vertical integration processes to better prepare them in understanding project developments in this dynamically changing IT industry.

5.9 Questions to think about: management perspectives

1. Think about how your organization identifies, prioritizes, and manages multiple IT projects. How does your organization allocate resources on projects? What are the key rationales for optimizing resource utilizations on projects? What are the major components of IT management? How do these components relate to your project environment? What is the perception of senior management of the need for a resource allocation model to guide the allocation of resources?
2. Think about how your organization estimates project costs. What are the three types of cost estimates and the level of accuracy? How do these three cost estimates relate to your project cost estimating approach? What is the preferred cost estimating technique for IT projects?
3. Think about how your organization manages data deliverables on projects. How are these deliverables documented and integrated? What are the major deliverables of the DA process? How do these DA deliverables and processes relate to your approach to the development of databases?
4. Think about how your organization manages applications deliverables on projects. How are these deliverables documented and integrated? What are the major deliverables of the AA process? How do these AA deliverables and processes relate to your approach to the development of applications?
5. Think about how your organization manages technology deliverables on projects. How are these deliverables documented and integrated? What are the major deliverables of the TA process? How do these TA deliverables and processes relate to your approach to the development of technology deliverables?
6. Think about how your organization supports applications during the warranty period after production migration. How are these support services communicated and executed? What are the major deliver-

ables of the SLAM? How is this model executed? How does this SLAM relate to your approach to supporting applications?

Selected bibliography

Arthur, L. J., *Improving Software Quality—An Insider's Guide to TQM*, New York: John Wiley & Sons, 1993.

CSC—Computer Sciences Corporation, *Catalyst Methodology*, CA CSC, 1999.

Hetzel, B., *The Complete Guide for Software Test Documentation*, New York: John Wiley & Sons/QED Press, 1998.

State of North Carolina, *North Carolina Technical Architecture*, 1997.

CHAPTER

6

Contents

Integrated IT Project Delivery Life-Cycle Model

It is not a question of how well each process works; the question is how well they all work together.
—Lloyd Dobens and Clare Crawford-Mason, Thinking about Quality

6.1 Introduction

In this technologically advancing world of software developments, we have to be able to adapt to the dynamic environment of parallel, recursive, prototyping, and iterative development approaches. A more model-centric[1] process to control and optimize this dynamic environment must replace the traditional waterfall or simple structured approach to software developments. The objective of this chapter is to present the deliverables, activities, and resources of the integrated IT PDLC from an IPM perspective, not to describe the prototyping and iterative approach to software development. There are many excellent books on this new and exciting development process, some of which is referenced in the selected bibliography at the end of this chapter.

The phases of the integrated IT PDLC connect the major processing components of business management, project management and IT management to show how the integrated nature of these components fit within the context of this dynamic environment. The IT PDLC model, presented in this chapter, shows the how business management, project management, and IT management components are integrated during each phase of the software development process. In most projects we undertake, the disciplined approach is to identify

1. Model-centric: An approach to software development that focuses on a deliverables-oriented approach to support the management and delivery of projects.

and define projects and to manage the delivery of these projects by dividing them into manageable components, based on the phases of the software development process.

The integrated PDLC model presented in this chapter represents further elaboration of the IT PDLC shown in Figure 6.1. It outlines the deliverables (what), activities (how), and human resources (who) for business management, project management, and IT management for each of the phases. The WBS, shown in Figure 6.2, is the technique used to structure the project deliverables and activities to effectively manage the projects and to optimize the utilization of resources. Appendix B shows a software tool implementation of this IPM-IT WBS, using the project management software tool Microsoft Project from Microsoft Corporation.

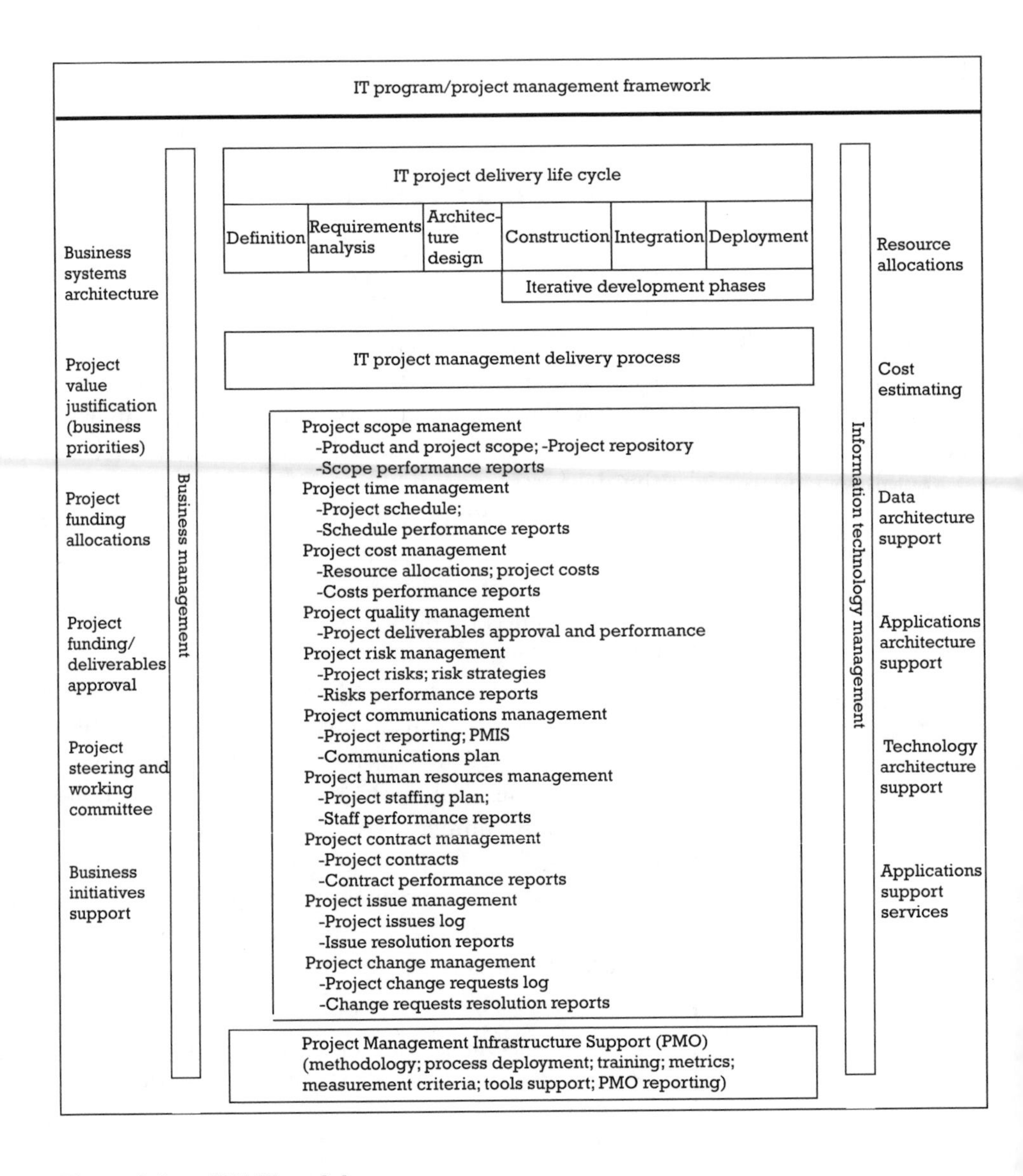

Figure 6.1 IPM-IT model.

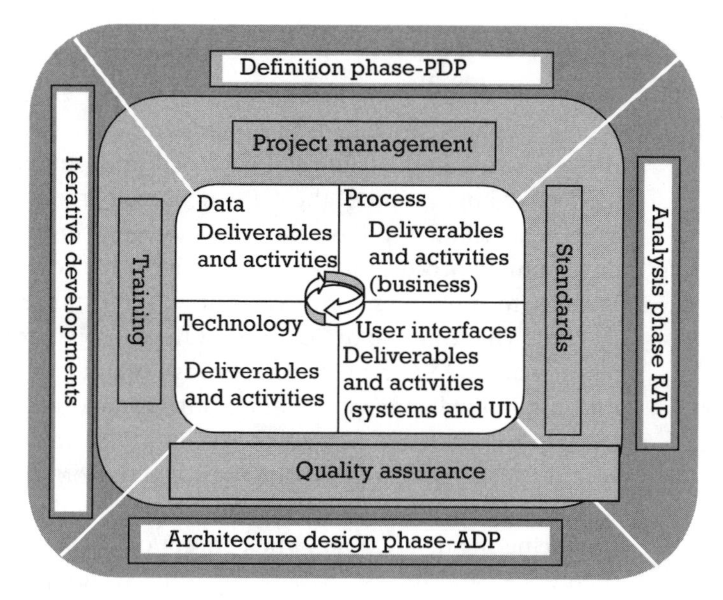

Figure 6.2 Master WBS.

6.2 IPM-IT framework

The IPM framework in Figure 6.1 represents the foundation principle to integrating business management, project management, and IT management components, using the phases of the IT PDLC.

6.3 Master WBS

The master WBS in Figure 6.2 shows a model-based approach to managing projects that supports the IPM-IT framework and PMI-PMBOK guiding processes and practices.

The PMBOK guiding definition of WBS is: "A deliverable-oriented grouping of project elements that organizes and defines the total work scope of the project. Each descending level represents an increasingly detailed definition of the project work." The core deliverables and supporting activities of IPM-IT have four distinguishing characteristics: they are model-driven, architecture-centric,[2] prototyping, and iterative. IPM-IT consists of demonstration and working prototypes during the requirements analysis and architecture phases and iterations during the IDPs. The core deliverables and supporting activities are managed based on these six phases: definition, requirements analysis, architecture, and the IDPs—construction, integration, and deployment.

2. Architecture-centric: An approach to software development that focuses on business, data, applications, and technology architecture constructs and that supports the model-centric approach to analyzing, designing, constructing, and deploying the software solution.

6.4 Real-world scenarios (WBS)

As a result of my 28 years of practical industry experiences as an IT professional, I finally concluded that one of the major reasons for failures of IT projects using both in-house-developed and purchased software, was the ineffective application of the software development processes. Project managers who seemed to have limited knowledge of the fundamental software development process may have contributed to these failures because of their inability to apply these processes during the management and delivery of IT projects. The thinking preferences of these project management professionals are narrowed to a procedural or task-oriented focus. They seem to demonstrate limited conceptualization, integration, and risk management skills and place more emphasis on the activities for each phase with little or no regard for the deliverables. The methodology or processes are applied with little consideration for why or how these activities should be integrated in producing the deliverables. The concept of WBS is unknown and treated as theoretical, and in cases where this WBS technique is applied, the focus is usually task oriented, not deliverables based. This is a typical example of the misuse of the WBS guidelines set forth in PMBOK-2000.

6.5 IT PDLC process model

If you can't describe what you are doing as a process, you don't know what you're doing.
—*W. Edwards Deming*

The IT PDLC process, represented in Figure 6.3, is a model-centric approach that provides a framework of phases, deliverables, and activities to support the IPM-IT framework through the horizontal integration of business management, IT management, and project management components. It is designed for new development projects, acquisition and installation of software packages, and application conversions or migration from one technology environment to another.

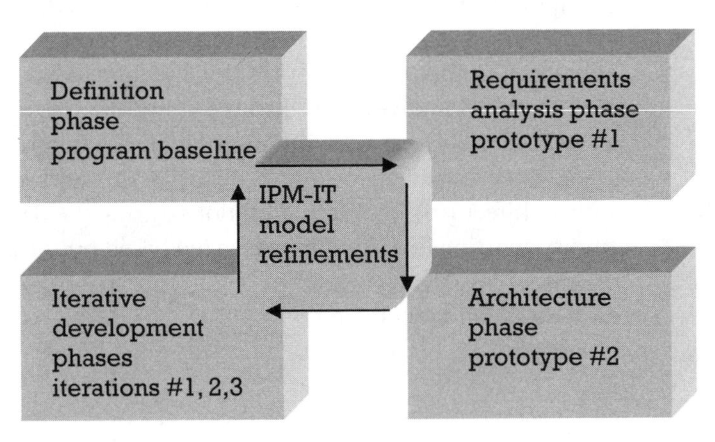

Figure 6.3 IT PDLC process.

The objective of this process model is to establish a framework that will assist business, IT, and project management staff to jointly implement cost-effective, integrated solutions that align with the company's business needs.

Specific objectives when applying this process are to do the following:

- Foster ongoing communications among the project team, business, IT, and project management support resources to ensure that project business requirements are clearly understood and satisfied.
- Provide consistent checkpoints for business, IT, and project management staff to review project progress, monitor project costs and schedule, and approve project continuation.
- Provide project managers with a framework of phases, deliverables, and activities.
- Provide a set of deliverables and activities that can be refined to meet the characteristics of a particular project.

6.6 Applying IT PDLC model

The effective application of this IT PDLC model produces a consistent and integrated approach to software development efforts based on prototyping and iterative development techniques. However, software development projects vary greatly in scope, objectives, and characteristics, and, as such, demands made of this model-centric approach should be proportional to the size and scope of the project.

Project managers should apply this process on major project initiatives. For small projects, the project manager may combine phases, deliverables, and activities to reflect the most appropriate development approach for that project. The integration skills of the project managers must now be applied in selecting and/or combining the deliverables and activities within each phase.

The demand for a prototyping approach is becoming increasingly important during the software development process to support this dynamic software development environment. Developing a prototype offers the opportunity to directly involve the main project stakeholders. A prototype can be a simple GUI, Windows or Web-based inputs/outputs or screen/report displays, or a business rule that requires further developments. Prototyping is the technique that promotes a natural adaptation of the IT PDLC model discussed in this chapter.

The prototype enhances the two-way communications among the business, project management and IT support resources, and the project team, and it simplifies the subsequent design/architecture process. The prototyping technique requires that accountability and control must continue to be applied throughout the project. This is usually achieved by establishing a prototyping strategy during the project definition part of the project plan document.

Varying degrees of prototyping can be used to help the business and IT support resources and the project team in determining specific analysis and design requirements. Prototypes can be throwaway—quick and dirty—or evolutionary—quick and clean. They can be repeatedly refined, especially during iterative developments. During each iteration of the prototype, appropriate deliverables and activities are repeated until the prototype has reached the required level of refinement, as determined by the project team.

Depending on the type of software development project and the reason for applying the prototyping techniques, prototypes can be one of the following: demonstration, evolutionary working model, or iterative development. Prototyping techniques for demonstration are usually addressed within the analysis phase. Evolutionary working model prototypes are usually produced during the design phase, and iterative development prototypes are refined during the construction, integration, and transition phases.

Iterative development results in the evolutionary working model prototype being refined or constructed, integrated, and deployed until the resulting model is transitioned into production. This type of iterative development prototype has a great impact on the software development process, as the need to distinguish the level of details between the analysis, design, and development phases is no longer required. The analysis-paralysis syndrome can be eliminated, or at least controlled. Emphasis is instead placed on each iteration of the working-model prototype and on prototyping the deliverables and activities, rather than on the traditional waterfall approach, which focuses on a rigid, stepwise, and sequential procedural-oriented approach to software development.

This traditional waterfall approach is not suitable for the evolving dynamic software development process. Tools of the 2000s require techniques of the 2000s not techniques of the 1970s and 1980s. During the earlier years of my extensive career in IT, I successfully implemented various projects using a combination of traditional waterfall and prototyping approaches. Even in those days, the 1970s, there existed elements of prototyping and iterative developments that were necessary for successful implementations. Technology advancements such as Internet, Web-based, client/server, object-oriented, *N*-tier, middleware, and the like, of the 1990s and 2000s resulted in the need for more parallel, recursive, or iterative approaches to software development.

6.7 Real-world scenarios: PDLC

Some IT consulting companies prefer to use their in-house waterfall-type software development methodologies to gain control of projects. Certain consulting firms use this process-control power to generate excessive billable hours by justifying the need for project extension via change requests, at the expense of the clients who may lack the required skills and knowledge to challenge the additional billable hours or change requests. Some project management consulting companies continue to defend the use of their rigid procedural-oriented waterfall-type software development methodologies on projects that use advancing technologies such as Internet Web-based, client/server, object-oriented, *N*-tier, middleware, and the like. Modern project management approaches require a more prototyping and iterative development approach to software development to effectively support this dynamically changing technological environment of the 2000s.

For IDSs to be deployed or transition effectively into the production environment, the following major deliverables must be produced during each phase:

- PDM-PDP;
- Project analysis/demonstration prototype[3] model–RAP;
- Project design/working prototype[4] solution–ADP;
- Project IDS[5]–IDPs;

Plans to use IDSs must include the following:

- Periodic reviews of functionality and funding;
- Cost and schedule estimates;
- Scope and quality measurements;
- Achievements of performance objectives;
- Deliverables and activities for project definition, analysis, and design phases.

Iterative developments are gaining widespread acceptance in this advancing technological world, mainly because the software development process must fit within the context of the new and changing dynamic IT environment. RUP from IBM Rational Corporation seems to be gaining wide acceptance in the IT community. Chapter 7 provides some foundation principles and processes on aligning PMI-PMBOK processes with RUP. I hope this introduction will provide guidance for IT professionals and prevent the type of methodology disasters that normally cause project failures.

3. Demonstration prototype solution: A solution that uses techniques to simulate on-line dialogues, transactions, and business rules, supported by user interfaces. There are no intentions of moving the simulated demonstration prototype into production. This type of prototyping is primarily conducted as part of the RAP, using currently available software tools.

4. Working prototype solution: A working solution of the system constructed during the ADP by elaborating on the demonstration prototype, using the appropriate software tools. This working model, or evolutionary prototype, uses tools and techniques that demonstrate a working model of requirements documented in the definition and RAPs. This working model is refined and transformed during the IDPs and usually becomes part of the final software product. It is synonymous with the evolutionary prototype solution.

5. IDS: A software solution constructed during on the IDPs by performing further refinements to the evolutionary prototype. These refinements are normally controlled in one to three iterations that are delivered during the construction, integration, and deployment subphases of the IDP. The software solution and the associated models, plans, and reports form the final system solution.

In order to effectively apply the integrated IT life cycle model presented in this chapter, the deliverables and activities that support this prototyping and iterative approach must be carefully monitored by executing project management reporting activities. The phases of the integrated IT PDLC are now highlighted to demonstrate how the components of business management, project management, and IT management are integrated horizontally within the context of the overall IPM-IT framework.

6.8 Project definition phase–project definition model

If the definition is wrong, you will be developing the right solution for the wrong problem.—James P. Lewis

The IT PDLC model, also referred to as the software development process, begins with an initial definition phase during which the need for the project is identified and defined, and the basic conceptual definition of the project evolves. This phase produces a PDM of the project to be developed. The PDM is a refinement the program planning model produced from the BSA, if completed during the business systems planning initiative.

The definition phase forms the basis for the following:

- The first formal initiation of the project;
- The first major project baseline–PDM;
- The analysis of the project.

This phase produces the following major deliverables, based on the IPM-IT framework (see Figure 6.1) and the IT PDLC process (see Figure 6.3). This is the project baseline document that is updated as the project progresses throughout the PDLC.

The PDM consists of the following:

- Business management deliverables:
  - BSA;
  - Project justification (benefits and priorities);
  - PFAs;
  - Project deliverables/funding approvals;
  - Program steering and working committee;
  - Project business support initiatives.
- Project management deliverables:
  - PSM management plan, requirements baseline, and reports:
    - Business requirements;
    - Data requirements;
    - Application requirements;
    - Technology requirements;
    - Scope management plan and reports.

- PTM management plan, schedule baseline, and reports;
- Project cost management plan, budget baseline, and reports;
- PQM management plan, quality baseline, and reports;
- Project change management plan, change requests, and reports;
- Project contract management plan, contract, and reports;
- PIM management plan, issues, and reports;
- PRM management plan, potential risks, and reports;
- PCM plan, progress status, and reports;
- Project human resources–staffing plan, RAM, and reports;
- PMO support services.

- IT management deliverables:
  - Resource (labor, technology, and facilities) allocations;
  - Cost estimates;
  - DA;
  - AA;
  - TA;
  - Application support services.

The definition phase starts as a result of the completion of a BSA or a BRS, in the absence of a BSA, and formally concludes with the project's first major deliverables checkpoint review–PDM. This review results in approval of the business, IT, and project management deliverables and formally declares the PDM as the first approved project baseline.

6.8.1 Project management problems: real-world issues

Problems in the traditional initiation and definition phases of the IT project delivery process or software development process, are mainly due to the difficulty in obtaining commitments from business, IT, and project management resources internal and external to the project. This is clearly the result of a lack of understanding and poor communication of the required deliverables (what), activities to produce the deliverables (how), and people resources required to deliver the deliverables (who). An atmosphere of confusion, disagreement, and conflicting views prevails among the people resources internal and external to the project. The most common and reoccurring problems that I have encountered during the initiation and definition phases of the traditional software development process include the following:

- *People problems—effort:* Disagreements, confusion, and conflicting views prevail over the contents of the staffing plan, organizational structure, and resource allocations, specifically as to who should be involved in the project development team. Conflicting views, as a result of power struggles, without proper understanding of the deliverables, often lead to emotional and irrational decisions in establishing the project development team.

- *Process problems—cost, schedule, quality:* Disagreements, confusion, and conflicting views prevail over the contents of the project economics—cost/benefits, priorities, cost estimates, funding approvals, schedule, and quality. Again, conflicting views, as a result of the striving for power, without proper understanding of the deliverables and supporting activities, often lead to emotional and irrational decisions in establishing the project cost-budget baseline, schedule baseline, quality measures, and the necessary approval levels.
- *Deliverables/requirements problems—scope:* Disagreements, confusion, and conflicting views prevail over the requirements or scope of the project–BRS, with an endless flow of requirements changes, making it difficult to complete the project requirements specifications.

  Again, conflicting views, as a result of the striving for power, lack of skilled and knowledgeable resources in developing and communicating the scope document, often lead to emotional and irrational decisions in establishing the project scope baseline, and the supporting project management plans—contract, change, issue, risk, and communications.
- *Technology problems—BSA:* Disagreements, confusion, and conflicting views prevail over the business, data, applications, and technology architecture, with a seemingly endless flow of architecture changes, making it difficult to complete the BSA. Again, conflicting views, as a result of the striving for power and a lack of skilled and knowledgeable resources in developing and communicating the BSA document, often lead to emotional and irrational decisions in developing the program delivery plan. This plan normally contains a list of development projects that defines the projects priorities, effort, scope, costs, and schedule baselines, and the supporting program management plans—contract, change, issue, risk and communications.

6.8.2 Recommended solutions: real-world scenarios

Identification of problems, without any recommended solutions, can be frustrating for readers, especially those individuals who may question the cause, effect, and impact of the problems. As a result, I decided to highlight some practical real-world solutions to the problems discussed above.

- *People problems—effort:* Determine the project organizational structure–project staffing plan only after the project scope–deliverables and activities are clearly understood by the key stakeholders. Chapter 2 provides a baseline project organizational structure with supporting roles and responsibilities, which can be easily modified to fit within the context of any existing economic, political, cultural, and technological environment.
- *Process problems—cost, schedule, quality:* In most cases, disagreements, confusion, and conflicting views on cost, schedule, and quality issues result from the lack of a sound project management principle. The

recommended solution in this case must originate from the famous project management triangle presented in Figure 6.4.

This triangle states that cost estimates, time/schedule estimates, and quality measures are dependent on or constrained by scope requirements. A clear understanding of the scope requirements to objectively determine the cost, schedule, and quality objectives will require an understanding of the project WBS. Chapter 2 provides a baseline WBS and supporting implementation schedule using the Microsoft Project tool, which can be easily modified to fit within the context of any existing economic, political, cultural, and technological environment. The project deliverables and supporting activities must form the basis in determining the cost, schedule, and quality objectives of the project.

- *Deliverables/requirements problems—scope:* During various assignments in my extensive IT career, I have had the amazing opportunity to met project managers who believed that WBS is a "theoretical" technique and, as such, is not applicable to their projects. Certain IT directors and other supporting project management staff responsible for project management services, sometimes share this view.

  The recommended solution is to employ project mangers who have gained business conceptualization skills, people, process, and technology integration skills, and excellent risks management skills through relevant professional qualifications and practical experience.

  Another recommended solution to resolving project-scope-related problems, such as commitment and obtaining consensus and approval from major stakeholders is to use techniques such as prototyping, use-case analysis, and model-based documentation to demonstrate, communicate, and test the requirements. The definition phase produces a view of the proposed business requirements and a conceptual solution, which is difficult to communicate, in many cases, using unstructured textual documentation.

  Techniques such as prototyping, use-case-analysis scenarios, and model-based documentation structures (later in this section) often

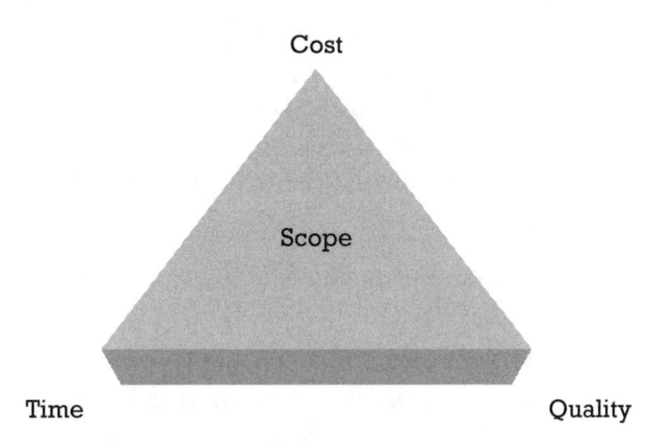

Figure 6.4 Project management triangle.

make the deliverables easier to communicate and integrate because models and scenarios are abstract representations of reality. Prototypes are concrete representations of reality, which often provides the required commitment and approval necessary for the next project phase–analysis.

- *Technology problems—BSA:* In many cases, disagreements, confusion, and conflicting views on architecture developments result from the lack of qualified business, IT, and project management resources involved as decision makers during the development of the BSA. This lack of qualified staff to advocate and disseminate the contents and usefulness of the BSA usually creates further confusion. The assignment of resources for this integrated activity should be decision makers who can communicate business and IT concepts from a logical or objective perspective. The recommended solution is to employ project mangers and systems architects who have gained business conceptualization skills, specialized people, process, and technology integration skills, and excellent risk management skills through relevant professional qualifications and practical experience.

6.8.3 Alignment with business management

The primary objective of this section is to demonstrate how the main deliverable of the definition phase, the PDM, aligns with the deliverables (what) from the business management component process, detailed in Chapter 3. The roles and responsibilities from the business management processes are further refined to determine the major activities (how) needed to produce the deliverables and the people resources (who) responsible for the creation, management, review, and approval of the deliverables.

6.8.3.1 Deliverables (what)

This section highlights the contents of the PDM in terms of business management, IT management, and project management deliverable components. Chapter 3 provides detailed definitions, deliverables, process flows, and checklist templates to be used as guidelines during the creation, management, review, and approval of these deliverables. The business management deliverables are replicated in this section to provide the reader with the necessary continuous and logical flow of information to enable a more analytical understanding of the deliverables produced during the PDP.

Business management deliverables include the following:

- BSA;
- Project justification (benefits and priorities);
- PFAs;
- Project deliverables/funding approvals;
- Program steering and working committee;
- Project business support initiatives.

6.8.3.2 Activities (how)

The roles and responsibilities of the business management component outlined in Chapter 3 are classified based on the creation, management, review, and approval responsibility scheme to determine the major activities required to support the deliverables during the PDP.

- Project BSA: Create, manage, review, approve–BSA;
- Project justification (benefits and priorities): Create, manage, review, approve–project justifications;
- PFAs: Create, manage, review, approve–funding allocations;
- Project deliverables/funding approvals: Create, manage, review, approve–deliverables approvals;
- Program steering and working committee: Create, manage, review, approve–steering committee;
- Project business support initiatives: Create, manage, review, approve–business initiatives.

6.8.3.3 Resources (who)

The labor resources needed to execute the activities to produce the deliverables are determined to show the type of business management resources responsible for the creation, management, review, and approval of the deliverables produced during the PDP.

- Project BSA (Create, manage, review, approve–BSA):
  - Program business manager–review;
  - Program delivery manager–manage;
  - Program IT manager–review;
  - Executive business manager–review and approve;
  - Executive project management manager–review;
  - Executive IT manager–review;
  - Project manager and team–create.
- Project justification (benefits and priorities) (Create, manage, review, approve–project justifications):
  - Program business manager–create;
  - Program delivery manager–manage;
  - Program IT manager–review;
  - Executive business manager–review and approve;
  - Executive project management manager–review;
  - Executive IT manager–review;
  - Project manager and team–review.
- PFAs (Create, manage, review, approve–funding allocations):
  - Program business manager–create;
  - Program delivery manager–manage;
  - Program IT manager–review;

- Executive business manager–review and approve;
- Executive project management manager–review;
- Executive IT manager–review;
- Project manager and team–review.

- Project deliverables/funding approvals (Create, manage, review, approve–deliverables approvals):
  - Program business manager–create;
  - Program delivery manager–manage;
  - Program IT manager–review;
  - Executive business manager–review and approve;
  - Executive project management manager–review;
  - Executive IT manager–review;
  - Project manager and team–review.
- Program steering and working committee (Create, manage, review, approve–steering committee):
  - Program business manager–create;
  - Program delivery manager–manage;
  - Program IT manager–review;
  - Executive business manager–review and approve;
  - Executive project management manager–review;
  - Executive IT manager–review;
  - Project manager and team–review.
- Project business support initiatives (Create, manage, review, approve–business initiatives):
  - Program business manager–create;
  - Program delivery manager–manage;
  - Program IT manager–review;
  - Executive business manager–review and approve;
  - Executive project management manager–review;
  - Executive IT manager–review;
  - Project manager and team–review.

6.8.4 Alignment with project management

The primary objective of this section is to demonstrate how the major deliverables of the PDP-PDM align with the deliverables (what) from the project management component, detailed in Chapter 4. The roles and responsibilities from the project management processes, presented in Chapter 4, are further refined to determine the major activities (how) needed to produce the deliverables and the people resources (who) responsible for the creation, management, review, and approval of the deliverables.

6.8.4.1 Deliverables (what)

This section highlights the contents of the PDM in terms of business management, IT management, and project management deliverables. Chapter 4 provides detailed definitions, deliverables, process flows, and checklist templates to be used as guidelines during the creation, management, review, and approval of these deliverables. The project management deliverables are replicated in this section to provide the reader with the necessary continuous and logical flow of information to enable a more analytical understanding of the deliverables produced during the PDP.

Project management deliverables include the following:

- PSM management plan, requirements baseline, and reports:
  - Business requirements;
  - Data requirements;
  - Application requirements;
  - Technology requirements;
  - Scope management plan and reports.
- PTM management plan, schedule baseline, and reports;
- Project cost management plan, budget baseline, and reports;
- PQM management plan, quality baseline, and reports;
- Project change management plan, change requests, and reports;
- Project contract management plan, contract, and reports;
- PIM management plan, issues, and reports;
- PRM management plan, potential risks, and reports;
- PCM plan, progress status, and reports;
- Project human resources–staffing plan, RAM, and reports;
- PMO support services.

6.8.4.2 Activities (how)

The roles and responsibilities of the project management component outlined in Chapter 4 are classified based on the creation-updating, management-integration, review-QA, and approval-commitment responsibility classifications to determine the major activities required to support the deliverables during the PDP. The integration of business management and IT management processes with project management processes is highlighted to demonstrate the integrated nature of project management during the IT PDLC processes. The manage/integrate responsibilities of project management are presented as similar tasks to demonstrate the integrated skills needed by the program/project manager.

- *PSM management plan, requirements baseline, and reports:* These scope management deliverables are integrated representations of the contents from business management–business architecture and IT management–data, applications, and technology architectures with project management–scope management (Create, manage/integrate, review, approve–scope management).

- *PTM management plan, schedule baseline, and reports:* The time management deliverables are integrated representations of the contents from business management–project deliverables/funding approvals and IT management–resource allocations with project management–time management (Create, manage/integrate, review, approve–time management).
- *Project cost management plan, budget baseline, and reports:* The costs management deliverables are integrated representations of the contents from business management–project justifications and PFAs and IT management–resource allocations and cost estimating with project management–cost management (Create, manage/integrate, review, approve–cost management).
- *PQM management plan, quality baseline, and reports:* The quality management deliverables are integrated representation of the contents from business management–project deliverables/funding approvals and IT management–data, applications, technology architectures, and applications support services with project management–quality management (Create, integrate, review, approve–quality management).
- *Project change management plan, change requests, and reports:* These are supporting deliverables for the core project management deliverables—cost, time, scope, and quality. They represent the integrated management of all changes made to the core project management deliverables (Create, manage, review, approve–change management).
- *Project contract management plan, contracts, and reports:* These are supporting deliverables for the core project management deliverables—cost, time, scope, and quality. They represent the integrated management of all contracts to support the core project management deliverables (Create, manage, review, approve–contract management).
- *PIM management plan, issues, and reports:* These are supporting deliverables for the core project management deliverables—cost, time, scope, and quality. They represent the integrated management of all issues that affect delivery of the core project management deliverables (Create, manage, review, approve–issue management).
- *PRM management plan, risks, and reports:* These are supporting deliverables for the core project management plans—cost, time, scope, and quality. They represent the integrated management of all risks that affect the delivery of the core project management deliverables (Create, integrate, review, approve–risk management).
- *PCM plan, progress status, and reports:* These are supporting deliverables for the core project management plans—cost, time, scope, and quality. They represent the integrated management of all communications to support the delivery of the core project management deliverables. Business management–project steering committee and IT management–resource allocation form an integral part of the communications

management plan (Create, integrate, review, approve–communications management).

- *PSM management plan, RAM, and reports:* These are the supporting deliverables for the core project management plans—cost, time, scope, and quality. They represent the integrated management of the staffing plans to support the delivery of the core project management deliverables. Business management–project steering committee and IT management–resource allocation form an integral part of the PSM management plan (Create, integrate, review, approve–human resources management).

6.8.4.3 Resources (who)

The labor resources needed to execute the activities to produce the deliverables are determined to show the type of project management resources responsible for the creation, management, review, approval, and support for the deliverables produced during the PDP.

- PSM management plan, requirements baseline, and reports (Create, manage/integrate, review, approve–scope management):
  - Program delivery manager–create, manage/integrate;
  - Program business manager–review;
  - Program IT manager–review;
  - Executive business manager–review;
  - Executive project management manager–review and approve;
  - Executive IT manager–review;
  - Project manager and team–review.
- PTM management plan, schedule baseline, and reports (Create, manage/integrate, review, approve–time management):
  - Program delivery manager–create, manage/integrate;
  - Program business manager–review;
  - Program IT manager–review;
  - Executive business manager–review;
  - Executive project management manager–review and approve;
  - Executive IT manager–review;
  - Project manager and team–review.
- Project cost management plan, budget baseline, and reports (Create, manage/integrate, review, approve–cost management):
  - Program delivery manager–create, manage/integrate;
  - Program business manager–review;
  - Program IT manager–review;
  - Executive business manager–review;
  - Executive project management manager–review and approve;
  - Executive IT manager–review;
  - Project manager and team–review.

- PQM management plan, quality baseline, and reports (Create, integrate, review, approve–quality management):
  - Program delivery manager–create, manage/integrate;
  - Program business manager–review;
  - Program IT manager–review;
  - Executive business manager–review;
  - Executive project management manager–review and approve;
  - Executive IT manager–review;
  - Project manager and team–review.
- Project change management plan, change requests, and reports (Create, manage, review, approve–change management):
  - Program delivery manager–create, manage/integrate;
  - Program business manager–review;
  - Program IT manager–review;
  - Executive business manager–review;
  - Executive project management manager–review and approve;
  - Executive IT manager–review;
  - Project manager and team–review.
- Project contract management plan, contracts, and reports (Create, manage, review, approve–contract management):
  - Program delivery manager–create, manage/integrate;
  - Program business manager–review;
  - Program IT manager–review;
  - Executive business manager–review;
  - Executive project management manager–review and approve;
  - Executive IT manager–review;
  - Project manager and team–review.
- PIM management plan, issues, and reports (Create, manage, review, approve–issue management):
  - Program delivery manager–create, manage/integrate;
  - Program business manager–review;
  - Program IT manager–review;
  - Executive business manager–review;
  - Executive project management manager–review and approve;
  - Executive IT manager–review;
  - Project manager and team–review.
- PRM management plan, risks, and reports (Create, integrate, review, approve–risk management):
  - Program delivery manager–create, manage/integrate;
  - Program business manager–review;
  - Program IT manager–review;
  - Executive business manager–review;
  - Executive project management manager–review and approve;

  - Executive IT manager–review;
  - Project manager and team–review.
- PCM plan, progress status, and reports (Create, integrate, review, approve–communications);
  - Program delivery manager–create, manage/integrate;
  - Program business manager–review;
  - Program IT manager–review;
  - Executive business manager–review;
  - Executive project management manager–review and approve;
  - Executive IT manager–review;
  - Project manager and team–review.
- PSM management plan, RAM, and reports (Create, integrate, review, approve–human resources management):
  - Program delivery manager–create, manage/integrate;
  - Program business manager–review;
  - Program IT manager–review;
  - Executive business manager–review;
  - Executive project management manager–review and approve;
  - Executive IT manager–review;
  - Project manager and team–review.

6.8.5 Alignment with IT management

The primary objective of this section is to demonstrate how the main deliverable of the definition phase, the PDM, aligns with the deliverables (what) from the technology management component detailed in Chapter 5. The roles and responsibilities from the it management processes, presented in Chapter 5, are further refined to determine the major activities (how) needed to produce the deliverables and the people resources (who) responsible for the creation, management, review, and approval of the deliverables.

6.8.5.1 Deliverables (what)

This section highlights the contents of the PDM, in terms of the business management, IT management, and project management deliverable components. Chapter 5 provides detailed definitions, deliverables, process flows, and checklist templates to be used as guidelines during the creation, management, review, and approval of these deliverables. The IT management deliverables are replicated in this section to provide readers with the necessary continuous and logical flow of information to enable a more analytical understanding of the deliverables produced during PDP.

IT management deliverables include the following:

- Resource allocations;

- Cost estimating;
- DA;
- AA;
- TA;
- Applications support services.

6.8.5.2 Activities (how)

The roles and responsibilities of IT management component outlined in Chapter 5 are classified based on the creation, management, review, and approval responsibility scheme to determine the major activities required to support the deliverables produced during the PDP.

- PRAs: Create, manage, review, approve–resource allocations;
- Project cost estimates: Create, manage, review, approve–cost estimates;
- Program DA: Create, manage, review, approve–DA;
- Program AA: Create, manage, review, approve–AA;
- Program TA: Create, manage, review, approve–TA;
- Program applications support services: Create, manage, review, approve–applications support.

6.8.5.3 Resources (who)

The labor resources needed to execute the activities to produce the deliverables are determined to show the type of IT management resources that are responsible for the creation, management, review, and approval of the deliverables produced during the PDP.

- PRAs (Create, manage, review, approve–resource allocations):
  - Program delivery manager–create, manage/integrate;
  - Program business manager–review;
  - Program IT manager–review;
  - Executive business manager–review;
  - Executive project management manager–review;
  - Executive IT manager–review and approve;
  - Project manager and team–review.
- Project cost estimates (Create, manage, review, approve–cost estimates):
  - Program delivery manager–create, manage/integrate;
  - Program business manager–review;
  - Program IT manager–review;
  - Executive business manager–review;
  - Executive project management manager–review and approve;
  - Executive IT manager–review;
  - Project manager and team–review.
- Program DA (Create, manage, review, approve–DA):

- Program delivery manager–manage;
- Program business manager–review;
- Program IT manager–create;
- Executive business manager–review;
- Executive project management manager–review;
- Executive IT manager–review and approve;
- Project manager and team–review.

- Program AA (Create, manage, review, approve–AA):
  - Program delivery manager–manage;
  - Program business manager–review;
  - Program IT manager–create;
  - Executive business manager–review;
  - Executive project management manager–review;
  - Executive IT manager–review and approve;
  - Project manager and team–review.
- Program TA (Create, manage, review, approve–TA):
  - Program delivery manager–manage;
  - Program business manager–review;
  - Program IT manager–create;
  - Executive business manager–review;
  - Executive project management manager–review;
  - Executive IT manager–review and approve;
  - Project manager and team–review.
- Program applications support services (Create, manage, review, approve–applications support):
  - Program delivery manager–manage;
  - Program business manager–review;
  - Program IT manager–create;
  - Executive business manager–review;
  - Executive project management manager–review;
  - Executive IT manager–review and approve;
  - Project manager and team–review.

6.8.6 Integration with business, IT, and project management

Tables 6.1 through 6.3[6] are an integrated representation to present how the deliverable from the RAP-RAM integrate with the deliverables (what), activities (how), and resources (who) from the business, IT, and project management components. The acronyms on the matrixes are abbreviations

6. Responsibility matrixes: A series of matrixes that show who is responsible for what in terms of creating/updating, managing/integrating, reviewing/assuring quality, and approving/committing responsibility classifications and that show the integrated nature of business management, IT management, and project management.

Table 6.1 Business Management Responsibility Matrix—PDP

Deliverables	*Roles* *EBM*	*EPM*	*EITM*	*PBM*	*PDM*	*PITM*	*PM/Team*
Business management							
BSA	RA	R	R	C	M/I	R	R
Project justifications	RA	R	R	C	M/I	R	R
PFAs	RA	R	R	C	M/I	R	R
Project deliverables approvals	RA	R	R	C	M/I	R	R
Program steering committee	RA	R	R	C	M/I	R	R
Business support initiatives	RA	R	R	C	M/I	R	R

Note: C = create; R = review; A = approve; M/I = manage/integrate

Table 6.2 Project Management Responsibility Matrix—PDP

Deliverables	*Roles* *EBM*	*EPM*	*EITM*	*PBM*	*PDM*	*PITM*	*PM/Team*
Project management							
Scope management	R	RA	R	R	CM	R	R
Time management	R	RA	R	R	CM	R	R
Quality management	R	RA	R	R	CM	R	R
Change management	R	RA	R	R	CM	R	R
Contract management	R	RA	R	R	CM	R	R
Issue management	R	RA	R	R	CM	R	R
Risk management	R	RA	R	R	CM	R	R
Communications management	R	RA	R	R	CM	R	R
Human resources management	R	RA	R	R	CM	R	R

Note: C = create; R = review; A = approve; M/I = manage/integrate

Table 6.3 IT Management Responsibility Matrix—PDP

Deliverables	*Roles* *EBM*	*EPM*	*EITM*	*PBM*	*PDM*	*PITM*	*PM/Team*
IT management							
Resource allocations	R	R	RA	R	CM	R	R
Cost estimating	R	RA	R	R	CM	R	R
DA	R	R	RA	R	M	C	R
AA	R	R	RA	R	M	C	R
TA	R	R	RA	R	M	C	R
Applications support services	R	R	RA	R	M	C	R

for executive managers, program managers, and project managers' responsibilities.

6.9 Project analysis: demonstration prototype

The purpose of models is not to fit the data but to sharpen the questions.
—R. A. Fisher, Royal Society

During the analysis phase, the basic conceptual definitions of the project in the definition phase—scope management deliverables—are analyzed, and the method for implementation—simulated demonstration prototype—is discussed with the objective of more clearly determining the business requirements, using the appropriate software tools. The business users review a demonstration of a simulated system to experiment with the GUIs. This prototype can be used to complement the requirements from the definition phase to further refine/develop the existing requirements.

This phase produces a demonstration prototype of the systems to be developed. It uses techniques to simulate on-line dialogues, transactions, and business rules, supported by user interfaces. There is no intention of moving the simulated demonstration prototype into production. This type of prototyping is primarily conducted as part of the RAP, using currently available software tools.

The RAP forms the basis for the following:

- The second major project baseline–RAM;
- A demonstration prototype to simulate on-line dialogues, transactions, and business rules, supported by user interfaces;
- Updated business, IT, and project management plans and deliverables;
- Design/architecture of the project (next phase–prototype #2);
- Approval and commitment to proceed to design/architecture—prototype #2.

During this phase, the business management, IT management, and project management deliverables of the PDM are updated, and the simulated demonstration prototype is developed. This is the software baseline solution, which may be refined, as the project progresses throughout the IT PDLC. In most cases, this software solution baseline will for the basis for the development of the evolutionary prototype.

The deliverables of the analysis phase focus mainly on updates to the business management and project management deliverables developed during the definition phase. Key IT management deliverables will be created during this phase and will be refined as the project progresses throughout the PDLC.

Project RAM will include updates to the following:

- Business management deliverables:
  - BSA;
  - Project justification (benefits and priorities);
  - PFAs;
  - Project deliverables/funding approvals;
  - Program steering and working committee;

  - Project business support initiatives.
- Project management deliverables:
  - PSM management plan, requirements baseline, and reports:
    - Business requirements;
    - Data requirements;
    - Application requirements;
    - Technology requirements;
    - Scope management plan and reports.
  - PTM management plan, schedule baseline, and reports;
  - Project cost management plan, budget baseline, and reports;
  - PQM management plan, quality baseline, and reports;
  - Project change management plan, change requests, and reports;
  - Project contract management plan, contract, and reports;
  - PIM management plan, issues, and reports;
  - PRM management plan, potential risks, and reports;
  - PCM plan, progress status, and reports;
  - Project human resources–staffing plan, RAM, and reports;
  - PMO support services.
- IT management deliverables:
  - Resource allocations;
  - Cost estimating;
  - DA–demonstration prototype level of detail: enterprise, business, and implementation data models;
  - AA–demonstration prototype level of detail:
    - Enterprise, business, and implementation object models;
    - Interface models (GUI and messaging models).
  - TA–demonstration prototype level of detail: enterprise, business, and implementation technology models;
  - Application support services–demonstration prototype level of detail: integrated systems testing and data conversion strategies.

The RAP starts as a result of completion of the PDM and formally concludes with the project's first major solution checkpoint review—project RAM. This review results in approval of the business, IT, and project management deliverables, and formally declares the project RAM and the supporting simulated demonstration prototype as the first approved project baseline solution.

6.9.1 Project management problems: real-world issues

Problems in the traditional analysis phases of the IT project delivery process or software development process are mainly due to difficulty in obtaining continuous commitments from business, IT, and project management resources internal and external to the project. Again, this is clearly the result of a lack of understanding and poor communications of the required deliverables (what), activities to produce the deliverables (how), and people resources required to deliver the deliverables (who). The analysis-paralysis starts, and an atmosphere of confusion, disagreements, and conflicting views continues to prevail among the resources internal and external to the project. The most common and reoccurring problems that I encountered during the analysis phases of the traditional software development process are the lack of continuous commitments as a result of the traditional analysis-paralysis syndrome. These problems are grouped based on people, process, deliverables, and technological conflicts to show the recurring similarities of conflicts between definition and analysis phases.

- *People problems—effort:* Disagreements, confusion, and conflicting views continue to prevail, but now over the execution of responsibilities defined in the staffing plan, organizational structure, and resource allocations. During the analysis phase, the reality of the organizational structure is more understood, and a sense of insecurity, resistance to change, and political maneuvers begin to evolve. Conflicting views, as a result of the striving for power and inconsistent understanding of how the deliverables will be produced, often lead to emotional and irrational decisions, and a restructuring of the previously committed organizational plan is produced, more often, based on hidden political motives.
- *Process problems—cost, schedule, quality:* Disagreements, confusion, and conflicting views continue to prevail, but now, over the processes involved in updating the cost, schedule, and quality deliverables, as a result of the inconsistent and subjective approach. Again, conflicting views, as a result of the striving for power, inconsistent processes, and improper understanding of the deliverables and supporting activities, often lead to emotional and irrational decisions. The result is a redevelopment of project management processes to manage the cost-budget baseline, schedule baseline, and quality measures with new approval levels.
- *Deliverables/requirements problems—scope:* This is the area where most of the disagreements, confusion, and conflicting views prevail over the requirements or scope of the project RAM, with an endless flow of requirements changes, making it difficult to complete the project requirements specifications. The analysis-paralysis syndrome starts, with unmanageable brainstorming sessions, unstructured and disjointed texts, nonproductive meetings, and disintegrated deliverables. The project team is buried in the information jungle, with each member seeking desperately for assistance amidst conflicting views. Suddenly, a hero appears who quickly produces a deliverable to the amazement and appreciation of the rest of the team members. In a majority of cases, the deliverable produced does not align with the deliverables of the previous phase. My assessment of such situations brings me back to the root

cause of IT projects failures—a lack of project managers with skills and knowledge in general business conceptualization to articulately visualize, lead, and communicate a conceptual baseline solution to support the business requirements.

- *Technology problems—BSA:* Disagreements, confusion, and conflicting views continue to prevail over the business, data, applications, and technology architecture, with a seemingly endless flow of architecture changes, as a result of the business and IT management input, making it difficult to update the BSA. Again, conflicting views, as a result of the striving for power for ownership of this document and a lack of skilled and knowledgeable resources to deliver and communicating the BSA solution, often lead to emotional and irrational decisions in updating and producing the program delivery plan. This plan normally contains a list of development projects, which define the projects priorities, effort, scope, costs, and schedule baselines and the supporting program management implementation plans—contract, change, issue, risk, and communications management plans.

6.9.2 Recommended solutions: real-world scenarios

The problems experienced during the traditional analysis phase focus on the lack of continuous commitment to people, process, deliverables/scope, and technology objectives, with major issues resulting from the lack of effectively communicating an understanding of the requirements–deliverables/scope. The business users, frustrated by the lack of understanding of the contents of the requirements analysis documentation, experience added frustration in trying to understand the processes and the approach to solving the business problems and, as a result, revert to trust, hoping for the magical silver bullet solution.

Here are highlights of some practical real-world solutions for the problems above:

- *People problems—effort:* Reconfirm the roles and responsibilities of the project, after the project scope, deliverables, and activities are clearly understood by the key stakeholders. Chapter 2 provides a baseline project organizational structure with supporting roles and responsibilities, which can be easily modified to fit within the context of any existing economic, political, cultural, and technological environment.
- *Process problems—cost, schedule, quality:* Communicate the project management processes involved in updating the cost, schedule, and quality constraints. In a majority of cases, disagreements, confusion, and conflicting views on updating cost, schedule, and quality deliverables result from the lack of effectively communicating the foundation project management processes.

  Reconfirm the team understanding of the project WBS and the need to clearly understand scope requirements to objectively determine the updates to cost, schedule, and quality objectives. Chapter 2 provides a baseline WBS with supporting implementation using the Microsoft Project tool, which can be easily modified to fit within the context of any existing economic, political, cultural, and technological environment. The project deliverables and supporting activities must form the basis for updating the cost, schedule, and quality objectives of the project.

- *Deliverables/requirements problems—scope:* The root cause of the analysis-paralysis syndrome in determining the appropriate level of details and the lack of commitment for the what and how of the analysis deliverables is the lack of a concrete deliverable that demonstrates or simulates a solution.

  The recommended solution to resolving project-scope-related problems, such as prolonged analysis paralysis and lack of commitment in obtaining consensus and approval from major stakeholders, is to build a simulated demonstration prototype–use-case analysis and model-based documentation to demonstrate, communicate, and test the requirements. The RAP produces a baseline view of a conceptual solution, which is easier to communicate in most cases, rather than using the traditional unstructured textual documentation.

  The simulated demonstration prototype, use-case-analysis scenarios, and model-based documentation structures (later in this section), often make the deliverables easier to communicate and integrate because models and scenarios are abstract representations of reality. The simulated demonstration prototype is a concrete representation of reality that often provides the required solution necessary to obtain commitment and approval for the next project phase—architecture.
- *Technology problems—BSA:* The communications and commitment problems experienced in this phase as a result of conflicting updates to the data, applications, and technology architectures are normally resolved by building the simulated demonstration prototype with supporting model-based documentation. The demonstration prototype confirms the need for experienced project managers with general business conceptualization skills, specialized people, process, and technology integration skills, and excellent risk management skills with professional qualifications and relevant practical experience.

6.9.3 Alignment with business management

The primary objective of this section is to demonstrate how the key deliverables of the RAP-RAM align with the deliverables (what) from business management component processes, detailed in Chapter 3. The roles and responsibilities from the business management processes, presented in Chapter 3, are further refined to determine the major activities (how) needed to produce the deliverables and the people resources (who) responsible for the update, integration, QA, and commitment for the deliverables.

6.9.3.1 Deliverables (what)

This section highlights the contents of the RAM in terms of business management, IT management, and project management deliverable components, based on refinements of the model produced in the definition phase. Chapter 3 provides detailed definitions, templates, process flows, and quality measures to be used as guidelines for the creation/update, management/integration, review/QA, and approval/commitment for these deliverables. The business management deliverables are replicated in this section to provide readers with the necessary continuous and logical flow of information to enable a continuous refinement of the deliverables produced during the project RAP.

Updated business management deliverables include the following:

- BSA updates;
- Project justification updates (benefits and priorities);
- PFA updates;
- Project deliverables/funding approvals updates;
- Program steering and working committee updates;
- Project business support initiatives updates.

6.9.3.2 Activities (how)

The roles and responsibilities of business management component outlined in Chapter 3 are classified based on the creation/update, integration, review, and approval responsibility scheme to determine the major activities required to support the deliverables during RAP.

- BSA updates: Update, integrate, review, approve–BSA;
- Project justification (benefits and priorities) updates: Update, integrate, review, approve–project justifications;
- PFA updates: Update, integrate, review, approve–funding allocations;
- Project deliverables/funding approvals updates: Update, integrate, review, approve–deliverables approvals;
- Program steering and working committee updates: Update, integrate, review, approve–steering committee;
- Project business support initiatives updates: Update, integrate, review, approve–business initiatives.

6.9.3.3 Resources (who)

The labor resources needed to execute the activities to produce the deliverables are determined to show the type of business management resources responsible to update, integrate, review, and approve the deliverables produced during the project RAP.

- Project BSA updates (Update, integrate, review, approve–BSA):
  - Program business manager–update;
  - Program delivery manager–integrate;

  - Program IT manager–review;
  - Executive business manager–review and approve;
  - Executive project management manager–review;
  - Executive IT manager–review;
  - Project manager and team–review.
- Project justification updates (benefits and priorities) (Update, integrate, review, approve–project justifications):
  - Program business manager–update;
  - Program delivery manager–integrate;
  - Program IT manager–review;
  - Executive business manager–review and approve;
  - Executive project management manager–review;
  - Executive IT manager–review;
  - Project manager and team–review.
- PFA updates (Update, integrate, review, approve–funding allocations):
  - Program business manager–update;
  - Program delivery manager–integrate;
  - Program IT manager–review;
  - Executive business manager–review and approve;
  - Executive project management manager–review;
  - Executive IT manager–review;
  - Project manager and team–review.
- Project deliverables/funding approvals updates (Update, integrate, review, approve–deliverables approvals):
  - Program business manager–update;
  - Program delivery manager–integrate;
  - Program IT manager–review;
  - Executive business manager–review and approve;
  - Executive project management manager–review;
  - Executive IT manager–review;
  - Project manager and team–review.
- Program steering and working committee updates (Update, integrate, review, approve–steering committee):
  - Program business manager–update;
  - Program delivery manager–integrate;
  - Program IT manager–review;
  - Executive business manager–review and approve;
  - Executive project management manager–review;
  - Executive IT manager–review;
  - Project manager and team–review.
- Project business support initiatives updates (Update, integrate, review, approve–business initiatives):
  - Program business manager–update;

- Program delivery manager–integrate;
- Program IT manager–review;
- Executive business manager–review and approve;
- Executive project management manager–review;
- Executive IT manager–review;
- Project manager and team–review.

6.9.4 Alignment with project management

The primary objective of this section is to demonstrate how the key deliverables of the RAP-RAM align with the deliverables (what) from the project management component detailed in Chapter 4. The roles and responsibilities from the project management processes presented in Chapter 4 are further refined to determine the major activities (how) needed to produce the deliverables and the people resources (who) responsible for the updates, integration, review, and approval of the deliverables.

6.9.4.1 Deliverables (what)

This section highlights the contents of the RAM in terms of business management, IT management, and project management deliverables. Chapter 4 provides detailed definitions, templates, process flows, and quality measures for the creation/updating, management/integration, approval/commitment, and review/QA of these deliverables. The project management deliverables are replicated in this section to provide readers with the necessary continuous and logical flow of information to enable a more analytical understanding and continuous refinement of the deliverables produced during the PDP.

Project management deliverables include the following:

- PSM management plan and requirements baseline updates:
  - Business requirements;
  - Data requirements;
  - Application requirements;
  - Technology requirements;
  - Scope management plan and reports updates.
- PTM management plan and schedule baseline updates;
- Project cost management plan, budget baseline, and reports updates;
- PQM management plan, quality baseline, and reports updates;
- Project change management plan, change requests, and reports updates;
- Project contract management plan, contract, and reports updates;
- PIM management plan, issues, and reports updates;
- PRM management plan, potential risks, and reports updates;
- PCM plan, progress status updates;
- Project human resources–staffing plan, RAM, and reports updates;
- PMO support services.

6.9.4.2 Activities (how)

The roles and responsibilities of project management component outlined in Chapter 4 are classified based on the creation/updating, management/integration, review/QA, and approval/commitment responsibility classifications to determine the major activities required to support the deliverables during the RAP. The integration of business management processes and IT management processes with project management processes is highlighted to demonstrate the integrated nature of project management during the IT PDLC processes. The manage and integrate responsibilities of project management are presented as similar tasks to demonstrate the integrated skills needed by the program/project manager.

- *PSM management plan, requirements baseline, and reports updates:* The scope management deliverables are integrated representation of the contents from business management–business architecture and IT management–data, applications, and technology architectures with project management–scope management (Update, integrate, review, approve–scope management).
- *PTM management plan, schedule baseline, and reports updates:* The time management deliverables are integrated representations of the contents from business management–project deliverables/funding approvals and IT management–resource allocations with project management–time management (Update, integrate, review, approve–time management).
- *Project cost management plan, budget baseline, and reports updates:* The cost management deliverables are integrated representations of the contents from business management–project justifications and PFAs and IT management–resource allocations and cost estimating with project management–cost management (Update, integrate, review, approve–cost management).
- *PQM management plan, quality baseline, and reports updates:* The quality management deliverables are integrated representations of the contents from business management–project deliverables/funding approvals and IT management–data, applications, TA, and applications support services with project management–quality management (Update, integrate, review, approve–quality management).
- *Project change management plan, change requests, and reports updates:* These are supporting deliverables for the core project management deliverables—cost, time, scope, and quality. They represent the integrated management of all changes made to the core project management deliverables (Update, integrate, review, approve–change management).
- *Project contract management plan, contracts, and reports updates:* These are supporting deliverables for the core project management deliverables—cost, time, scope, and quality. They represent the integrated management of all contracts to support the core project management

deliverables (Update, integrate, review, approve–contract management).

- *PIM management plan, issues, and reports updates:* These are supporting deliverables for the core project management deliverables—cost, time, scope, and quality. They represent the integrated management of all issues that affect delivery of the core project management deliverables (Update, integrate, review, approve–issue management).
- *PRM management plan, risks, and reports updates:* These are supporting deliverables for the core project management plans—cost, time, scope, and quality. They represent the integrated management of all risks that affect the delivery of the core project management deliverables (Update, integrate, review, approve–risk management).
- *PCM plan, progress status updates:* These are supporting deliverables for the core project management plans—cost, time, scope, and quality. They represent the integrated management of all communications to support the delivery of the core project management deliverables. Business management–project steering committee and IT management–resource allocation form an integral part of the communications management plan (Update, integrate, review, approve–communications).
- *Project human resources management plan, RAM, and reports updates:* These are the supporting deliverables for the core project management plans—cost, time, scope, and quality. They represent the integrated management of the staffing plans to support the delivery of the core project management deliverables. Business management–project steering committee and IT management–resource allocation form an integral part of the PSM management plan (Update, integrate, review, approve–human resources management).

6.9.4.3 Resources (who)

The labor resources needed to execute the activities to produce the deliverables are determined to show the type of project management resources responsible for the updating, integration, review, and approval of the deliverables during the RAP.

- PSM management plan, requirements baseline, and reports updates (Update, integrate, review, approve–scope management):
  - Program delivery manager–update, integrate;
  - Program business manager–review;
  - Program IT manager–review;
  - Executive business manager–review;
  - Executive project management manager–review and approve;
  - Executive IT manager–review;
  - Project manager and team–review.
- PTM management plan, schedule baseline, and reports updates:

  - Update, manage/integrate, review, approve–time management;
  - Program delivery manager–update, integrate;
  - Program business manager–review;
  - Program IT manager–review;
  - Executive business manager–review;
  - Executive project management manager–review and approve;
  - Executive IT manager–review;
  - Project manager and team–review.
- Project cost management plan, budget baseline, and reports updates (Update, manage/integrate, review, approve–cost management):
  - Program delivery manager–update, integrate;
  - Program business manager–review;
  - Program IT manager–review;
  - Executive business manager–review;
  - Executive project management manager–review and approve;
  - Executive IT manager–review;
  - Project manager and team–review.
- PQM management plan, quality baseline, and reports updates (Update, integrate, review, approve–quality management):
  - Program delivery manager–update, integrate;
  - Program business manager–review;
  - Program IT manager–review;
  - Executive business manager–review;
  - Executive project management manager–review and approve;
  - Executive IT manager–review;
  - Project manager and team–review.
- Project change management plan, change requests, and reports updates (Update, manage, review, approve–change management):
  - Program delivery manager–update, integrate;
  - Program business manager–review;
  - Program IT manager–review;
  - Executive business manager–review;
  - Executive project management manager–review and approve;
  - Executive IT manager–review;
  - Project manager and team–review.
- Project contract management plan, contracts, and reports updates (Update, manage, review, approve–contract management):
  - Program delivery manager–update, integrate;
  - Program business manager–review;
  - Program IT manager–review;
  - Executive business manager–review;
  - Executive project management manager–review and approve;
  - Executive IT manager–review;

    - Project manager and team–review.
- PIM management plan, issues, and reports updates (Update, manage, review, approve–issue management):
    - Program delivery manager–update, integrate;
    - Program business manager–review;
    - Program IT manager–review;
    - Executive business manager–review;
    - Executive project management manager–review and approve;
    - Executive IT manager–review;
    - Project manager and team–review.
- PRM management plan, risks, and reports updates (Update, integrate, review, approve–risk management):
    - Program delivery manager–update, integrate;
    - Program business manager–review;
    - Program IT manager–review;
    - Executive business manager–review;
    - Executive project management manager–review and approve;
    - Executive IT manager–review;
    - Project manager and team–review.
- PCM plan, and progress status updates (Update, integrate, review, approve–communications):
    - Program delivery manager–update, integrate;
    - Program business manager–review;
    - Program IT manager–review;
    - Executive business manager–review;
    - Executive project management manager–review and approve;
    - Executive IT manager–review;
    - Project manager and team–review.
- Project human resources staffing plan, RAM, and reports updates (Update, integrate, review, approve–human resources management):
    - Program delivery manager–update, integrate;
    - Program business manager–review;
    - Program IT manager–review;
    - Executive business manager–review;
    - Executive project management manager–review and approve;
    - Executive IT manager–review;
    - Project manager and team–review.

6.9.5 Alignment with IT management

The primary objective of this section is to demonstrate how the main deliverable of the RAP, the RAM, aligns with the deliverables (what) from the technology management component detailed in Chapter 5. The roles and responsibilities from the IT management processes presented in Chapter 5 are further refined to determine the major activities (how) needed to produce the deliverables and the people resources (who) responsible for the creation/updating, management/integration, approval/commitment, and review/QA of the deliverables.

6.9.5.1 Deliverables (what)

This section highlights the contents of the project RAM in terms of business management, IT management, and project management deliverable components. Chapter 5 provides detailed definitions, templates, process flows, and quality measures for the creation, management, approval, and review of these deliverables. The IT management deliverables are replicated in this section to provide readers with the necessary continuous and logical flow of information to enable a more analytical and continuous refinement understanding of the deliverables produced during the project RAP.

IT management deliverables include the following:

- Resource allocations updates;
- Cost estimates updates;
- DA updates–demonstration prototype level of detail: enterprise, business, and implementation data models;
- AA updates–demonstration prototype:
  - Enterprise, business, and implementation object (data and process) models;
  - Interface models (GUI and messaging models).
- TA updates–demonstration prototype: Enterprise, business, and implementation technology models
- Application support services updates–demonstration prototype: integrated systems testing and data conversion strategies updates.

6.9.5.2 Activities (how)

The roles and responsibilities of the IT management component outlined in Chapter 5 are classified based on the creation/updating, management/integration, approval/commitment, and review/QA responsibility scheme to determine the major activities required to support the deliverables during the RAP.

- PRA updates: Update, integrate, review, approve–resource allocations;
- Project cost estimates updates: Update, integrate, review, approve–cost estimates;
- Program DA updates: Update, integrate, review, approve–DA;
- Program AA updates: Update, integrate, review, approve–AA;
- Program TA updates: Update, integrate, review, approve–TA;
- Program applications support services updates: Update, integrate, review, approve–applications support.

6.9.5.3 Resources (who)

The labor resources needed to execute the activities to produce the deliverables are determined to show the type of IT management resources that are responsible for the creation, management, review, and approval of the deliverables produced during the PDP.

- PRA updates (Update, integrate, review, approve–resource allocations):
  - Program delivery manager–update, manage/integrate;
  - Program business manager–review;
  - Program IT manager–review;
  - Executive business manager–review;
  - Executive project management manager–review;
  - Executive IT manager–review and approve;
  - Project manager and team–review.
- Project cost estimates updates (Update, integrate, review, approve–cost estimates):
  - Program delivery manager–update, manage/integrate
  - Program business manager–review;
  - Program IT manager–review;
  - Executive business manager–review;
  - Executive project management manager–review and approve;
  - Executive IT manager–review;
  - Project manager and team–review.
- Program DA updates (Update, integrate, review, approve–DA):
  - Program delivery manager–integrate;
  - Program business manager–review;
  - Program IT manager–update;
  - Executive business manager–review;
  - Executive project management manager–review;
  - Executive IT manager–review and approve;
  - Project manager and team–create/update.
- Program AA updates (Update, manage, review, approve–AA):
  - Program delivery manager–integrate;
  - Program business manager–review;
  - Program IT manager–update;
  - Executive business manager–review;
  - Executive project management manager–review;
  - Executive IT manager–review and approve;
  - Project manager and team–create/update.
- Program TA updates (Update, manage, review, approve–TA):
  - Program delivery manager–integrate;
  - Program business manager–review;

- Program IT manager–update;
- Executive business manager–review;
- Executive project management manager–review;
- Executive IT manager–review and approve;
- Project manager and team–create/update.

- Program applications support services updates (Update, manage, review, approve–applications support):
  - Program delivery manager–manage;
  - Program business manager–review;
  - Program IT manager–update;
  - Executive business manager–review;
  - Executive project management manager–review;
  - Executive IT manager–review and approve;
  - Project manager and team–create/update.

6.9.6 Integration with business, IT, and project management

Tables 6.4 through 6.6 show integrated representations to present how the deliverables from the RAP-RAM, integrate with the deliverables (what), activities (how), and resources (who) from the business, IT, and project management components. The acronyms on the matrixes are abbreviations for executive managers, program managers, and project managers' responsibilities.

6.10 Project architecture: evolutionary prototype

A model is an abstraction of reality constructed to explore particular aspects or properties of a system.
—NAAIDT Computer Conference, 2003

During the ADP, the RAM and simulated demonstration prototype is further elaborated with the objective of building a working prototype, using the

Table 6.4 Business Management Responsibility Matrix—RAP

Deliverables	*Roles* *EBM*	*EPM*	*EITM*	*PBM*	*PDM*	*PITM*	*PM/Team*
Business management							
BSA	RA	R	R	U	M/I	R	R
Project justifications	RA	R	R	U	M/I	R	R
PFAs	RA	R	R	U	M/I	R	R
Project deliverables approvals	RA	R	R	U	M/I	R	R
Program steering committee	RA	R	R	U	M/I	R	R
Business support initiatives	RA	R	R	U	M/I	R	R

Note: U= update; R = review; A = approve; M/I = manage/integrate.

Table 6.5 Project Management Responsibility Matrix—RAP

Roles Deliverables	*EBM*	*EPM*	*EITM*	*PBM*	*PDM*	*PITM*	*PM/Team*
Project management							
Scope management	R	RA	R	R	UI	R	R
Time management	R	RA	R	R	UI	R	R
Quality management	R	RA	R	R	UI	R	R
Change management	R	RA	R	R	UI	R	R
Contract management	R	RA	R	R	UI	R	R
Issue management	R	RA	R	R	UI	R	R
Risk management	R	RA	R	R	UI	R	R
Communications management	R	RA	R	R	UI	R	R
Human resources management	R	RA	R	R	UI	R	R

Note: U = update; R = review; A = approve; M/I = manage/integrate.

Table 6.6 IT Management Responsibility Matrix—RAP

Roles Deliverables	*EBM*	*EPM*	*EITM*	*PBM*	*PDM*	*PITM*	*PM/Team*
IT management							
Resource allocations	R	R	RA	R	UI	R	R
Cost estimating	R	RA	R	R	UI	R	R
DA	R	R	RA	R	M/I	U	CU
AA	R	R	RA	R	M/I	U	CU
TA	R	R	RA	R	M/I	U	CU
Applications support services	R	R	RA	R	M/I	U	CU

Note: C = create; U = update; R = review; A = approve; M/I = manage/integrate.

appropriate software tools. This working model or evolutionary prototype uses tools and techniques that demonstrate a working model of requirements documented in the definition and RAPs. This working model is refined and transformed during the IDPs and usually becomes part of the final software product. This type of prototyping reduces the information, confusion, conflict, and design issues that usually occur during the traditional design specifications process. It produces a working concrete model of reality, using the recommended future-state software tools and techniques.

The ADP forms the basis for the following:

- The first major working model solution–PAS;
- A working model evolutionary prototype that demonstrates a continuous transformation from analysis to design for production transformation during the IDPs;
- Updated business, IT, and project management plans and deliverables;
- IDPs–construction, integration, and deployment;
- Approval and commitment to proceed to IPD.

During this phase, the business management, IT management, and project management deliverables of the project RAM are updated, and the simulated demonstration prototype is refined, which results in the working model evolutionary prototype. This is another software baseline solution that is refined as the project progresses throughout the PDLC. In most cases, this software solution baseline will form the basis for the development of the final solution during the IDPs—construction, integration, and transition.

The deliverables of the ADP focus mainly on updates to the business management, project management, and IT management deliverables developed during the RAP. IT management deliverables will be elaborated during this phase and will be further refined as the project progresses throughout the IT PDLC.

PAS will include further updates to the following:

- Business management deliverables:
  - BSA;
  - Project justification (benefits and priorities);
  - PFAs;
  - Project deliverables/funding approvals;
  - Program steering and working committee;
  - Project business support initiatives.
- Project management deliverables:
  - PSM management plan, requirements baseline, and reports:
    Business requirements;
    Data requirements;
    Application requirements;
    Technology requirements;
    Scope management plan and reports.
  - PTM management plan, schedule baseline, and reports;
  - Project cost management plan, budget baseline, and reports;
  - PQM management plan, quality baseline, and reports;
  - Project change management plan, change requests, and reports;
  - Project contract management plan, contract, and reports;
  - PIM management plan, issues, and reports;
  - PRM management plan, potential risks, and reports;
  - PCM plan, progress status, and reports;
  - Project human resources–staffing plan, RAM, and reports;
  - PMO support services.
- IT management deliverables:
  - Resource allocations;
  - Cost estimating;
  - DA–demonstration prototype level of detail: enterprise, business, and implementation data models;
  - AA—demonstration prototype level of detail:
    Enterprise, business, and implementation object models;
    Interface models (GUI and messaging models).

- TA—demonstration prototype level of detail: enterprise, business, and implementation technology models;
- Application support services–demonstration prototype level of detail: integrated systems testing and data conversion strategies.

The ADP starts as a result of the completion of a RAM and the simulated demonstration prototype and formally concludes with the project's second major solution checkpoint review–PAS. This review results in approval of the business, IT, and project management deliverables and formally declares the PAS and the supporting working model evolutionary prototype as the second approved project baseline solution.

6.10.1 Project management problems: real-world issues

Problems in the traditional design phase of the IT project delivery process or software development process are mainly due to the difficulty in obtaining technical resources or the procurement of technical resources in defining a technical design solution that aligns with the analysis and definition requirements. Again, this is clearly the result of a lack of understanding and poor communication of the required deliverables (what), activities to produce the deliverables using continuous refinements from the previous phases (how), technical skills, and knowledge of people resources required to deliver the deliverables (who). The design process starts with technical jargons, and an atmosphere of confusion, disagreement, and conflicting views continues to prevail among the people resources internal and external to the project. The most common and reoccurring problem that I have encountered during the design phase of the traditional software development process is the lack of continuous and stepwise refinements to show how the analysis deliverables map to the design deliverables, resulting in redevelopment of the deliverables from the analysis phase. Again, these problems are grouped based on people, process, deliverables, and technological conflicts to show the recurring similarities of conflicts between definition, analysis, and design phases. This is a classical case of failure to address the root cause of the problem early in the project life cycle.

- *People problems—effort:* Disagreements, confusion, and conflicting views continue to prevail, but now, over the difficulties in technical staffing defined in the staffing plan, organizational structure, and resource allocations. During the design phase, the procurement of technical development resources becomes a matter of subjective judgment for the nontechnical project manager, and political manoeuvres begin to evolve. Conflicting views, as a result of the striving for power between the contracted resources and the client and inconsistent understanding of how the deliverables will be produced as a refinement of the previous deliverables, often lead to emotional and irrational decisions. These decisions often result in "firing" the contracted resources or sometimes the client project managers, based on hidden political motives.

- *Process problems—cost, schedule, quality, procurement:* Disagreements, confusion, and conflicting views continue to prevail, but now over the processes involved in contract negotiations and award. After contract award, conflicts occur between the contracting resources and the client on updates to cost, schedule, and quality deliverables as a result of the inconsistent and subjective procurement process. Again, conflicting views, as a result of the striving for power between contractors and client, inconsistent processes, and improper understanding of the deliverables and supporting activities, often lead to emotional and irrational decisions. These decisions often result in redevelopment of project management processes to manage costs/budget, schedule, quality, contracts, and the necessary approval levels.
- *Deliverables/requirements problems—scope:* This is an area where disagreements, confusion, and conflicting views continue to prevail, although approval was given for previous deliverables. Endless flows of requirements changes occur, making it difficult to complete the design specifications, resulting in redeveloping the analysis specifications. The creative technical design starts, with little or no references to the previous deliverables, unmanageable technical brainstorming sessions, unstructured and disjointed texts, nonproductive meetings, and disintegrated deliverables. Similar scenarios occur during analysis and design; the project team is buried in the information jungle with each member desperately seeking assistance amidst conflicting views. Suddenly, a hero appears who quickly produces a deliverable to the amazement and appreciation of the rest of the team members. In a majority of cases, the deliverable produced does not align with the deliverables of the previous phase. My assessment of such situation brings me back to the root cause of IT projects failures—a lack of project managers with skills and knowledge in general business conceptualization to visualize, lead, and communicate a conceptual baseline solution to support the business requirements.
- *Technology problems—technical design issues:* This is the area where most of the disagreements, confusion, and conflicting views prevail over the detailed data, applications, and technology architectures with a seemingly endless flow of architecture changes as a result of IT management input, making it difficult to properly design a technical solution, much less update the BSA. Again, conflicting views between the contracting resources and the client technical support staff, the striving for power based on technical competency or incompetence, and the lack of skilled and knowledgeable resources for delivering and communicating the technical design solution, often result in technical design specifications that align poorly with the requirements and analysis specifications. A design document is produced at the mercy of

the construction team that now has to painfully weed through a series of unstructured documents with the hope that the design team can provide clarity. This is the stage where the finger-pointing and politics start, the innocent get punished, and a hero who normally plays the political game appears. The project normally gets canceled, rescoped, or approved for the next phase, based on illogical, irrational, and emotional decisions.

6.10.2 Recommended solutions: real-world scenarios

The problems experienced during the traditional design phase arise because of the lack of continuous commitment to people, process, deliverables/scope, and technology objectives, with major issues resulting from the lack of effectively communicating and understanding of the technical solution to support the analysis and project definition requirements. The business users, frustrated by the lack of understanding of the contents of the technical design solution, experience added frustration in trying to understand the technical jargon and the approach to solving the business problems, and as a result, revert to trust—hoping for the magical silver bullet solution.

Here are highlights of some practical real-world solutions to the problems above:

- *People problems—effort:* Project managers should ensure that the deliverables from the analysis phase include the required technical staffing levels with stated knowledge, skills, and experiences to deliver a technical solution that meets the requirements analysis. In the case of procurement of technical resources, the contract should be awarded based on a thorough understanding of the proposed solution from the contracting resources. Chapter 2 provides a baseline project organizational structure with supporting roles and responsibilities, which can easily be modified to fit within the context of any existing economic, political, cultural, and technological environment.
- *Process problems—cost, schedule, quality, procurement:* Communicate the project management processes involved in updating the cost, schedule, and quality constraints during contract negotiations and award. In the majority of cases, disagreements, confusion, and conflicting views on updating cost, schedule, and quality deliverables result from failure in effectively communicating the foundation project management processes, including contract management processes.

  Reconfirm the contracting firm's understanding of the project WBS and the need to clearly understand scope and analysis requirements to objectively determine the updates to cost, schedule, and quality objectives. Clearly understand how the contracting firm's WBS (CWBS) fits within the context of the client WBS. From experience, this level of understanding will eliminate change requests that normally cause emotional discomfort for both parties. Chapter 2 provides a baseline WBS with supporting implementation, using the Microsoft Project tool, which can easily be modified to fit within the context of any existing economic, political, cultural, and technological environment. The project deliverables and supporting activities must form the basis for updating the cost, schedule, and quality objectives of the project and supporting contractual arrangements.

- *Deliverables/requirements problems—scope:* The root cause of the technical design issues—the inability to determine the appropriate level of detail and a lack of commitment for which design solution is produced and how—is the failure to design a concrete deliverable that demonstrates a working-model.

  The recommended solution to resolving technical design issues, such as reanalysis or analysis paralysis, lack of commitment, or difficulty obtaining consensus and approval from major stakeholders, is to build a working model evolutionary prototype from the simulated demonstration prototype. Deploy use-case analysis and model-based documentation to demonstrate, communicate, and test the requirements. The evolutionary prototype produces a second baseline view of a real solution, which is easier to communicate in most cases, rather than using the traditional unstructured textual documentation.

  The working prototype, use-case analysis scenarios, and model-based documentation structures (later in this section) often make the deliverables easier to communicate and integrate because models and scenarios are abstract representation of reality. The working evolutionary prototype represents a concrete representation of reality that often provides the required commitment and approval necessary for the next project phases, or IPDs.

- *Technology problems—technical design issues:* The technical communications and commitment problems experienced in this phase, as a result of conflicting updates to the data, applications, and technology architectures, are normally resolved by building a working evolutionary prototype, with supporting model-based documentation. This evolutionary prototyping solution confirms the need for experienced project managers with general business conceptualization skills, specialized people, process, and technology integration skills, and excellent risk management skills with professional qualifications and relevant practical experience.

6.10.3 Alignment with business management

The primary objective of this section is to demonstrate how the main deliverable of the project architecture phase, the PAP, aligns with the deliverables (what) from the business management component process, detailed in Chapter 3. The roles and responsibilities from the business management processes, presented in Chapter 3, are further refined to determine the major activities (how) needed to produce the deliverables and the people resources (who) responsible for the updating, integration, review/QA, and approval/commitment for the deliverables.

6.10.3.1 Deliverables (what)

This section highlights the contents of the project architecture model in terms of business management, IT management, and project management components, based on refinements to the model produced in the analysis phase. Chapter 3 provides detailed definitions, templates, process flows, and quality measures for the creation/updating, management/integration, review/QA, and approval/commitment of these deliverables. The business management deliverables are replicated in this section to provide readers with the necessary continuous and logical flow of information to enable a continuous refinement of the deliverables produced during the project RAP.

Updated business management deliverables include the following:

- BSA updates;
- Project justification updates (benefits and priorities);
- PFA updates;
- Project deliverables/funding approvals updates;
- Program steering and working committee updates;
- Project business support initiatives updates.

6.10.3.2 Activities (how)

The roles and responsibilities of the business management component outlined in Chapter 3 are classified based on the update, integration, review, and approve responsibility scheme to determine the major activities required to support the deliverables during the ADP.

- BSA updates: Update, integrate, review, approve–BSA;
- Project justification (benefits and priorities) updates: Update, integrate, review, approve–project justifications;
- PFA updates: Update, integrate, review, approve–funding allocations;
- Project deliverables/funding approvals updates: Update, integrate, review, approve–deliverables approvals;
- Program steering and working committee updates: Update, integrate, review, approve–steering committee;
- Project business support initiatives updates: Update, integrate, review, approve–business initiatives.

6.10.3.3 Resources (who)

The labor resources needed to execute the activities to produce the deliverables are determined to show the type of business management resources responsible for the updating, integration, review, and approval of the deliverables produced during the ADP.

- Project BSA updates (Update, integrate, review, approve–BSA):

  - Program business manager–update;
  - Program delivery manager–integrate;
  - Program IT manager–review;
  - Executive business manager–review and approve;
  - Executive project management manager–review;
  - Executive IT manager–review;
  - Project manager and team–review.
- Project justification updates (benefits and priorities) (Update, integrate, review, approve–project justifications):
  - Program business manager–update;
  - Program delivery manager–integrate;
  - Program IT manager–review;
  - Executive business manager–review and approve;
  - Executive project management manager–review;
  - Executive IT manager–review;
  - Project manager and team–review.
- PFA updates (Update, integrate, review, approve–funding allocations):
  - Program business manager–update;
  - Program delivery manager–integrate;
  - Program IT manager–review;
  - Executive business manager–review and approve;
  - Executive project management manager–review;
  - Executive IT manager–review;
  - Project manager and team–review.
- Project deliverables/funding approvals updates (Update, integrate, review, approve–deliverables approvals):
  - Program business manager–update;
  - Program delivery manager–integrate;
  - Program IT manager–review;
  - Executive business manager–review and approve;
  - Executive project management manager–review;
  - Executive IT manager–review;
  - Project manager and team–review.
- Program steering and working committee updates (Update, integrate, review, approve–steering committee):
  - Program business manager–update;
  - Program delivery manager–integrate;
  - Program IT manager–review;
  - Executive business manager–review and approve;
  - Executive project management manager–review;
  - Executive IT manager–review;
  - Project manager and team–review.

- Project business support initiatives updates (Update, integrate, review, approve–business initiatives):
  - Program business manager–update;
  - Program delivery manager–integrate;
  - Program IT manager–review;
  - Executive business manager–review and approve;
  - Executive project management manager–review;
  - Executive IT manager–review;
  - Project manager and team–review.

6.10.4 Alignment with project management

The primary objective of this section is to demonstrate how the key deliverables of the ADP, the project architecture model, align with the deliverables (what) from the project management component detailed in Chapter 4. The roles and responsibilities from the project management processes presented in Chapter 4 are further refined to determine the major activities (how) needed to produce the deliverables and the people resources (who) responsible for the updating, integration, review, and approval of the deliverables.

6.10.4.1 Deliverables (what)

This section highlights the contents of the project architecture model in terms of business management, IT management, and project management deliverables. Chapter 4 provides detailed definitions, templates, process flows, and quality measures for the creation/updates, management/integration, approval/commitment, and review/QA of these deliverables. The project management deliverables are replicated in this section to provide the reader the necessary continuous and logical flow of information to enable a more analytical understanding and continuous refinement of the deliverables produced during the project architecture phase.

Project management deliverables include the following:

- PSM management plan and requirements baseline updates:
  - Business requirements;
  - Data requirements;
  - Application requirements;
  - Technology requirements;
  - Scope management plan, and reports updates.
- PTM management plan and schedule baseline updates;
- Project cost management plan, budget baseline, and reports updates;
- PQM management plan, quality baseline, and reports updates;
- Project change management plan, change requests, and reports updates;
- Project contract management plan, contract, and reports updates;
- PIM management plan, issues, and reports updates;

- PRM management plan, potential risks, and reports updates;
- PCM plan, progress status updates;
- Project human resources–staffing plan, RAM, and reports updates;
- PMO support services.

6.10.4.2 Activities (how)

The roles and responsibilities of the project management component outlined in Chapter 4 are classified based on the creation-updating, management-integration, review-QA, and approval-commitment responsibility classifications to determine the major activities required to support the deliverables during the project architecture phase. The integration of business management processes and IT management processes with project management processes is highlighted to demonstrate the integrated nature of project management during the IT PDLC processes. The manage and integrate responsibilities of project management are presented as similar tasks to demonstrate the integration skills needed by the program/project manager.

- *PSM management plan, requirements baseline, and reports updates:* The scope management deliverables are integrated representations of the contents from business management–business architecture and IT management–data, applications, and technology architectures with project management–scope management (Update, integrate, review, approve–scope management).
- *PTM management plan, schedule baseline, and reports updates:* The time management deliverables are integrated representations of the contents from business management–project deliverables/funding approvals and IT management–resource allocations with project management–time management (Update, integrate, review, approve–time management).
- *Project cost management plan, budget baseline, and reports updates:* The costs management deliverables are integrated representations of the contents from business management–project justifications and PFAs and IT management–resource allocations and cost estimating with project management–cost management (Update, integrate, review, approve–cost management).
- *PQM management plan, quality baseline, and reports updates:* The quality management deliverables are integrated representations of the contents from business management–project deliverables/funding approvals and IT management–data, applications, TA, and applications support services with project management–quality management (Update, integrate, review, approve–quality management).
- *Project change management plan, change requests, and reports updates:* These are supporting deliverables for the core project management deliverables—cost, time, scope, and quality. They represent the integrated management of all changes made to the core project management

deliverables (Update, integrate, review, approve–change management).

- *Project contract management plan, contracts, and reports updates:* These are supporting deliverables for the core project management deliverables—cost, time, scope, and quality. They represent the integrated management of all contracts to support the core project management deliverables (Update, integrate, review, approve–contract management).
- *PIM management plan, issues, and reports updates:* These are supporting deliverables for the core project management deliverables—cost, time, scope, and quality. They represent the integrated management of all issues that affect delivery of the core project management deliverables (Update, integrate, review, approve–issue management).
- *PRM management plan, risks, and reports updates:* These are supporting deliverables for the core project management plans—cost, time, scope, and quality. They represent the integrated management of all risks that affect the delivery of the core project management deliverables (Update, integrate, review, approve–risk management).
- *PCM plan, progress status updates:* These are supporting deliverables for the core project management plans—cost, time, scope, and quality. They represent the integrated management of all communications to support the delivery of the core project management deliverables. Business management–project steering committee and IT management–resource allocation form an integral part of the communications management plan (Update, integrate, review, approve–communications).
- *Project human resources management plan, RAM, and reports updates:* These are the supporting deliverables for the core project management plans—cost, time, scope, and quality. They represent the integrated management of the staffing plans to support the delivery of the core project management deliverables. Business management–project steering committee and IT management–resource allocation form an integral part of the PSM management plan (Update, integrate, review, approve–human resources management).

6.10.4.3 Resources (who)

The labor resources needed to execute the activities to produce the deliverables are determined, to show the type of project management resources responsible for the updating, integration, review, and approval of the deliverables produced during the project architecture phase.

- PSM management plan, requirements baseline, and reports updates (Update, integrate, review, approve–scope management):
  - Program delivery manager–update, integrate;
  - Program business manager–review;

- Program IT manager–review;
- Executive business manager–review;
- Executive project management manager–review and approve;
- Executive IT manager–review;
- Project manager and team–review.

- PTM management plan, schedule baseline, and reports updates (Update, manage/integrate, review, approve–time management):
  - Program delivery manager–update, integrate;
  - Program business manager–review;
  - Program IT manager–review;
  - Executive business manager–review;
  - Executive project management manager–review and approve;
  - Executive IT manager–review;
  - Project manager and team–review.
- Project cost management plan, budget baseline, and reports updates (Update, manage/integrate, review, approve–cost management):
  - Program delivery manager–update, integrate;
  - Program business manager–review;
  - Program IT manager–review;
  - Executive business manager–review;
  - Executive project management manager–review and approve;
  - Executive IT manager–review;
  - Project manager and team–review.
- PQM management plan, quality baseline, and reports updates (Update, integrate, review, approve–quality management):
  - Program delivery manager–update, integrate;
  - Program business manager–review;
  - Program IT manager–review;
  - Executive business manager–review;
  - Executive project management manager–review and approve;
  - Executive IT manager–review;
  - Project manager and team–review.
- Project change management plan, change requests, and reports updates (Update, manage, review, approve–change management):
  - Program delivery manager–update, integrate;
  - Program business manager–review;
  - Program IT manager–review;
  - Executive business manager–review;
  - Executive project management manager–review and approve;
  - Executive IT manager–review;
  - Project manager and team–review.
- Project contract management plan, contracts, and reports updates (Update, manage, review, approve–contract management):

- Program delivery manager–update, integrate;
- Program business manager–review;
- Program IT manager–review;
- Executive business manager–review;
- Executive project management manager–review and approve;
- Executive IT manager–review;
- Project manager and team–review.

- PIM management plan, issues, and reports updates (Update, manage, review, approve–issue management):
  - Program delivery manager–update, integrate;
  - Program business manager–review;
  - Program IT manager–review;
  - Executive business manager–review;
  - Executive project management manager–review and approve;
  - Executive IT manager–review;
  - Project manager and team–review.
- PRM management plan, risks, and reports updates (Update, integrate, review, approve–risk management):
  - Program delivery manager–update, integrate;
  - Program business manager–review;
  - Program IT manager–review;
  - Executive business manager–review;
  - Executive project management manager–review and approve;
  - Executive IT manager–review;
  - Project manager and team–review.
- PCM plan, and progress status updates (Update, integrate, review, approve–communications):
  - Program delivery manager–update, integrate;
  - Program business manager–review;
  - Program IT manager–review;
  - Executive business manager–review;
  - Executive project management manager–review and approve;
  - Executive IT manager–review;
  - Project manager and team–review.
- Project human resources staffing plan, RAM, and reports updates (Update, integrate, review, approve–human resources management):
  - Program delivery manager–update, integrate;
  - Program business manager–review;
  - Program IT manager–review;
  - Executive business manager–review;
  - Executive project management manager–review and approve;
  - Executive IT manager–review;
  - Project manager and team–review.

6.10.5 Alignment with IT management

The primary objective of this section is to demonstrate how the key deliverables of the architecture phase–PAS, aligns with the deliverables (what) from technology management component detailed in Chapter 5. The roles and responsibilities from IT management processes presented in Chapter 5 are further refined to determine the major activities (how) needed to produce the deliverables and the people resources (who) responsible for the creation/updating, management/integration, review, and approval of the deliverables.

6.10.5.1 Deliverables (what)

This section highlights the contents of the project architecture model solution, in terms of business management, IT management, and project management deliverable components. Chapter 5 provides detailed definitions, templates, process flows, and quality measures for the creation, management, approval, and review of these deliverables. The IT management deliverables are replicated in this section to provide readers with the necessary continuous and logical flow of information to enable a more analytical and continuous refinement understanding of the deliverables produced during the project architecture phase.

Updated IT Management deliverables include the following:

- Resource allocations updates;
- Cost estimates updates
- DA updates–demonstration prototype level of detail: enterprise, business, and implementation data models;
- AA updates–demonstration prototype:
  - Enterprise, business, and implementation object (data and process) models;
  - Interface models (GUI and messaging models).
- TA updates–demonstration prototype: enterprise, business, and implementation technology models;
- Application support services updates–demonstration prototype: integrated systems testing and data conversion strategies updates

6.10.5.2 Activities (how)

The roles and responsibilities of IT management component outlined in Chapter 5 are classified based on the creation/updating, management/integration, approval/commitment, and review/QA responsibility scheme to determine the major activities required to support the deliverables during the architecture phase.

- PRA updates: Update, integrate, review, approve–resource allocations;
- Project cost estimates updates: Update, integrate, review, approve–cost estimates;

- Program DA updates: Update, integrate, review, approve–DA;
- Program AA updates: Update, integrate, review, approve–AA;
- Program TA updates: Update, integrate, review, approve–TA;
- Program applications support services updates: Update, integrate, review, approve–applications support.

6.10.5.3 Resources (who)

The labor resources needed to execute the activities to produce the deliverables are determined, to show the type of IT management resources responsible for the creation, management, review, and approval of deliverables during project architecture phase.

- PRA updates (Update, integrate, review, approve–resource allocations):
  - Program delivery manager–update, manage/integrate;
  - Program business manager–review;
  - Program IT manager–review;
  - Executive business manager–review;
  - Executive project management manager–review;
  - Executive IT manager–review and approve;
  - Project manager and team–review.
- Project cost estimates updates (Update, integrate, review, approve–cost estimates):
  - Program delivery manager–update, manage/integrate;
  - Program business manager–review;
  - Program IT manager–review;
  - Executive business manager–review;
  - Executive project management manager–review and approve;
  - Executive IT manager–review;
  - Project manager and team–review.
- Program DA updates (Update, integrate, review, approve–DA):
  - Program delivery manager–integrate;
  - Program business manager–review;
  - Program IT manager–update;
  - Executive business manager–review;
  - Executive project management manager–review;
  - Executive IT manager–review and approve;
  - Project manager and team–create/update.
- Program AA updates (Update, manage, review, approve–AA):
  - Program delivery manager–integrate;
  - Program business manager–review;
  - Program IT manager–update;
  - Executive business manager–review;
  - Executive project management manager–review;

- Executive IT manager–review and approve;
- Project manager and team–create/update.

- Program TA updates (Update, manage, review, approve–TA):
  - Program delivery manager–integrate;
  - Program business manager–review;
  - Program IT manager–update;
  - Executive business manager–review;
  - Executive project management manager–review;
  - Executive IT manager–review and approve;
  - Project manager and team–create/update.
- Program applications support services updates (Update, manage, review, approve–applications support):
  - Program delivery manager–manage;
  - Program business manager–review;
  - Program IT manager–update
  - Executive business manager–review
  - Executive project management manager–review
  - Executive IT manager–review and approve
  - Project manager and team–create/update

6.10.6 Integration with business, IT, and project management

Tables 6.7 through 6.9 are integrated representations to present how the deliverables from the project architecture phase–project architecture model solution, integrate with the deliverables (what), activities (how), and resources (who) from the business, IT, and project management components. The acronyms on the matrixes are abbreviations for executive managers, program managers, and project managers' responsibilities.

Table 6.7 Business Management Responsibility Matrix—ADP

Deliverables	*Roles* *EBM*	*EPM*	*EITM*	*PBM*	*PDM*	*PITM*	*PM/Team*
Business management							
BSA	RA	R	R	U	M/I	R	R
Project justifications	RA	R	R	U	M/I	R	R
PFAs	RA	R	R	U	M/I	R	R
Project deliverables approvals	RA	R	R	U	M/I	R	R
Program steering committee	RA	R	R	U	M/I	R	R
Business support initiatives	RA	R	R	U	M/I	R	R

Note: U = update; R = review; A = approve; M/I = manage/integrate.

Table 6.8 Project Management Responsibility Matrix—ADP

Deliverables	*Roles* *EBM*	*EPM*	*EITM*	*PBM*	*PDM*	*PITM*	*PM/Team*
Project management							
Scope management	R	RA	R	R	UI	R	R
Time management	R	RA	R	R	UI	R	R
Scope management	R	RA	R	R	UI	R	R
Quality management	R	RA	R	R	UI	R	R
Change management	R	RA	R	R	UI	R	R
Contract management	R	RA	R	R	UI	R	R
Issue management	R	RA	R	R	UI	R	R
Risk management	R	RA	R	R	UI	R	R
Communications management	R	RA	R	R	UI	R	R
Human resources management	R	RA	R	R	UI	R	R

Note: C = create; U = update; R = review; A = approve; M/I = manage/integrate.

Table 6.9 IT Management Responsibility Matrix—ADP

Deliverables	*Roles* *EBM*	*EPM*	*EITM*	*PBM*	*PDM*	*PITM*	*PM/Team*
IT management							
Resource allocations	R	R	RA	R	UI	R	R
Cost estimating	R	RA	R	R	UI	R	R
DA	R	R	RA	R	M/I	U	CU
AA	R	R	RA	R	M/I	U	CU
TA	R	R	RA	R	M/I	U	CU
Applications support services	R	R	RA	R	M/I	U	CU

Note: C = create; U = update; R = review; A = approve; M/I = manage/integrate.

6.11 Project IDPs: Iterations #1 to #3

To improve is to change; to be perfect is to change often.
—Sir Winston Churchill

During the architecture phase, executive and senior management approve the recommendation for either a custom-based or packaged-based development solution. The IDP consists of three major subphases: construction/build, integration/conversion, and deployment/transition. The iterative development process starts during this phase, and I recommend three deliverables-focused iterations to prevent the endless uncontrollable iterative cycles that can result in project disasters. The IDP further develops the models, plans and evolutionary prototype from the architecture phase, plans and executes the integration of these components, and plans and deploys these components, using three iterations and subphase approvals. Iterative #1 represents refinements to the evolutionary prototype that incorporates execution of integration components and deployment of these components to production and applications support. Iterative #2 represents further refinements to the evolutionary prototype development and further refinements to the integration and deployment of these components to production and applications support. Iterative #3 represents final refinements to the evolutionary prototype development and final refinements to the integration and transition of these components to production and applications support. Each of these iterations must be approved by the appropriate business, IT, and project management users prior to continuing with the next iteration.

This type of iterative development process reduces the amount of unstructured information, confusion, and conflict and the number of implementation issues that usually occur during the traditional implementation process. It produces three major iterations of working concrete models of reality that incorporate integration and deployment components during each iteration, using the recommended future-state software tools and techniques. This iterative development process addresses the scope management, integration management, and risk management issues of traditional implementation processes to resolving the complex communications and commitment problems that continue to haunt us during software development.

The IDP forms the basis for the following:

- Refinements to the working-model solution baseline–PAS that incorporates integration and deployment components;
- Software development iterations that demonstrate a continuous transformation from analysis to design to construction to integration to deployment for production transformation and applications support;
- Updated business, IT, and project management plans and deliverables;
- IDPs—construction, integration and transition;
- Approval and commitment to proceed with each iterative solution.

During this phase, the business management, IT management, and project management deliverables of the project architecture model is updated, and the working evolutionary prototype is further developed, integrated, and deployed resulting in an integrated solution, Iteration #1. This integrated solution–Iteration #1 software baseline solution must be refined as the project progresses throughout IPD phase. In most cases, this software solution baseline will form the basis for the development of the final solution during the IDPs—construction, integration and transition.

The deliverables of the IDP focus on updates to the business management, IT management, and project management deliverables developed during the architecture phase. IT management deliverables will be further elaborated during this phase and will be further refined as the project progresses throughout the iterative development subphases—construction, integration, and transition.

Project IDS will include iterations and further updates that incorporate construction, integration, and transition components.

- Business management deliverables:

    - BSA;
    - Project justification (benefits and priorities);
    - PFAs;
    - Project deliverables/funding approvals;
    - Program steering and working committee;
    - Project business support initiatives.
- Project management deliverables:
    - PSM management plan, requirements baseline, and reports:
        - Business requirements;
        - Data requirements;
        - Application requirements;
        - Technology requirements;
        - Scope management plan, and reports.
    - PTM management plan, schedule baseline, and reports;
    - Project cost management plan, budget baseline, and reports;
    - PQM management plan, quality baseline, and reports;
    - Project change management plan, change requests, and reports;
    - Project contract management plan, contract, and reports;
    - PIM management plan, issues, and reports;
    - PRM management plan, potential risks, and reports;
    - PCM plan, progress status, and reports;
    - Project human resources–staffing plan, RAM, and reports;
    - PMO support services.
- IT management deliverables:
    - Resource allocations;
    - Cost estimating;
    - DA—Iterations #1, #2, and #3 levels of detail: business, system, and technical data models;
    - AA—Iterations #1, #2, and #3 levels of detail:
        - Enterprise, business, and implementation object models;
        - Interface models (GUI and messaging models).
    - TA—Iterations #1, #2, and #3 levels of detail: Enterprise, business, and implementation technology models;
    - Application support services—Iterations #1, #2, and #3 levels of detail:
        - Integrated systems testing and data conversion strategies;
        - Applications support model.

The IDP starts as a result of refinements to the architecture/design model and the working evolutionary prototype, and formally concludes with the project's #1, #2, and #3 integrated iteration solution checkpoint reviews—Iterations #1, #2, and #3. These reviews result in approval of the business, IT, and project management deliverables, and software iterations, and formally declare the project complete for transition to production and applications support.

6.11.1 Project management problems: real-world issues

Problems during the traditional implementation phase of the IT project delivery cycle or software development process, are mainly due to the difficulties in resolving last minute design failures and controlling last minute changes, inability to effectively control and monitor contractor and vendor deliverables, ineffective management of budget, schedule, quality and people issues, and finally, project acceptance problems. Again, this is clearly the result of a lack of understanding and poor communications of the required deliverables (what), activities or tasks required to construct, integrate, and deploy the deliverables using continuous refinements from the previous phases and iterative developments (how) and technical skills/knowledge of people resources required to coordinate the delivery of the final product (who). The traditional implementation process usually starts with constructing/installing the software code, databases, user interfaces and technical components, with little or no consideration for integration and deployment requirements.

This lack of focus on integration and deployment-production readiness requirements, usually create an atmosphere of confusion, disagreements and conflicting views. Project acceptance problems begin to manifest themselves among the people resources internal and external to the project. The most common and reoccurring problems that I encountered during the implementation phase of the traditional software development process are the lack of understanding of the integration and deployment deliverables and processes required to migrate the software solution to production and applications maintenance. This lack of understanding of integration and deployment processes and requirements is the root cause of the budget, schedule, quality, people and project acceptance problems that usually occur during the traditional software implementation phase.

A major corporation decided to address these integration and deployment issues, by creating three separate project team having individual project managers for each phase-construction, integration, and deployments. Senior management, on the advice of a consulting firm, awarded contracts to three different consulting companies to manage and deliver the three projects. The end result was the delivery of three separate project solutions with little or no considerations of how these three projects are integrated.

Integration was never achieved and this company continued to develop projects based on the decision to assign separate project managers to manage each of the construction, integration, and deployment phase. This scenario is a classical case of failure to recognize the root cause of the software implementation problems or lack of understanding of integration and deployment objectives and strategies, that usually result in communications, commitment and project acceptance problems.

The problems experienced during traditional implementation phases, are also grouped based on people, process, deliverables, and technological categories, to show the similarities in conflicts that usually reoccur throughout the traditional applications development process.

- *People problems—effort:* Disagreements, confusion, and conflicting views continue to prevail, but now, over the difficulties in handling technical design failures and last minute changes, controlling and monitoring contractor and vendor progress, integrated testing and data conversion problems, and production readiness or deployment problems. During the traditional implementation phase, tracking of project progress of the team members, including contractors and vendors, is based on trust and completion of impressive-looking set of activities, rather than being deliverables based. The coordination and assessment of the technical development resources, becomes a matter of subjective judgment for the nontechnical project manager, and political manoeuvres begin to evolve. Conflicting views, as a result of the striving for power between the contracted resources and the client, inconsistent understanding of how the deliverables will be constructed, integrated, and deployed, often lead to emotional and irrational decisions. These decisions may result in "firing" the contracted resources or sometimes the client project managers, sometimes based on hidden political motives.
- *Process problems—cost, schedule, quality, people:* Disagreements, confusion, and conflicting views continue to prevail, but now over problems with budget overrun, schedule changes, quality measures, and staff motivation. Conflict occurs between the contracting resources and the client in determining the final cost, schedule, and quality deliverables, as a result of the inconsistent and subjective construction, integration, and deployment processes, tools, and techniques. Now, the search for the guilty starts, amidst confusion and frustration, and the secrecy style of management begins to take precedence. This in conjunction with inconsistent processes, tools, and techniques, without proper understanding of the deliverables and supporting activities, often leads to emotional and irrational decisions. These decisions may result in reestablishing project management processes and practices to determine the final costs/budget, schedule, quality measures, contract management, and the necessary final approval levels.
- *Deliverables/requirements and technical design failures—scope changes:* This is an area where disagreements, confusion, and conflicting views continue to prevail, although approval was previously given for produced deliverables. An endless flow of requirements and technical design changes results, making it difficult to complete the construction, integration, and deployment deliverables, and further resulting in redeveloping the analysis and technical design specifications. The programmers start coding, with little or no references to the previous

deliverables, resulting in unmanageable code-review sessions, and spaghetti-type code with meaningless documentation, nonproductive meetings, and disintegrated deliverables. Similar scenarios usually occur during the integration and deployment phases, with the project team under pressure to get something working and to complete the project. Finally, the business users or customers see an initial version of the end product. At this point, different interpretations of the requirements and technical design emerge, which often need to be resolved at the senior management level. My assessment of such situations brings me back to the root cause of IT projects failures—the lack of project manager skills and knowledge of scope management, integration management, and risk management to lead the construction, integration, and deployment of the software product and to resolve the communications, commitment, and acceptance problems.

- *Technology problems—integration/conversion and deployment issues:* Another critical phase where most projects fail is during the conversion and deployment activities. Figure 6.5 is a real-world example of a deployment strategy that can be used as a guideline to prevent similar problems to those discussed in this section from reoccurring.

This is the area where most of the disagreements, confusion, and conflicting views over the construction, integration, and deployment of the data, applications, and technology architectures prevail. This usually results

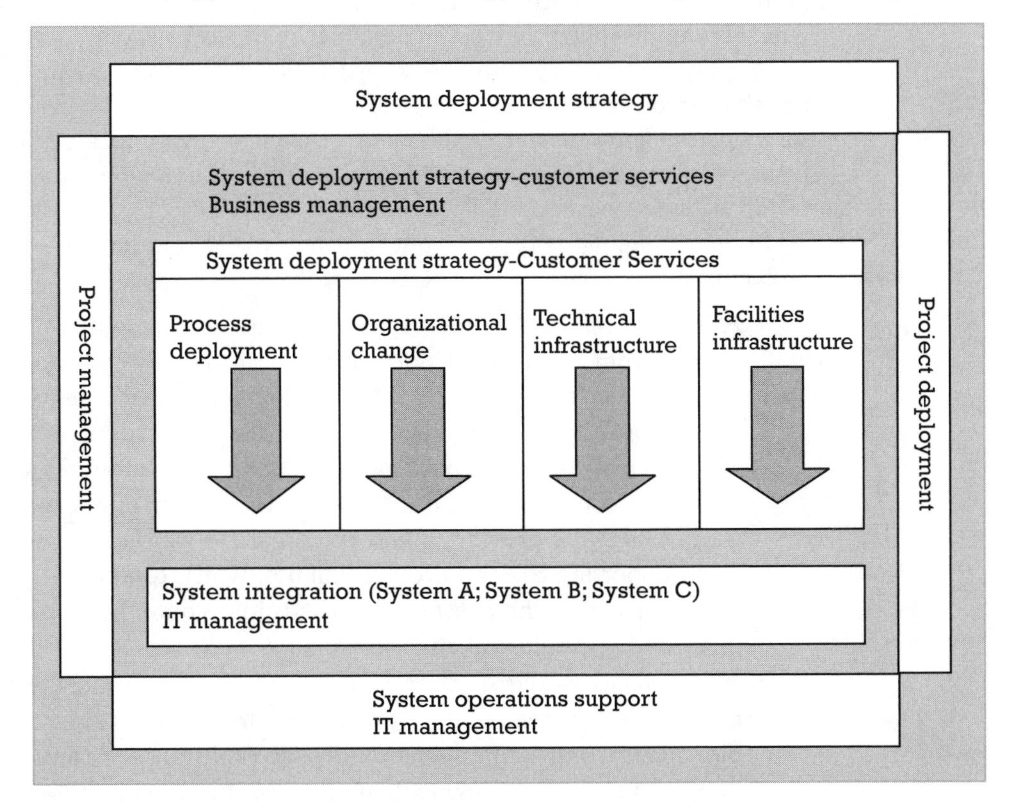

Figure 6.5 System deployment model process for customer services.

in a seemingly endless flow of architecture changes because of the technical resources involved, making it difficult to properly construct a technical solution, much less update the design documentation. Again, conflicting views between the contracting resources and the client technical support staff, the striving for power based on technical competence or incompetence, and the lack of skilled and knowledgeable resources for constructing and communicating the technical solution often lead to emotional and irrational decisions. The result of this building/coding process is a technical solution that aligns poorly with the requirements, analysis, and design specifications. The project team, under pressure to get something working suddenly builds something, and finally the business users or customers see an initial version of the end product. The business users or customers usually express dissatisfaction and frustration over integration testing and conversion problems and deployment issues. This is the stage where the finger-pointing gradually emerges, the political game starts, the innocent get punished, and a hero who plays the political game suddenly appears. The project gets canceled, rescoped, or approved pending completion of future changes based, in most cases, on illogical, irrational, and emotional decisions.

6.11.2 Recommended solutions: real-world scenarios

The problems experienced during the traditional implementation phase center on the lack of continuous commitment and acceptance of people, process, deliverables/scope, and technology objectives, mainly because of poor communications and understanding of the constructed, integrated, and deployed software solution to support the design, analysis, and project definition requirements. The business users, frustrated by the constructed solution, integration, and deployment problems, revert to accepting the final product, pending delivery of future changes, still hoping for the magical silver bullet solution.

Here are highlights of some practical real-world solutions to solve the problems above:

- *People problems—effort:* Project managers should ensure that the deliverables from the construction subphase are reviewed and accepted based on meeting integration and deployment requirements. In the case of procurement of technical resources, the contract should be awarded based on a thorough understanding of the proposed solution from the contracting resources with integration and deployment skills. Chapter 2 provides a baseline project organizational structure with supporting roles and responsibilities, which can be easily modified to fit within the context of any existing economic, political, cultural, and technological environment.
- *Process problems—cost, schedule, quality, people:* Communicate the project management processes involved in determining the final cost, schedule, quality, and staff motivation objectives during the construction, integration, and deployment subphases. In a majority of cases,

disagreements, confusion, and conflicting views on the final cost, schedule, quality, and staff assessments, result from the lack of effectively communicating the foundation project management processes during these final stages of developments.

Reconfirm the contracting firm's understanding of the project WBS and the need to clearly understand construction, integration, and deployment requirements to objectively determine the final cost, schedule, and quality objectives. Clearly understand how the CWBS fits within the context of the client WBS. From experience, this level of understanding eliminates change requests that manifest themselves during this stage of development and usually cause emotional discomfort for both parties. Chapter 2 provides a baseline WBS with supporting implementation, using Microsoft Project tool, which can be easily modified to fit within the context of any existing economic, political, cultural, and technological environment. The project deliverables and supporting activities must form the basis for determining the final cost, schedule, and quality objectives of the project and supporting contractual arrangements for acceptance.

- *Deliverables/requirements and technical design failures—scope changes:* The root cause of the technical design failures, last minute changes, and commitment, communications, and acceptance problems is the result of the lack of a concrete deliverable that demonstrates a working product—IDSs (Iterations #1, #2, and #3).

  The recommended solution to resolving technical design failures, handling last minute changes, and commitment, communications, and stakeholder acceptance problems is to construct, integrate, and deploy an IDS. Iteration #1–development completion of the IDS produces a constructed, integrated, and deployed software solution, which is easier to communicate, rather than using the traditional unstructured textual documentation. Iteration #2–production readiness represents refinements to Iteration #1. Iteration #3–production support delivers the constructed, integrated, and deployed software solution, migrated to production, ready for applications support services.

- *Technology problems—integration/conversion and deployment issues:* The technical communications, commitment, and stakeholder acceptance problems experienced in these phases are the result of integration problems (testing and data conversion), deployment impact issues (business changes, organizational changes, training, technological infrastructure changes, facilities changes, and operational support changes). These problems are usually resolved by building the IDS with supporting model-based documentation. Figure 6.5 is a real-world example of a deployment strategy that can be used as a guideline to prevent similar deployment problems discussed in this section from reoccurring. The IDS iterations confirm the need for experienced project managers with general business conceptualization skills, specialized people, process, and technology integration skills, and excellent risk management skills, who have attained the relevant professional qualifications and practical industry experiences.

6.11.3 Alignment with business management

The primary objective of this section is to demonstrate how the key deliverables of the IDP-IDS align with the deliverables (what) from the business management component process, detailed in Chapter 3. The roles and responsibilities from the business management processes, presented in Chapter 3, are further refined to determine the major activities (how) needed to produce the deliverables and the people resources (who) responsible for the update, integration, review, and approval of the deliverables.

6.11.3.1 Deliverables (what)

This section highlights the contents of the IDS, in terms of business management, IT management, and project management deliverable components, based on refinements to the PAS, produced in the ADP. Chapter 3 provides detailed definitions, templates, process flows, and quality measures for the creation/updating, management/integration, review/QA and approval/commitment for these deliverables. The business management deliverables are replicated in this section to provide readers with the necessary continuous and logical flow of information to show the continuous refinement of the deliverables produced during the ADP of the project.

Final business management deliverables include the following:

- BSA updates;
- Project justification updates (benefits and priorities);
- PFA updates;
- Project deliverables/funding approvals updates;
- Program steering and working committee updates;
- Project business support initiatives updates.

6.11.3.2 Activities (how)

The roles and responsibilities of the business management component outlined in Chapter 3 are classified based on the update, integration, review, and approval responsibility scheme to determine the major activities required to support the deliverables during this IDP.

- BSA updates: Update, integrate, review, approve–BSA;
- Project justification (benefits and priorities) updates: Update, integrate, review, approve–project justifications;
- PFA updates: Update, integrate, review, approve–funding allocations;
- Project deliverables/funding approvals updates: Update, integrate, review, approve–deliverables approvals;
- Program steering and working committee updates: Update, integrate, review, approve–steering committee;

- Project business support initiatives updates: Update, integrate, review, approve–business initiatives.

6.11.3.3 Resources (who)

The labor resources needed to execute the activities to produce the deliverables are determined to show the type of business management resources responsible for the updating, integration, review, and approval of the deliverables produced during the IDP.

- Project BSA updates (Update, integrate, review, approve–BSA):
  - Program business manager–update;
  - Program delivery manager–integrate;
  - Program IT manager–review;
  - Executive business manager–review and approve;
  - Executive project management manager–review;
  - Executive IT manager–review;
  - Project manager and team–review.
- Project justification updates (benefits and priorities) (Update, integrate, review, approve–project justifications):
  - Program business manager–update;
  - Program delivery manager–integrate;
  - Program IT manager–review;
  - Executive business manager–review and approve;
  - Executive project management manager–review;
  - Executive IT manager–review;
  - Project manager and team–review.
- PFA updates (Update, integrate, review, approve–funding allocations):
  - Program business manager–update;
  - Program delivery manager–integrate;
  - Program IT manager–review;
  - Executive business manager–review and approve;
  - Executive project management manager–review;
  - Executive IT manager–review;
  - Project manager and team–review.
- Project deliverables/funding approvals updates (Update, integrate, review, approve–deliverables approvals):
  - Program business manager–update;
  - Program delivery manager–integrate;
  - Program IT manager–review;
  - Executive business manager–review and approve;
  - Executive project management manager–review;
  - Executive IT manager–review;
  - Project manager and team–review.

- Program steering and working committee updates (Update, integrate, review, approve–steering committee):
  - Program business manager–update;
  - Program delivery manager–integrate;
  - Program IT manager–review;
  - Executive business manager–review and approve;
  - Executive project management manager–review;
  - Executive IT manager–review;
  - Project manager and team–review.
- Project business support initiatives updates (Update, integrate, review, approve–business initiatives):
  - Program business manager–update;
  - Program delivery manager–integrate;
  - Program IT manager–review;
  - Executive business manager–review and approve;
  - Executive project management manager–review;
  - Executive IT manager–review;
  - Project manager and team–review.

6.11.4 Alignment with project management

The primary objective of this section is to demonstrate how the key deliverables of the IDP-IDS (Iterations #1, #2, and #3) align with the deliverables (what) from the project management component, detailed in Chapter 4. The roles and responsibilities from the project management processes, presented in Chapter 4, are further refined to determine the major activities (how) needed to produce the deliverables and the people resources (who) responsible for the updating, integration, review, and approval of the deliverables.

6.11.4.1 Deliverables (what)

This section highlights the contents of the IDS, in terms of business management, IT management, and project management deliverables. Chapter 4 provides detailed definitions, templates, process flows, and quality measures for the creation/updating, management/integration, approval/commitment and review/QA for these deliverables. The project management deliverables are replicated in this section to provide readers with the necessary continuous and logical flow of information to enable a more analytical understanding and continuous refinement of the deliverables produced during the project ADP.

Final project management deliverables include the following:

- PSM management plan and requirements baseline updates:
  - Business requirements;
  - Data requirements;

- Application requirements;
- Technology requirements;
- Scope management plan and reports updates.

- PTM management plan and schedule baseline updates;
- Project cost management plan, budget baseline, and reports updates;
- PQM management plan, quality baseline, and reports updates;
- Project change management plan, change requests, and reports updates;
- Project contract management plan, contract, and reports updates;
- PIM management plan, issues, and reports updates;
- PRM management plan, potential risks, and reports updates;
- PCM plan, progress status updates;
- Project human resources–staffing plan, RAM, and reports updates;
- PMO support services.

6.11.4.2 Activities (how)

The roles and responsibilities of the project management component outlined in Chapter 4 are classified based on the creation/updating, management/integration, review/QA, and approval/commitment responsibility classifications to determine the major activities required to support the deliverables during the IDP. The integration of business management processes and IT management processes with project management processes is highlighted to demonstrate the integrated nature of project management during the IT PDLC processes. The manage and integrate responsibilities of project management are presented as similar tasks to demonstrate the integrated skills needed by the program/project manager.

- *PSM management plan, requirements baseline, and reports updates:* The scope management deliverables are integrated representations of the contents from business management–business architecture and IT management–data, applications, and technology architectures with project management–scope management (Update, integrate, review, approve–scope management).
- *PTM management plan, schedule baseline, and reports updates:* The time management deliverables are integrated representations of the contents from business management–project deliverables/funding approvals and IT management–resource allocations with project management–time management (Update, integrate, review, approve– time management).
- *Project cost management plan, budget baseline, and reports updates:* The costs management deliverables are integrated representations of the contents from business management–project justifications and PFAs and IT management–resource allocations and cost estimating with project management–cost management (Update, integrate, review, approve–cost management).

- *PQM management plan, quality baseline, and reports updates:* The quality management deliverables are integrated representations of the contents from business management–project deliverables/funding approvals and IT management–data, applications, TA, and applications support services with project management–quality management (Update, integrate, review, approve–quality management).
- *Project change management plan, change requests, and reports updates:* These are supporting deliverables for the core project management deliverables—cost, time, scope, and quality. They represent the integrated management of all changes made to the core project management deliverables (Update, integrate, review, approve–change management).
- *Project contract management plan, contracts, and reports updates:* These are supporting deliverables for the core project management deliverables—cost, time, scope, and quality. They represent the integrated management of all contracts to support the core project management deliverables (Update, integrate, review, approve–contract management).
- *PIM management plan, issues, and reports updates:* These are supporting deliverables for the core project management deliverables—cost, time, scope, and quality. They represent the integrated management of all issues that affect delivery of the core project management deliverables (Update, integrate, review, approve–issue management).
- *PRM management plan, risks, and reports updates:* These are supporting deliverables for the core project management plans—cost, time, scope, and quality. They represent the integrated management of all risks that affect the delivery of the core project management deliverables (Update, integrate, review, approve–risk management).
- *PCM plan, progress status updates:* These are supporting deliverables for the core project management plans—cost, time, scope, and quality. They represent the integrated management of all communications to support the delivery of the core project management deliverables. Business management–project steering committee and IT management–resource allocation form an integral part of the communications management plan (Update, integrate, review, approve–communications).
- *Project human resources management plan, RAM, and reports updates:* These are the supporting deliverables for the core project management plans—cost, time, scope, and quality. They represent the integrated management of the staffing plans to support delivery of the core project management deliverables. Business management–project steering committee and IT management–resource allocation form an integral part of the PSM management plan (Update, integrate, review, approve–human resources management).

6.11.4.3 Resources (who)

The labor resources needed to execute the activities to produce the deliverables are determined, to show the type of project management resources responsible for the updated, integration, QA and commitment of the deliverables during the IDP.

- PSM management plan, requirements baseline, and reports updates (Update, integrate, review, approve–scope management):
  - Program delivery manager–update, integrate;
  - Program business manager–review;
  - Program IT manager–review;
  - Executive business manager–review;
  - Executive project management manager–review and approve;
  - Executive IT manager–review;
  - Project manager and team–review.
- PTM management plan, schedule baseline, and reports updates (Update, manage/integrate, review, approve-time management):
  - Program delivery manager–update, integrate;
  - Program business manager–review;
  - Program IT manager–review;
  - Executive business manager–review;
  - Executive project management manager–review and approve;
  - Executive IT manager–review;
  - Project manager and team–review.
- Project cost management plan, budget baseline, and reports updates (Update, manage/integrate, review, approve–cost management):
  - Program delivery manager–update, integrate;
  - Program business manager–review;
  - Program IT manager–review;
  - Executive business manager–review;
  - Executive project management manager–review and approve;
  - Executive IT manager–review;
  - Project manager and team–review.
- PQM management plan, quality baseline, and reports updates (Update, integrate, review, approve–quality management):
  - Program delivery manager–update, integrate;
  - Program business manager–review;
  - Program IT manager–review;
  - Executive business manager–review;
  - Executive project management manager–review and approve;
  - Executive IT manager–review;
  - Project manager and team–review.
- Project change management plan, change requests, and reports updates (Update, manage, review, approve–change management):

- Program delivery manager–update, integrate;
- Program business manager–review;
- Program IT manager–review;
- Executive business manager–review;
- Executive project management manager–review and approve;
- Executive IT manager–review;
- Project manager and team–review.

- Project contract management plan, contracts, and reports updates (Update, manage, review, approve–contract management):
  - Program delivery manager–update, integrate;
  - Program business manager–review;
  - Program IT manager–review;
  - Executive business manager–review;
  - Executive project management manager–review and approve;
  - Executive IT manager–review;
  - Project manager and team–review.
- PIM management plan, issues, and reports updates (Update, manage, review, approve–issue management):
  - Program delivery manager–update, integrate;
  - Program business manager–review;
  - Program IT manager–review;
  - Executive business manager–review;
  - Executive project management manager–review and approve;
  - Executive IT manager–review;
  - Project manager and team–review,
- PRM management plan, risks, and reports updates (Update, integrate, review, approve–risk management):
  - Program delivery manager–update, integrate;
  - Program business manager–review;
  - Program IT manager–review;
  - Executive business manager–review;
  - Executive project management manager–review and approve;
  - Executive IT manager–review;
  - Project manager and team–review.
- PCM plan, and progress status updates (Update, integrate, review, approve–communications):
  - Program delivery manager–update, integrate;
  - Program business manager–review;
  - Program IT manager–review;
  - Executive business manager–review;
  - Executive project management manager–review and approve;
  - Executive IT manager–review;
  - Project manager and team–review.

- Project human resources staffing plan, RAM, and reports updates (Update, integrate, review, approve–human resources management):
  - Program delivery manager–update, integrate;
  - Program business manager–review;
  - Program IT manager–review;
  - Executive business manager–review
  - Executive project management manager–review and approve;
  - Executive IT manager–review;
  - Project manager and team–review.

6.11.5 Alignment with IT management

The primary objective of this section is to demonstrate how the main deliverable of the IDPs, the IDS, aligns with the deliverables (what) from technology management component detailed in Chapter 5. The roles and responsibilities from IT management processes, presented in Chapter 5, are further refined to determine the major activities (how) needed to produce the deliverables and the people resources (who) responsible for the creation/updating, management/integration, approval/commitment and review/QA of the deliverables.

6.11.5.1 Deliverables (what)

This section highlights the contents of the IDS, in terms of business management, IT management, and project management deliverable components. Chapter 5 provides detailed definitions, templates, process flows, and quality measures for the creation, management, approval, and review of these deliverables. The IT management deliverables are replicated in this section to provide readers with the necessary continuous and logical flow of information to enable a more analytical understanding and continuous refinement of the deliverables produced during the project IDPs.

IT management deliverables include the following:

- Resource allocations;
- Cost estimating
- DA—Iterations #1, #2, and #3 levels of detail: business, system, and technical data models;
- AA—Iterations #1, #2, and #3 levels of detail:
  - Enterprise, business, and implementation object models;
  - Interface models (GUI and messaging models).
- TA—Iterations #1, #2, and #3 levels of detail: enterprise, business, and implementation technology models;
- Application support services—Iterations #1, #2, and #3 levels of detail:
  - Integrated systems testing and data conversion strategies;
  - Applications support model.

6.11.5.2 Activities (how)

The roles and responsibilities of the IT management component outlined in Chapter 5 are classified based on the creation/updating, management/integration, approval/commitment, and review/QA responsibility scheme to determine the major activities required to support the deliverables during the IDP.

- PRA updates: Update, integrate, review, approve–resource allocations;
- Project cost estimates updates: Update, integrate, review, approve–cost estimates;
- Program DA updates: Update, integrate, review, approve–DA;
- Program AA updates: Update, integrate, review, approve–AA;
- Program TA updates: Update, integrate, review, approve–TA;
- Program applications support services updates: Update, integrate, review, approve–applications support.

6.11.5.3 Resources (who)

The labor resources needed to execute the activities to produce the deliverables are determined to show the type of IT management resources responsible for the creation, management, review, and approval of the deliverables during the project IDP.

- PRA updates (Update, integrate, review, approve–resource allocations):
  - Program delivery manager–update, manage/integrate;
  - Program business manager–review;
  - Program IT manager–review;
  - Executive business manager–review;
  - Executive project management manager–review;
  - Executive IT manager–review and approve;
  - Project manager and team–review.
- Project cost estimates updates (Update, integrate, review, approve–cost estimates):
  - Program delivery manager–update, manage/integrate;
  - Program business manager–review;
  - Program IT manager–review;
  - Executive business manager–review;
  - Executive project management manager–review and approve;
  - Executive IT manager–review;
  - Project manager and team–review.
- Program DA updates (Update, integrate, review, approve–DA):
  - Program delivery manager–integrate;
  - Program business manager–review;
  - Program IT manager–update;
  - Executive business manager–review;
  - Executive project management manager–review;

  - Executive IT manager–review and approve;
  - Project manager and team–create/update.
- Program AA updates (Update, manage, review, approve–AA):
  - Program delivery manager–integrate;
  - Program business manager–review;
  - Program IT manager–update;
  - Executive business manager–review;
  - Executive project management manager–review;
  - Executive IT manager–review and approve;
  - Project manager and team–create/update.
- Program TA updates (Update, manage, review, approve–TA):
  - Program delivery manager–integrate;
  - Program business manager–review;
  - Program IT manager–update;
- Executive business manager–review;
  - Executive project management manager–review;
  - Executive IT manager–review and approve;
  - Project manager and team–create/update.
- Program applications support services updates (Update, manage, review, approve–applications support):
  - Program delivery manager–manage;
  - Program business manager–review;
  - Program IT manager–update;
  - Executive business manager–review;
  - Executive project management manager–review;
  - Executive IT manager–review and approve;
  - Project manager and team–create/update.

6.11.6 Integration with business, IT, and project management

Tables 6.10 through 6.12 show integrated representations to present how the deliverable from the ADP project architecture model solution integrate with the deliverables (what), activities (how), and resources (who) from business, IT, and project management components. The acronyms on the matrixes are abbreviations for executive managers, program managers, and project managers' responsibilities.

Another critical phase during which most projects fail is the conversion and deployment activities phase. Figure 6.5 is a real-world example of a deployment strategy[7] that can be used as a guideline to prevent the deployment problems discussed in this section from reoccurring.

System deployment process descriptions

The system deployment strategy consists of:

Table 6.10 Business Management Responsibility Matrix—IDP

Deliverables	*Roles* *EBM*	*EPM*	*EITM*	*PBM*	*PDM*	*PITM*	*PM/Team*
Business management							
BSA	RA	R	R	U	M/I	R	R
Project justifications	RA	R	R	U	M/I	R	R
PFAs	RA	R	R	U	M/I	R	R
Project deliverable approval	RA	R	R	U	M/I	R	R
Program steering committee	RA	R	R	U	M/I	R	R
Business support initiatives	RA	R	R	U	M/I	R	R

Note: U = update; R = review; A = approve; M/I = manage/integrate.

Table 6.11 Project Management Responsibility Matrix—IDP

Deliverables	*Roles* *EBM*	*EPM*	*EITM*	*PBM*	*PDM*	*PITM*	*PM/Team*
Project management							
Scope management	R	RA	R	R	UI	R	R
Time management	R	RA	R	R	UI	R	R
Scope management	R	RA	R	R	UI	R	R
Quality management	R	RA	R	R	UI	R	R
Change management	R	RA	R	R	UI	R	R
Contract management	R	RA	R	R	UI	R	R
Issue management	R	RA	R	R	UI	R	R
Risk management	R	RA	R	R	UI	R	R
Communications management	R	RA	R	R	UI	R	R
Human resources management	R	RA	R	R	UI	R	R

Note: U = update; R = review; A = approve; M/I = manage/integrate.

Table 6.12 IT Management Responsibility Matrix—IDP

Deliverables	*Roles* *EBM*	*EPM*	*EITM*	*PBM*	*PDM*	*PITM*	*PM/Team*
IT management							
Resource allocations	R	R	RA	R	UI	R	R
Cost estimating	R	RA	R	R	UI	R	R
DA	R	R	RA	R	M/I	U	CU
AA	R	R	RA	R	M/I	U	CU
TA	R	R	RA	R	M/I	U	CU
Applications support services	R	R	RA	R	M/I	U	CU

Note: C = create; U = update; R = review; A = approve; M/I = manage/integrate.

7. Deployment strategy: A strategy that describes how the components of the deployment subphase of the IDP should be implemented. This deployment strategy is based on process deployment, organizational change, training, technology, and facilities infrastructure support.

- System end-user training plan;
- System communications plan;
- System help desk support;
- System business transition plan.

In the system deployment strategy–customer services business management, system customer services processes deployment:

- Customer services business change is announced and communicated.
- Fundamental system principles and policies are presented to customer services account management team.
- Customer services account management teams are taught how to apply system principles and policies during business processes deployment.
- Customer services account management teams are given computer-based training on how to use the first release of system.
- On-the-job support (help-desk) is provided to customer services account management teams.

In the system deployment strategy–customer services business management, the system customer services organizational change:

- Establishes procedures for implementing the new organization infrastructure.
- Establishes procedures and facilities for storing, maintaining, and distributing training materials.
- Establishes procedures and facilities for creating, maintaining, and distributing communication materials.

In the system deployment strategy–customer services business management, the system technical infrastructure determines applications, data, and technology (hardware, system software, network) configuration for customer services deployment site.

In the system deployment strategy–customer services business management, the system facilities infrastructure determines facilities (telephone, CTI) configuration for customer services deployment site.

The system integration (legacy systems)–IT management:

- Integrates technical infrastructure (hardware; system software; network).
- Integrate system with technical infrastructure.
- Integrate system with legacy applications.
- Create integrated release of system deployment.

The project deployment team:

- Executes deployment plan/strategy.
- Installs system hardware, system software, and applications.
- Conducts deployment–site readiness testing.
- Makes deployment–site cutover go/no-go decision.

- Completes deployment–site cutover.
- Monitors and tunes production system.
- Conducts deployment–site acceptance review.
- Turns over to deployment–site operations support.

The system operations support–IT management supports deployment at site operations.

Project management:

- Establishes and executes site deployment plan.
- Determines system deployment teams.
- Reports, tracks, and reports problems.
- Manages problem resolution.

6.12 PMO support services

The quality of a leader is reflected in the standards they set for themselves.
—Ray Kroc, founder, McDonald's

The primary objective of this section is to demonstrate how the PMO supports the project team during each phase of the PDLC. The PMO is responsible for deploying and supporting a set of consistent and IPM processes within the company, including policies, roles and responsibilities, procedures—deliverables templates, process flows, and checklists—and best practices. These project management processes and program deliverables are normally stored in an integrated project management information repository (IPMIR),[8] which is accessed, maintained, and reported by an integrated project management information system (IPMIS). Chapter 4 provides a structure of the IPMIR directory and a systems flow diagram of the IPMIS. PMO is not a one-time event, but a broad initiative that could cover a number of years to ensure the consistency, integrity, and reusability of the project deliverables and processes. The program delivery manager, project managers, and supporting staff are usually part of this group.

6.12.1 Project management problems: real-world issues

Companies involved in the development and deployment of many projects are usually confronted with the standard project management problems of budget overrun, schedule slippage, unsatisfactory quality, and inefficient utilization of people resources. These companies make various attempts at resolving these issues, using general management principles for leading, communicating, negotiating, problem solving, and influencing without

8. IPMIR: The integrated project management information repository is an integrated database containing all of the business management, project management, and IT management deliverables. It is accessed, manipulated, managed, and reported on using the integrated project management system (IPMS). It is synonymous with project management information repository (PMIR)

applying any form of standardized, consistent, or integrated process. In most cases, these typical project management problems remain unresolved, with compounded communications, commitment, and acceptance problems. The search for the magical silver bullet solution continues, with certain companies reverting to project management consulting firms to provide that silver bullet solution.

6.12.2 Recommended solution: real-world scenarios

A major part of the solution to executing projects better, more quickly, more cheaply, and with high quality is the organization's ability and commitment to guide and implement common and integrated processes and practices across the entire organization. If implemented properly and with the right project management culture, there may be a very short learning curve for the project manager and the team members as they transition from one project to another.

Most PMOs should be responsible for deploying and supporting a set of consistent IPM processes within the company, including policies, roles and responsibilities, procedures—deliverables templates, process flows, and checklists—and best practices. Some critics may argue that the PMO demands additional resources, which can add overhead to the project. As result of my extensive IT project management experiences, I strongly believe that if implemented properly, investment in the PMO will be more than recouped by implementing consistent and integrated processes and practices that will allow every project within the organization to be completed better, more quickly and cheaply, and with higher quality.

An effective PMO can offer many potential products and services, depending on the project management needs and culture of the organization and the vision of the PMO sponsor (the person who is generally responsible for the PMO funding). Before the PMO can be successful, it must gain agreement from executive and senior management and the project management team on its overall role and the general expectations it needs to achieve.

This book provides a framework to assist in the resolution of the typical project management problems identified above using integrating business management, IT management, and project management process components and applying standard, consistent, and integrated policies, processes, deliverables, and practices.

6.12.3 PMO alignment with IPM-IT

The primary objective of this section is to demonstrate how the key deliverables of the PMO align with the deliverables (what) of IPM-IT presented in Chapters 3, 4, and 5. The roles and responsibilities of the project management–PMO process presented in Chapter 4 are further refined to determine the major activities (how) needed to produce the deliverables and the people resources (who) responsible for the creation/delivery/

support, management/integration, review/QA, and approval/commitment of the PMO deliverables.

6.12.3.1 Deliverables (what)

This section highlights the products and services of the PMO to support the IPM-IT processes—the business management, IT management, and project management deliverable components. Chapter 4 provides detailed definitions, templates, process flows, and quality measures for the creation, management, approval, and review of these deliverables. The project management products and services are replicated in this section to provide readers with the necessary continuous and logical flow of information to enable a more analytical and continuously refined understanding of the value of the PMO during implementation of the IPM-IT model.

PMO products and services include the following:

- Project methodology support;
- Project management processes and templates deployment;
- Project management training and coaching;
- Project metrics and measurement criteria support;
- Project management tools support;
- Project progress reporting support.

6.12.3.2 Activities (how)

The roles and responsibilities of the project management–PMO component outlined in Chapter 4 are classified based on the creation/updating, management/integration, approval/commitment, and review/QA responsibility scheme to determine the major activities required to support PMO support services during the IT PDLC.

- Project methodology support: create, manage, review, approve–methodology support;
- Project management processes and template deployment: create, manage, review, approve–project management processes;
- Project management training and coaching: create, manage, review, approve–project management training;
- Project metrics and measurement criteria support: create, manage, review, approve–project metrics;
- Project management tools support: create, manage, review, approve–project management tools;
- Project progress reporting support: create, manage, review, approve–project reporting support.

6.12.3.3 Resources (who)

The labor resources needed to execute the activities to deliver the PMO services are determined to show the type of PMO resources responsible for the creation, management, review, and approval of the services during the IT PDLC to enhance project team productivity.

- Project methodology support (Create, manage, review, approve–methodology support):
  - Program delivery manager–manage;
  - Program business manager–review;
  - Program IT manager–review;
  - Executive business manager–review;
  - Executive project management manager–review and approve;
  - Executive IT manager–approve;
  - Project manager and PMO team–create/update.
- Project management processes and template deployment (Create, manage, review, approve–project management processes):
  - Program delivery manager–manage;
  - Program business manager–review;
  - Program IT manager–review;
  - Executive business manager–review;
  - Executive project management manager–review and approve;
  - Executive IT manager–approve;
  - Project manager and PMO team–create/update.
- Project management training and coaching (Create, manage, review, approve–project management training):
  - Program delivery manager–manage;
  - Program business manager–review;
  - Program IT manager–review;
  - Executive business manager–review;
  - Executive project management manager–review and approve;
  - Executive IT manager–approve;
  - Project manager and PMO team–create/update.
- Project metrics and measurement criteria support (Create, manage, review, approve–project metrics):
  - Program delivery manager–manage;
  - Program business manager–review;
  - Program IT manager–review;
  - Executive business manager–review;
  - Executive project management manager–review and approve;
  - Executive IT manager–approve;
  - Project manager and PMO team–create/update.
- Project management tools support (Create, manage, review, approve–project management tools):
  - Program delivery manager–manage;
  - Program business manager–review;
  - Program IT manager–review;

- Executive business manager–review;
- Executive project management manager–review and approve;
- Executive IT manager–approve;
- Project manager and PMO team–create/update.

- Project progress reporting support (Create, manage, review, approve–project reporting support):
  - Program delivery manager–manage;
  - Program business manager–review;
  - Program IT manager–review;
  - Executive business manager–review;
  - Executive project management manager–review and approve;
  - Executive IT manager–approve;
  - Project manager and PMO team–create/update.

6.12.4 PMO integration: IPM-IT

Table 6.13 is an integrated representation to show how the services from the PMO support the IPM-IT model during the IT PDLC.

Figure 6.6 is a process flow diagram that represents the buyer's methodology or processes during the acquisition or purchase of another company. Various companies are constantly consolidating their assets by purchasing other companies in order to survive in this competitive marketplace. Since IT assets are a major resource at most companies, most buyers are experiencing difficulties in assessing the current state of the purchased company and in determining how to consolidate and integrate IT assets effectively and efficiently. The intent of Figure 6.6 is to address the buyers' consolidation and integration issues of IT assets by providing framework to guide the assessment, recommendation, acquisition, and transition (ARAT) processes.[9]

6.13 Summary

The integrated IT PDLC model presented in this chapter describes how the components of business management, project management, and IT management are horizontally integrated, based on the phases of the IT PDLC model. For each phase, the alignment with business management, project management, and IT management is demonstrated based on the deliverables (what), activities (how), and key responsibilities of the resources (who) needed to meet the milestone. Real-world problems and recommended solutions for each phase are also discussed to provide readers with scenarios for proactive actions. The business management, project management, and IT

9. ARAT: Assessment, recommendation, acquisition and transition is a methodology or process that Buyers can use to evaluate, manage and control the IT resources or assets of another company during companies' acquisition or takeover bidding processes.

Table 6.13 PMO Integrated Services: IT PDLC

Deliverables	*Roles* Executive Business	Executive PM	Executive IT	Program Business Manager	Program Delivery Manager	Program IT Manager	Project Manager/ PMO Team
PMO							
Project methodology	R	RA	R	R	M/I	R	CS
PM processes	R	RA	R	R	M/I	R	CS
PM training	R	RA	R	R	M/I	R	CS
PM metrics	R	RA	R	R	M/I	R	CS
PM tools	R	RA	R	R	M/I	R	CS
PM reporting	R	RA	R	R	M/I	R	CS

Note: C = create/update; R = review; A = approve; M/I = manage/integrate; S = support.

management deliverables (what), activities (how), and roles and responsibilities (who, what) matrixes provide an excellent representation of how these components are horizontally integrated during the IT PDLC.

IT organizations in the process of managing and delivering multiple IT projects with the goal of integrated or enterprise project management should consider the following recommendations as a framework to guide them towards the successful management and delivery of multiple IT projects during the IT PDLC.

- Develop a PDM that consists of key business management, IT management, and project management deliverables, similar to the model presented in Section 6.8. This PDM will form the baseline to better manage the scope–deliverables, effort, and costs, schedule, and quality objectives of the development projects during the PDP.
- Develop a demonstration prototype, similar to the model presented in Section 6.9. This demonstration prototype will form the baseline for obtaining business and management commitment for the requirements or scope of the project by providing a concrete solution that the business users can easily understand and refine as the project progresses. The major causes of most project failures occur during the analysis phase.
- Develop an evolutionary prototype, similar to the model presented in Section 6.10. This evolutionary prototype will form the baseline to ensure a smooth transition that clearly maps the business requirements to achieve a working automated solution. Traditional software development processes have failed miserably to provide any form of meaningful transition to meet the expectations of both the business requirements and the IT design solution. In most cases, the architects or designers have been forced to rework the analysis and deemed previous analysis to be futile.
- Develop one to three iterations similar to the IDS presented in Section 6.11. These iterations will ensure commitment and acceptance of the final product by providing concrete solutions that business and IT staff

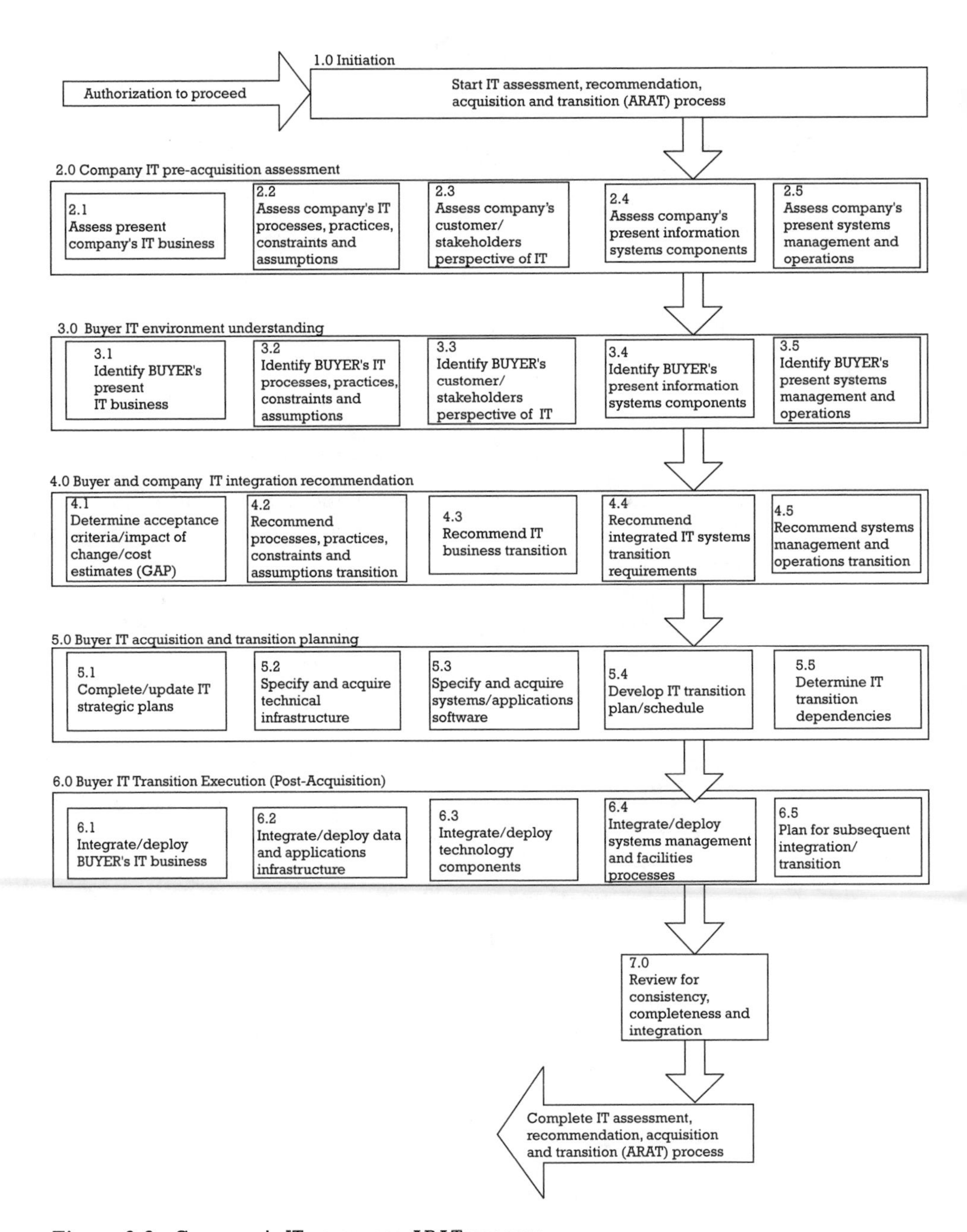

Figure 6.6 Company's IT resources ARAT process.

can understand and refine prior to production migration. One of the critical root causes of most project failures usually occurs during the deployment-conversion phase.

- Establish a PMO function that provides the services, similar to those discussed in Section 6.12. This will ensure the completeness, consistency, integrity, and reusability of project deliverables to improve project team productivity.

6.14 Questions to think about: management perspectives

1. Think about how your organization identifies, prioritizes, and manages multiple IT projects. How does your organization control project deliverables? What are the key rationales for obtaining business users' commitments to requirements? What are the major components of the IT PDLC? How do these components relate to your project environment? What is the perception of senior management of the need for a PDLC model to manage the project deliverables?
2. Think about how your organization defines "project." What are the three components of a business definition model? How do these three major deliverables relate to your project definition approach? What is the major problem that occurs during project definition?
3. Think about how your organization determines project requirements on projects. What is the major deliverable of the RAP? How does this deliverable relate to your organization? What is the most frequent problem that occurs during the RAP? What is the root cause? What is the recommended solution?
4. Think about how your organization designs project solutions. What is the major deliverable of the ADP? How does this deliverable relate to your organization? What is the most frequent problem that occurs during the RAP? What is the root cause? What is the recommended solution?
5. Think about how your organization implements projects. What are the major phases and deliverables of the IDP? How do these phases and deliverables relate to your organization? What is the most frequent problem that occurs during the IDP? What is the root cause? What is the recommended solution? What are the deliverables of the deployment subphase? How do these deliverables relate to your organization?
6. Think about how your organization ensures consistency, completeness, and integration of project deliverables. What are the services of the PMO? How do these services relate to your organization? What is the perception of senior management of the value of the PMO at your organization? What are the major causes of PMO failures? What is the key recommended solution? What are the six processing components of ARAT?

Selected bibliography

Arthur, L. J., *Improving Software Quality: An Insider's Guide to TQM*, New York: John Wiley & Sons, 1993.

CSC—Computer Sciences Corporation, *Catalyst Methodology*, CA CSC, 1999.

Hetzel, B., *The Complete Guide for Software Test Documentation*, New York: John Wiley & Sons/QED Press, 1998.

Kerzner, H., *Project Management: A Systems Approach to Planning, Scheduling, and Controlling*, 7th ed., New York: John Wiley & Sons, 2001.

Lewis, J. P., *Project Planning Scheduling and Control*, 3rd ed., New York: McGraw-Hill, 2001.

Muller, R. J., *Productive Objects: An Applied Software Project Management Framework*, San Francisco CA: Morgan Kaufmann, 1998.

PMI, P*roject Management Institute: A Guide to Project Management Body of Knowledge*, 2000 Edition, Newton Square, PA: PMI, 2000.

Royce, W., Software Project Management—A Unified Framework, Reading, MA: Addison-Wesley, 1998.

State of North Carolina, *North Carolina Technical Architecture*, 1997.

CHAPTER

7

Contents

Aligning PMI-PMBOK with IBM Rational Corporation RUP

You got to be careful if you don't know where you're going because you might not get there.
—Yogi Berra

7.1 Introduction

One of the software development methodologies that is gaining widespread acceptance in the IT industry is RUP[1] from IBM Rational Corporation. The main focus of this approach, based on the objective of effectively controlling and optimizing the dynamic IT environment, is movement toward an IDS. The RUP approach is similar to the software development process solution discussed in Chapter 6 with respect to the iterative strategy. However, the RUP approach differs slightly from the prototyping approach discussed in Chapter 6. As mentioned above, a model-centric process to control and optimize this dynamic environment must replace the traditional waterfall or tedious structured approach to software development. RUP from IBM Rational Corporation is one of the software development methodologies that is gaining widespread acceptance in meeting this need. The objective of this chapter is to discuss the deliverables, activities, and resources of RUP and to show the alignment with PMI-PMBOK project management processes from an IPM perspective, not to describe the RUP iterative approach to software development. There are many excellent books on this new and exciting development process, some of which are referenced in the selected bibliography at the end of this chapter.

1. RUP: The rational unified process is a use-case-driven, architecture-centric, iterative, and incremental software engineering process developed by Rational Corporation.

The phases of RUP—inception, elaboration, construction, and transition—connect the major processing components of business management, project management, and IT RUP workflow management to show how the integrated nature of these components fits within the context of IPM-IT,[2] the theme of this book. The integrated RUP-PMBOK project management model presented in Figure 7.1 shows how business management, PMBOK project management, and RUP workflow management[3] components are integrated during each phase of the IBM Rational Corporation RUP software development process. This RUP integration is based on three distinguishing

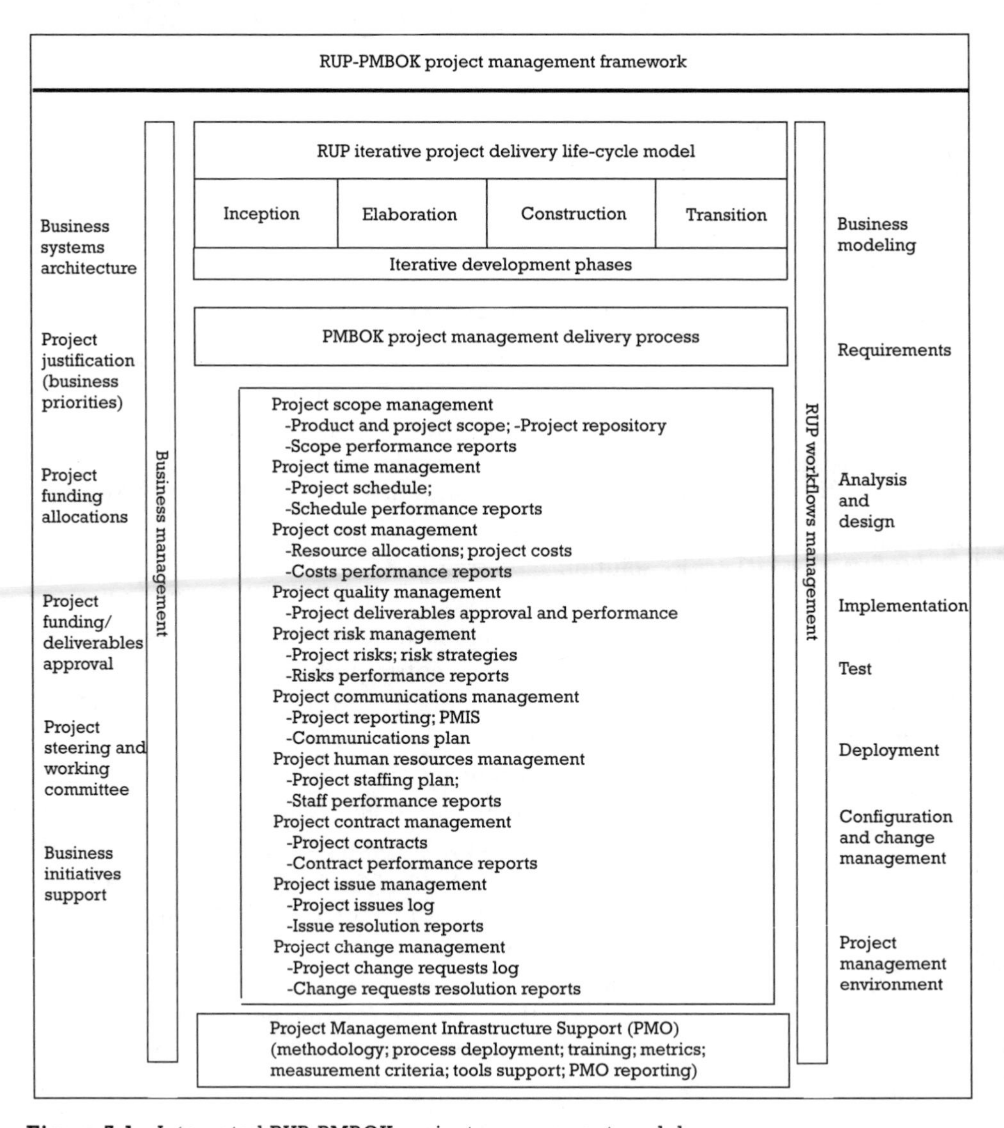

Figure 7.1 Integrated RUP-PMBOK project management model.

2. IPM-IT: IPM-IT is a model-centric framework that integrates business management and IT management with Project management. It is the major theme of this book.
3. RUP workflow management: RUP workflow management consists of six core workflows and three core supporting workflows that produce results of value to the project.

characteristics: It is use-case driven,[4] architecture-centric,[5] and iterative and incremental.

The RUP iterative PDLC model[6] (or phases) presented in this chapter represents the deliverables (what), activities (how), and human resources (who) for each of the phases shown in Figure 7.1. The WBS, shown in Figure 7.2, is the technique used to structure the project deliverables and activities to effectively manage the projects and to optimize the utilization of resources based on aligning RUP with PMBOK processes. Appendix C is a software tool implementation of this WBS, using Microsoft's project management software, Microsoft Project.

7.2 RUP-PMBOK project management framework

This RUP-PMBOK project management framework represents the foundation principle to demonstrating the alignment of PMI-PMBOK with RUP.

7.3 RUP-PMBOK WBS

This WBS shows a model-based approach to managing projects that enforce RUP and PMI-PMBOK guiding processes and practices.

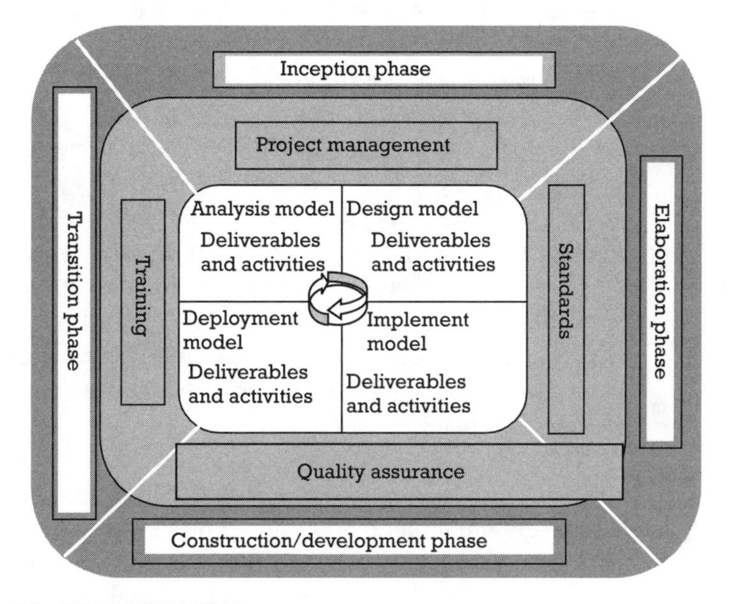

Figure 7.2 RUP-PMBOK WBS.

4. Use-case driven: RUP uses use cases as the foundation of the RUP. A use case is a sequence of actions a system performs that yields an observable result to a particular actor.
5. Architecture-centric: RUP is based on an architecture baseline. This architecture baseline is produced at the end of the elaboration phase, at which time the foundation structure and behavior of the system are stabilized.
6. RUP iterative PDLC model: The phases of RUP—inception, elaboration, construction, and transition—that support an iterative approach to project delivery.

PMBOK defines WBS as: "A deliverable-oriented grouping of project elements that organizes and defines the total work scope of the project. Each descending level represents an increasingly detailed definition of the project work." The core deliverables and supporting activities of RUP have three distinguishing characteristics: They are use-case driven, architecture-centric, and iterative and incremental. RUP consists of cycles that may iterate during software development with an iteration consisting of core artifacts or deliverables and supporting management deliverables or artifacts for the phases—inception, elaboration, construction, and transition. The core deliverables and supporting activities for each phase are grouped as the analysis model, design model, implementation model, and deployment model. The supporting management deliverables and activities for each of the core deliverables for each phase are grouped as project management, standards, QA, and training components.

7.4 The merits of understanding RUP and PMBOK processes

Incompetents invariably make trouble for people other than themselves.
—Larry McMurtry, Lonesome Dove

During my extensive IT career, I have witnessed frequent misunderstandings by IT professionals of the foundation processes needed to effectively apply tools to produce the required value and benefits during the management of IT projects. A major cause of these misunderstandings was the result of the limited skill levels of some IT professionals, including project managers. They seemed to lack the necessary business conceptualization skills, specialized people, process, and technology integration skills, and excellent risk management skills to properly apply these foundation concepts to support tools and technology implementation. The symptoms—not root causes—of the failures were usually attributed to the technology and the supporting process, not to project management inefficiencies. The inefficient project management process prevailed, as managers searched for the nonexistent magical solution.

There are many project management tools, some of which are listed in the selected bibliography at the end of this chapter, and they cannot be effectively implemented without a proper understanding of the basic WBS concepts as defined in PMBOK. There are some project managers who manage projects using project management tools and who totally ignore vital concepts such as WBS. These project managers hold the firm belief that WBS concepts are theoretical and have no place in the practical world.

RUP is supported by various tools, many of which are also listed in the selected bibliography at the end of this chapter. It would be extremely difficult to effectively use RUP software tools without proper training and understanding of the basic RUP software development process. Some IT professionals and project managers have managed and developed projects using software development tools and have totally ignored concepts such as models. They believe that these concepts are theoretical and have no place in the practical world. Certain senior IT managers pronounced their non-commitment to models, and software development processes and approaches with zero maturity level still continued to progressively function in this technologically advancing world.

My main objective in writing this chapter is to eliminate or a least prevent the massive proliferation of ineffective software tools application and implementation by those IT professionals and project managers who have unconsciously contributed to software development failures of the past. These IT professionals and project managers continued to unconsciously execute ineffective traditional methods in the IT industry, some even disguised themselves as project management consultants. I hope that these practicing IT professionals and project managers will fully accept the reality of technology advancements and applicability and improve their knowledge, skills, and practices to support the changing roles and responsibilities of IT professionals and project managers.

The moral of this situation can be expressed using a simple but effective statement: *A fool with a tool is still a fool.*

7.5 RUP-PMBOK PDLC process model

The RUP PDLC is iterative and incremental, use-case driven, and architecture-centric. It provides a framework of phases, deliverables, and activities to support an iterative and integrated RUP-PMBOK project management framework, through horizontal integration of business management, IT RUP workflows management, and PMBOK–project management components. It is designed for new development projects, acquisition and installation of software packages, and application conversions or migrations from one technical environment to another.

The objectives of this RUP model are to highlight the phases of IBM Rational Corporation RUP during software development and to discuss the integrated alignment with business management, IT RUP workflow management, and PMBOK–project management processes based on the RUP phases.

Specific objectives when aligning RUP with PMBOK processes are to do the following:

- Enhance ongoing communications between the project management team and the business and RUP workflow support resources to ensure that project business, IT, and project management requirements are clearly understood and satisfied.
- Provide consistent checkpoints for business and RUP workflow management to review project progress, monitor project costs and schedule, assess risk, and approve project continuation.

- Provide project managers with a framework of RUP phases, iterative deliverables and activities, and PMI-PMBOK project management deliverables and processes.
- Provide a set of RUP iterative deliverables and activities that align with PMBOK processes are and refined during each phase to meet the characteristics of the project.

Figure 7.3 is a representation of the core iterative deliverables that are refined during each of the RUP phases—inception, elaboration, construction, and transition.

7.6 Applying the RUP-PMBOK project delivery process

The effective application of the RUP-PMBOK project delivery process produces an iterative integrated approach to software development efforts based on use-case driven, architecture-centric, and iterative and incremental development techniques. However, software development projects vary greatly in scope, objectives, and characteristics, and as such demands of this iterative approach should be proportional to the size, scope, and number of iterations of the project.

The project managers should apply this iterative process on major project initiatives. However, for small projects, the project manager may combine phases, deliverables, and activities for each iteration to reflect the most appropriate iterative development approach for that project. The conceptualization, integration, and risk management skills of the project managers must now be applied in selecting and/or combining the deliverables, activities, and number of iterations within each phase.

The implementation of RUP is based on three key concepts—build, integration, and prototypes. In order to effectively manage the implementation of these three constructs, a structured and disciplined project management philosophy must be realized. The key modeling concept presented in this

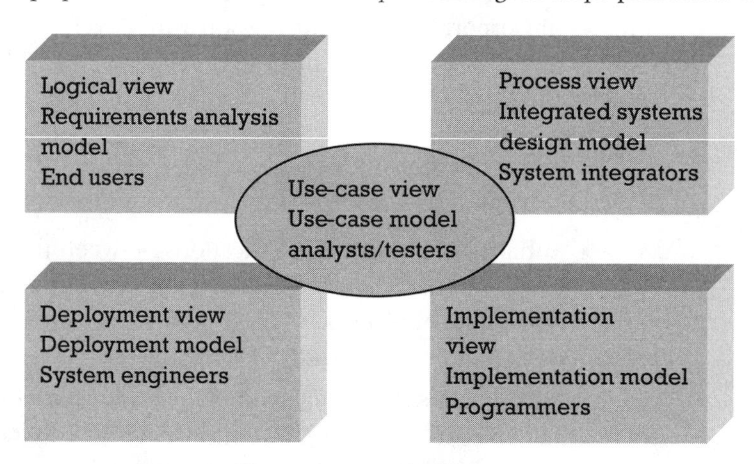

Figure 7.3 The RUP 4 + 1 view model of architecture.

chapter is based on PMBOK process models that are aligned with business management and RUP workflow management process to support the RUP phases as represented in Figure 7.1.

RUP uses prototypes in an iterative way to reduce risk. Prototypes can reduce uncertainty surrounding the following issues:

- Business value of the software product being developed;
- Stability or performance of key technology;
- Project commitment or funding (proof-of-concept prototype);
- Understanding of requirements;
- Look and feel and ultimately the usability of the product.

A prototype can help gain support and commitment for the product by demonstrating something concrete and executable to project stakeholders. However, the nature of prototype development can lead to chaotic, endless iterations, in the absence of clear objectives of the prototype. PMI-PMBOK processes provide project management guidelines to reduce the risk of the prototypes reaching that chaotic state.

RUP iterative software developments view prototypes in two ways—as behavioral prototypes and structural prototypes.

- A behavioral prototype focuses on exploring the specific behavior of the system.
- A structural prototype explores architectural or technological concerns. The result of this structural view is two kinds of prototypes:
  - An exploratory prototype, also called a throwaway prototype;
  - An evolutionary prototype, which evolves to become the final system.

For the RUP iterative development approach to be effectively deployed or transitioned into a production environment, the following major deliverables or artifacts[7] must be produced, with incremental iterations, as shown in Figure 7.3, during each of the RUP phases:

- Project management, configuration and change management, and environment artifact set–management plans and reports;
- Business modeling and requirements artifact set–RAM;
- Analysis and design artifact set–design model;
- Implementation and test artifact set–implementation model;
- Deployment artifact set–deployment model.

RUP project management, configuration and change management, and environment artifact set–management plans, and reports, as provided in Philippe Kruchten's book *The Rational Unified Process: An Introduction, Second Edition,* consists of the following:

- RUP project management artifact set:

7. Artifact: RUP uses the term artifact to represent a model, a model element, or a document deliverable.

- Business case;
- Software development plan:
  - Iteration plan;
  - Problem resolution plan;
  - Risk management plan;
  - Product acceptance plan;
  - Measurement plan.
- Iteration assessment;
- Status assessment;
- Work order;
- Project measurements;
- Review record.

- RUP Configuration and change management artifact set:
  - Configuration management plan;
  - Project repository;
  - Configuration audit findings;
  - Change request.
- Environment artifact set:
  - QA plan;
  - Development organization assessment;
  - Product-specific templates;
  - Development case: guidelines (design, test);
  - Supporting environment;
  - Tool support assessment;
  - Tools.

The project management delivery process model (PMBOK) project structure to support RUP iterative development prototyping-product structure consists of the following:

- PSM management plan, requirements baseline, and reports:
  - Business requirements;
  - Data requirements;
  - Application requirements;
  - Technology requirements;
  - Scope management plans, and reports.
- PTM management plan, schedule baseline, and reports;
- Project cost management plan, budget baseline, and reports;
- PQM management plan, quality baseline, and reports;
- Project change management plan, change requests, and reports;
- Project contract management plan, contract, and reports;
- PIM management plan, issues, and reports;
- PRM management plan, potential risks, and reports;
- PCM plan, progress status, and reports;

- Project human resources–staffing plan, RAM, and reports;
- PMO support services.

In order to effective apply this RUP-PMBOK project delivery framework,[8] which focuses on a prototyping and iterative development approach, deliverables and activities must be closely examined to eliminate unnecessary duplicated processes and deliverables to ensure optimum integration. The phases of the RUP iterative delivery life-cycle model are now discussed to demonstrate how the horizontal integrated components of business management, project management, and RUP workflow management fit within the context of the RUP-PMBOK project management framework, as presented in Figure 7.1.

7.7 Project inception phase: Iteration #1

By failing to prepare, you are preparing to fail.
—Benjamin Franklin

During the inception phase,[9] the core idea is developed into a software product vision. In this phase, an understanding of the core business drivers is reviewed and confirmed. A business case is developed to fully understand why the software product is required. The focus is to gain an understanding of the overall requirements and to determine the feasibility and scope of the software product. This phase produces a basic conceptual definition of the project—Iteration #1, consisting of management, requirements, design, implementation and deployment artifact sets that are incrementally iterated as the project progresses throughout the RUP-PMBOK delivery life-cycle model or phases. Iteration # 1 produces one or several prototypes and supporting models and reports as the major deliverables of this phase.

The inception phase forms the basis for the following:

- The first formal iteration of the project;
- The first major product baseline: Iteration #1–business modeling and requirements artifact set–RAM and foundation architectural prototype;
- The conceptual definition of the project;
- The first major project baseline plan;
- Iteration #2: next incremental iteration.

This phase produces the following major deliverables, introduced above, in accordance with the model discussed in Figure 7.1. The product baseline models, prototype, and supporting project plans are incrementally iterated

8. RUP-PMBOK project delivery framework: A framework that integrates business management and RUP workflow management with PMBOK–project management processes.

9. Inception phase: The RUP inception phase specifies the end-product vision and its business case and defines the scope of the product. The inception phase concludes by delivering the life-cycle objectives (LCO) milestone.

as the project progresses throughout the PDLC. The RUP artifact sets, provided in Philippe Kruchten's book *The Rational Unified Process: An Introduction, Second Edition,* are integrated with the model presented in Figure 7.1 to show the alignment with business management, PMBOK–project management, and RUP workflow management components.

The deliverables of this inception phase focus mainly on creating the business management, project management, and RUP support workflow artifact set deliverables. The RUP core workflow deliverables—business modeling and requirements artifact set—will also be developed during this phase to be further refined as the project progresses throughout the RUP iterative PDLC.

The major deliverables from this inception phase include the following:

Iteration #1: One or several foundation architectural prototypes and supporting documents:

- Business management deliverables:
  - BSA;
  - Project justification (benefits and priorities);
  - PFAs;
  - Project deliverables/funding approvals;
  - Program steering and working committee;
  - Project business support initiatives.
- PMBOK–project management deliverables:
  - PSM management plan, requirements baseline, and reports:
    Business requirements;
    Data requirements
    Application requirements;
    Technology requirements;
    Scope management plan and reports.
  - PTM management plan, schedule baseline, and reports;
  - Project cost management plan, budget baseline, and reports;
  - PQM management plan, quality baseline, and reports;
  - Project change management plan, change requests, and reports;
  - Project contract management plan, contract, and reports;
  - PIM management plan, issues, and reports;
  - PRM management plan, potential risks, and reports;
  - PCM plan, progress status, and reports;
  - Project human resources–staffing plan, RAM, and reports;
  - PMO support services.
- RUP workflow management deliverables:
  - Business modeling artifact set;
  - Requirements artifact set;
  - Analysis and design artifact set;
  - Implementation artifact set;
  - Test artifact set;

- Deployment artifact set;
- Environment artifact set;
- Project management artifact set;
- Configuration and change management artifact set.

The inception phase starts as a result of completing certain major business management and project management deliverables and RUP supporting deliverables—project management workflows, configuration and change management workflows, and environment workflows. This phase delivers a RUP business modeling and requirements artifact set and a foundation architectural prototype. It formally concludes with the project's first major deliverables checkpoint review, Iteration #1. This review results in approval of the business, RUP workflow, and project management deliverables and formally declares Iteration #1, as the first approved project baseline.

7.7.1 RUP alignment with PMBOK project management

The primary objective of this section is to demonstrate how the main deliverable of the RUP inception phase, Iteration #1, aligns with the deliverables (what) from PMBOK–project management component, RUP workflow components, and business management components. The roles and responsibilities from the project management processes, presented in Chapter 4, are summarized to determine the major activities (how) needed to produce the deliverables and the people resources (who) responsible to create/update, manage/integrate, review/test, approve/commit, and support/maintain the deliverables.

7.7.1.1 Deliverables (what)

This section highlights the contents of Iteration #1–foundation architectural prototype and supporting business modeling and requirements artifacts set, in terms of business management, RUP workflows, and project management deliverables. Chapter 4 provides detailed definitions, templates, process flows, and quality measures for the creation, management, review, approval, and support of these deliverables. The project management deliverables are replicated in this section to provide readers with the necessary continuous and logical flow of information to enable a more analytical understanding of the deliverables produced during RUP–inception phase.

PMBOK–project management deliverables include the following:

- PSM management plan, requirements baseline, and reports:
  - Business requirements;
  - Data requirements;
  - Application requirements;
  - Technology requirements;
  - Scope management plan and reports.

- PTM management plan, schedule baseline, and reports;
- Project cost management plan, budget baseline, and reports;
- PQM management plan, quality baseline, and reports;
- Project change management plan, change requests, and reports;
- Project contract management plan, contract, and reports;
- PIM management plan, issues, and reports;
- PRM management plan, potential risks, and reports;
- PCM plan, progress status, and reports;
- Project human resources–staffing plan, RAM, and reports;
- PMO support services.

7.7.1.2 Activities (how)

The roles and responsibilities of project management component outlined in Chapter 4 are classified based on the creation/updating, management/integration, review/QA, approval/commitment, and support/maintenance responsibility classifications to determine the major activities required to produce the deliverables during the RUP inception phase. The integration of business management processes and RUP workflows with project management processes is highlighted to demonstrate the integrated nature of project management during the IT PDLC processes. The manage and integrate responsibilities of project management are presented as similar tasks to demonstrate the integrated skills needed by the program/project manager.

- *PSM management plan:* This plan is an integrated representation of the contents from business management–business architecture with the contents from RUP workflow management–business modeling, requirements, analysis and design, implementation, test, deployment, project management artifact sets (Create, manage/integrate, review, approve, support–scope management).
- *PTM management plan:* This plan is an integrated representation of the contents from business management–project deliverables approval with the contents from RUP workflow management–project management artifact set (Create, manage/integrate, review, approve, support–time management).
- *Project cost management plan:* This plan is an integrated representation of the contents from business management–project justifications and PFAs with the contents from RUP workflow management–project management artifact set (Create, manage/integrate, review, approve, support–cost management).
- *PQM management plan:* This plan is an integrated representation of the contents from business management–project deliverables approval with the contents from RUP workflow management–business modeling, requirements, analysis and design, implementation, test, deployment, project management artifact sets (Create, manage, review, approve, support–quality management).

- *Project change management plan:* This is one of the supportive plans for the core project management plans—cost, time, scope, and quality. It represents the integrated management of all the changes made to the core project management deliverables, with the contents from RUP workflow management–configuration and change management (Create, manage, review, approve, support–change management).
- *Project contract management plan:* This is one of the supportive plans for the core project management plans—cost, time, scope, and quality. It represents the integrated management of all the contractual arrangements to support the core project management deliverables, with the contents from RUP workflow management–project management artifact set (Create, manage, review, approve, support–contract management).
- *PIM management plan:* This is one of the supportive plans for the core project management plans—cost, time, scope, and quality. It represents the integrated management of all issues that affect the delivery of the core project management deliverables, with the contents from RUP workflow management–project management artifact set (Create, manage/integrate, review, approve, support–issue management).
- *PRM management plan:* This is one of the supportive plans for the core project management plans—cost, time, scope, and quality. It represents the integrated management of all risks that affect the delivery of the core project management deliverables, with the contents from RUP workflow management–project management artifact set (Create, manage/integrate, review, approve, support–risk management).
- *PCM plan:* This is one of the supportive plans for the core project management plans—cost, time, scope, and quality. It represents the integrated management of all communications to support the delivery of the core project management deliverables. The contents from business management–program steering committee process form an integral part of this communications management plan, with the contents from RUP workflow management–project management artifact set (Create, manage, review, approve, support–communications).
- *PSM management plan:* This is one of the supportive plans for the core project management plans—cost, time, scope, and quality. It represents the integrated management of the staffing plans to support the delivery of the core project management deliverables. The contents from business management–program steering committee process also form an integral part of this communications management plan, with the contents from RUP workflow management–project management artifact set (Create, manage/integrate, review, approve, support–human resources management).

7.7.1.3 Resources (who)

The labor resources needed to execute the activities to produce the deliverables are determined to show the type of project management resources responsible for the creation, management, review, approval, and support of the deliverables during the RUP project inception phase. It makes reference to the project managers roles, provided in Philippe Kruchten's book *The Rational Unified Process: An Introduction, Second Edition.*

- PSM management plan, requirements baseline, and reports (Create, manage/integrate, review, approve, support–scope management):
  - Program delivery manager/RUP program manager–create, manage;
  - Program business manager–support;
  - Program IT manager/RUP reviewer–support;
  - Executive business manager–review;
  - Executive PM manager/PM reviewer–review and approve;
  - Executive IT manager/RUP reviewer–review;
  - RUP project manager and team–support.
- PTM management plan, requirements baseline, and reports (Create, manage/integrate, review, approve, support–time management):
  - Program delivery manager/RUP program manager–create, manage;
  - Program business manager–support;
  - Program IT manager/RUP reviewer–support;
  - Executive business manager–review;
  - Executive PM manager/RUP reviewer–review and approve;
  - Executive IT manager/RUP reviewer–review;
  - RUP project manager and team–support.
- Project cost management plan, requirements baseline, and reports (Create, manage/integrate, review, approve, support–cost management):
  - Program delivery manager/RUP program manager–create, manage;
  - Program business manager–support;
  - Program IT manager/RUP reviewer–support;
  - Executive business manager–review;
  - Executive PM manager/RUP reviewer–review and approve;
  - Executive IT manager/RUP reviewer–review;
  - RUP project manager and team–support;
- PQM management plan, requirements baseline, and reports (Create, manage/integrate, review, approve, support–quality management):
  - Program delivery manager/RUP program manager–create, manage;
  - Program business manager–support;
  - Program IT manager/RUP reviewer–support;
  - Executive business manager–review;
  - Executive PM manager/RUP reviewer–review and approve;
  - Executive IT manager/RUP reviewer–review;
  - RUP project manager and team–support.

- Project change management plan, requirements baseline, and reports (Create, manage/integrate, review, approve, support–change management):
  - Program delivery manager/RUP program manager–create, manage;
  - Program business manager–support;
  - Program IT manager/RUP reviewer–support;
  - Executive business manager–review;
  - Executive PM manager/RUP reviewer–review and approve;
  - Executive IT manager/RUP reviewer–review;
  - RUP project manager and team–support.
- Project contract management plan, requirements baseline, and reports (Create, manage/integrate, review, approve, support–contract management):
  - Program delivery manager/RUP program manager–create, manage;
  - Program business manager–support;
  - Program IT manager/RUP reviewer–support;
  - Executive business manager–review;
  - Executive PM manager/RUP reviewer–review and approve;
  - Executive IT manager/RUP reviewer–review;
  - RUP project manager and team–support.
- PIM management plan, requirements baseline, and reports (Create, manage/integrate, review, approve, support–issue management):
  - Program delivery manager/RUP program manager–create, manage;
  - Program business manager–support;
  - Program IT manager/RUP reviewer–support;
  - Executive business manager–review;
  - Executive PM manager/RUP reviewer–review and approve;
  - Executive IT manager/RUP reviewer–review;
  - RUP project manager and team–support.
- PRM management plan, requirements baseline, and reports (Create, manage/integrate, review, approve, support–risk management):
  - Program delivery manager/RUP program manager–create, manage;
  - Program business manager–support;
  - Program IT manager/RUP reviewer–support;
  - Executive business manager–review;
  - Executive PM manager/RUP reviewer–review and approve;
  - Executive IT manager/RUP reviewer–review;
  - RUP project manager and team–support.
- PCM plan, requirements baseline, and reports (Create, manage/integrate, review, approve, support–communications):
  - Program delivery manager/RUP program manager–create, manage;
  - Program business manager–support;
  - Program IT manager/RUP reviewer–support;

- Executive business manager–review;
- Executive PM manager/RUP reviewer–review and approve;
- Executive IT manager/RUP reviewer–review;
- Project manager and team–support.

- Project human resources management plan, requirements baseline, and reports (Create, manage/integrate, review, approve, support–human resources management):
  - Program delivery manager/RUP program manager–create, manage;
  - Program business manager–support;
  - Program IT manager/RUP reviewer–support;
  - Executive business manager–review;
  - Executive PM manager/RUP reviewer–review and approve;
  - Executive IT manager–RUP reviewer–review;
  - RUP project manager and team–support.

7.7.2 Integration with business, RUP, and project management

Table 7.1 is an integrated representation of how the deliverable from the inception phase, Iteration #1, integrates with the deliverables (what), activities (how), and resources (who) from business, RUP workflow, and the PMBOK–project management component. The acronyms on the matrixes are abbreviations for executive managers, program managers, and project managers' responsibilities.

7.8 Elaboration phase: Iterations #2 and #3

High achievement always takes place in the framework of high expectation.
—Jack Kinder

During the elaboration phase,[10] the majority of the use cases are defined at a sufficient level of detail to establish a sound systems architectural design foundation. The project management plans, as defined in PMBOK, are refined for the entire project. In this phase, an understanding of the systems architecture specifications–product scope and the project management plans–project/work scope are reviewed and confirmed. This phase is the most critical of the four RUP phases because it established the foundation for both product scope and project/work scope. The focus is to gain a complete understanding of the business, data, applications, and technology architectural design and to determine the project/work scope, prior to committing to the construction and transition phases. This phase produces the executable prototype, Iterations #2 and #3, depending on the scope, costs,

10. Elaboration phase: This RUP phase plans the necessary activities and required resources by specifying the features and designing the architecture. The elaboration phase concludes by delivering the life-cycle architecture (LCA) milestone.

Table 7.1 Integrated Business, RUP Inception, and PM Responsibility Matrix[11]

	Roles						
Deliverables	*EBM*	*EPM RUP*	*EITM RUP*	*PBM*	*PDM RUP*	*PITM RUP*	*PM/Team RUP*
Business							
Project economics	RA	R	R	C	M/I	S	S
PFAs	RA	R	R	C	M/I	S	S
Project funding approvals	RA	R	R	C	M/I	S	S
Project organizational structure	RA	R	R	C	M/I	S	S
Project business architecture	RA	R	R	C	M/I	S	S
Project business support initiatives	RA	R	R	C	M/I	S	S
RUP workflows							
Business modeling artifact set	RA	R	R	C	M/I	S	S
Requirements artifact set	RA	R	R	S	M/I	S	C
Analysis/design artifact set	R	R	RA	S	M/I	S	C
Implementation artifact set	R	R	RA	S	M/I	S	C
Test artifact set	RA	R	R	S	M/I	S	C
Deployment artifact set	RA	R	R	S	M/I	S	C
Configuration and change management artifact set	R	RA	R	S	C	S	S
Project management artifact set	R	RA	R	S	C	S	S
Environment artifact set	R	RA	R	S	C	S	S
Project management							
Scope management plan	R	RA	R	S	CM	S	S
Time management plan	R	RA	R	S	CM	S	S
Cost management plan	R	RA	R	S	CM	S	S

11. Responsibility matrixes: A series of matrixes that show who is responsible for what in terms of creating/updating, managing/integrating, reviewing/assuring quality, approving/committing, and supporting responsibility classifications and that show the integrated nature of business management, RUP management, and project management for each phase of the project.

Table 7.1 (continued)

	Roles						
Deliverables	*EBM*	*EPM RUP*	*EITM RUP*	*PBM*	*PDM RUP*	*PITM RUP*	*PM/Team RUP*
Quality management plan	R	RA	R	S	CM	S	S
Contract management plan	R	RA	R	S	CM	S	S
Issue management plan	R	RA	R	S	CM	S	S
Change management plan	R	RA	R	S	CM	S	S
Risk management plan	R	RA	R	S	CM	S	S
Communications management plan	R	RA	R	S	CM	S	S
Human resources management plan	R	RA	R	S	CM	S	S

Note: C = create/update; R = review; A = approve; M/I = manage/integrate; S = support.
The shaded areas represent the major project management deliverables produced during this RUP phase.

time, quality, and risk factors. It also produces management, requirements, design, implementation, and deployment artifact sets that are incrementally iterated as the project progresses throughout the RUP-PMBOK delivery life cycle model or phases. Iterations # 2 and #3 produce one or several executable prototypes and supporting models and reports as the major deliverables of this phase.

The elaboration phase forms the basis for the following:

- The second major project baseline: Iterations #2 and #3;
- An executable architectural prototype to decide whether or not to commit to the construction and transition phases;
- Updated business, RUP workflow–IT, and project management plans and deliverables;
- Systems design/architecture foundation for the next two project phases;
- Approval and commitment to proceed with construction and transition, Iterations #4 . . . *n*.

During this phase, the business management, RUP workflow–IT management, and project management deliverables of Iteration #1 are updated, and the executable architectural prototype and supporting models and reports are produced, based on Iterations #2 and #3. This is the software baseline solution, which may be refined as the project progresses throughout the PDLC. In most cases, this executable architectural prototype software solution baseline will form the basis for the development of the production quality evolutionary prototype.

The deliverables of this elaboration phase focus mainly on updates to the business management, project management, and RUP support workflow artifact set deliverables developed during the inception phase. The RUP core workflow deliverables–analysis and design artifact set will be defined during this phase to be further refined as the project progresses throughout the RUP iterative PDLC.

Iterations #2 and #3 entail one or several executable architectural prototypes and supporting documents, which include updates to the following:

- Business management deliverables:
  - BSA;
  - Project justification (benefits and priorities);
  - PFAs;
  - Project deliverables/funding approvals;
  - Program steering and working committee;
  - Project business support initiatives.
- PMBOK–project management deliverables:
  - PSM management plan, requirements baseline, and reports:
    - Business requirements;
    - Data requirements;
    - Application requirements;
    - Technology requirements;
    - Scope management plan and reports.
  - PTM management plan, schedule baseline, and reports;
  - Project cost management plan, budget baseline, and reports;
  - PQM management plan, quality baseline, and reports;
  - Project change management plan, change requests, and reports;
  - Project contract management plan, contract, and reports;
  - PIM management plan, issues, and reports;
  - PRM management plan, potential risks, and reports;
  - PCM plan, progress status, and reports;
  - Project human resources–staffing plan, RAM, and reports;
  - PMO support services.
- RUP workflow management deliverables:
  - Business modeling artifact set:
    - Logical business view–business model;
    - Use-case view–business use-case model.
  - Requirements artifact set:
    - Logical system view–systems model;
    - Use-case view–system requirements use-case model.
  - Analysis and design artifact set:
    - Process view–design model;
    - Use-case view–system design use-case model.
  - Implementation artifact set:
    - Implementation view–implementation model;

Use-case view–system implementation use-case model.

- Test artifact set:

  Implementation testing view–testing model;

  Use-case view–system testing use-case model.

- Deployment artifact set:

  Deployment view–deployment model;

  Use-case view–deployment use-case model.

- Environment artifact set: management view–project work model;
- Project management artifact set: management view–project work model;
- Configuration and change management artifact set: management view–project work model.

The elaboration phase starts as a result of completion of the business management and project management deliverables and RUP supporting deliverables–project management workflows, configuration and change management workflows, and environment workflows. This phase delivers RUP analysis and design artifact set and an executable architectural prototype. It formally concludes with the project's first major software solution baseline—the executable architectural prototype as a result of Iterations #2 and #3. This review results in approval of the business, RUP workflow, and project management deliverables and formally declares the executable architectural prototype as the deciding factor in determining whether or not to commit to the construction and transition phases.

7.8.1 RUP alignment with PMBOK: project management

The primary objective of this section is to demonstrate how the main deliverable of the RUP elaboration phase, Iterations #2 and #3, aligns with the deliverables (what) from the PMBOK–project management component, RUP workflow components, and business management components. The roles and responsibilities from the project management processes, presented in Chapter 4, are summarized to determine the major activities (how) needed to produce the deliverables and the people resources (who) responsible to create/update, manage/integrate, review/test, approve/commit, and support/maintain the deliverables.

7.8.1.1 Deliverables (what)

This section highlights the contents of Iterations #2 and #3—executable architectural prototype and supporting analysis and design artifacts set in terms of business management, RUP workflows, and project management deliverables. Chapter 4 provides detailed definitions, templates, process flows, and quality measures for the creation, management, review, approval, and support of these deliverables. The project management deliverables are replicated in this section to provide readers with the necessary continuous and logical flow of information to enable a more analytical understanding of the deliverables produced during RUP–elaboration phase.

PMBOK–project management deliverables include the following:

- PSM management plan and requirements baseline updates:
  - Business requirements;
  - Data requirements;
  - Application requirements;
  - Technology requirements;
  - Scope management plan, and reports updates.
- PTM management plan and schedule baseline updates;
- Project cost management plan, budget baseline, and reports updates:
  - PQM management plan, quality baseline, and reports updates;
  - Project change management plan, change requests, and reports updates;
  - Project contract management plan, contract, and reports updates;
  - PIM management plan, issues, and reports updates;
  - PRM management plan, potential risks, and reports updates;
  - PCM plan and progress status updates;
  - Project human resources–staffing plan, RAM, and reports updates;
  - PMO support services.

7.8.1.2 Activities (how)

The roles and responsibilities of the project management component outlined in Chapter 4 are classified based on the creation/updating, management/integration, review/QA, approval/commitment, and support/maintenance responsibility classifications to determine the major activities required to produce the deliverables during the RUP elaboration phase. The integration of business management processes and RUP workflows with project management processes is highlighted to demonstrate the integrated nature of project management during the IT PDLC processes. The manage and integrate responsibilities of project management are presented as similar tasks to demonstrate the integrated skills needed by the program/project manager.

- *PSM management plan, requirements baseline, and reports updates:* This plan is an integrated representation of the contents from business management–business architecture with the contents from RUP workflow management–business modeling, requirements, analysis and design, implementation, test, deployment, and project management artifact sets (Update, integrate, review, approve, support–scope management).
- *PTM management plan, schedule baseline, and reports updates:* This plan is an integrated representation of the contents from business management–project deliverables approval with the contents from RUP

workflow management–project management artifact set (Update, integrate, review, approve, support–time management).

- *Project cost management plan, budget baseline, and reports updates:* This plan is an integrated representation of the contents from business management–project justifications and PFAs with the contents from RUP workflow management–project management artifact set (Update, integrate, review, approve, support–cost management).
- *PQM management plan, quality baseline, and reports updates:* This plan is an integrated representation of the contents from business management–project deliverables approval with the contents from RUP workflow management–business modeling, requirements, analysis and design, implementation, test, deployment, and project management artifact sets (Update, integrate, review, approve, support–quality management).
- *Project change management plan, change requests, and reports updates:* This is one of the supportive plans for the core project management plans—cost, time, scope, and quality. It represents the integrated management of all the changes made to the core project management deliverables, with the contents from RUP workflow management–configuration and change management (Update, integrate, review, approve, support–change management).
- *Project contract management plan, contracts, and reports:* This is one of the supportive plans for the core project management plans—cost, time, scope, and quality. It represents the integrated management of all the contractual arrangements to support the core project management deliverables with the contents from RUP workflow management–project management artifact (Update, integrate, review, approve, support–contract management).
- *PIM management plan, issues, and reports updates:* This is one of the supportive plans for the core project management plans—cost, time, scope, and quality. It represents the integrated management of all issues that affect the delivery of the core project management deliverables, with the contents from RUP workflow management–project management artifact set (Update, integrate, review, approve, support–issues management).
- *PRM management plan, risk, and reports updates:* This is one of the supportive plans for the core project management plans—cost, time, scope, and quality. It represents the integrated management of all risks that affect the delivery of the core project management deliverables, with the contents from RUP workflow management–project management artifact set (Update, integrate, review, approve, support–risk management).
- *PCM plan, progress status updates:* This is one of the supportive plans for the core project management plans—cost, time, scope, and quality. It represents the integrated management of all communications to support the

delivery of the core project management deliverables. The contents from business management–program steering committee process form an integral part of this communications management plan with the contents from RUP workflow management–project management artifact set (Update, integrate, review, approve, support–communications).

- *PSM management plan, RAM, and reports updates:* This is one of the supportive plans for the core project management plans—cost, time, scope, and quality. It represents the integrated management of the staffing plans to support the delivery of the core project management deliverables. The contents from the business management–program steering committee process also form an integral part of this communications management plan with the contents from RUP workflow management–project management artifact set (Update, integrate, review, approve, support–human resources management).

7.8.1.3 Resources (who)

The labor resources needed to execute the activities to produce the deliverables are determined to show the type of project management resources responsible for the updating, management/integration, review/QA, approval/commitment, and support of the deliverables during the RUP project elaboration phase. It makes reference to the project manager roles, provided in Philippe Kruchten's book *The Rational Unified Process: An Introduction, Second Edition.*

- PSM management plan, requirements baseline, and reports updates (Create, manage/integrate, review, approve, support–scope management):
  - Program delivery manager/RUP program manager–update/integrate;
  - Program business manager–support;
  - Program IT manager/RUP reviewer–support;
  - Executive business manager–review;
  - Executive PM manager/RUP reviewer–review and approve;
  - Executive IT manager–RUP reviewer–review;
  - RUP project manager and team–support.
- PTM management plan, requirements baseline, and reports (Create, manage/integrate, review, approve, support–time management):
  - Program delivery manager/RUP program manager–update/integrate;
  - Program business manager–support;
  - Program IT manager/RUP reviewer–support;
  - Executive business manager–review;
  - Executive PM manager/RUP reviewer–review and approve;
  - Executive IT manager–RUP reviewer–review;
  - RUP project manager and team–support.

- Project cost management plan, requirements baseline, and reports (Create, manage/integrate, review, approve, support–cost management):
  - Program delivery manager/RUP program manager–update/integrate;
  - Program business manager–support;
  - Program IT manager/RUP reviewer–support;
  - Executive business manager–review;
  - Executive PM manager/RUP reviewer–review and approve;
  - Executive IT manager–RUP reviewer–review;
  - RUP project manager and team–support.
- PQM management plan, requirements baseline, and reports (Create, manage/integrate, review, approve, support–quality management)
  - Program delivery manager/RUP program manager–update/integrate;
  - Program business manager–support;
  - Program IT manager/RUP reviewer–support;
  - Executive business manager–review;
  - Executive PM manager/RUP reviewer–review and approve;
  - Executive IT manager–RUP reviewer–review;
  - RUP project manager and team–support.
- Project change management plan, requirements baseline, and reports (Create, manage/integrate, review, approve, support–change management):
  - Program delivery manager/RUP program manager–update/integrate;
  - Program business manager–support;
  - Program IT manager/RUP reviewer–support;
  - Executive business manager–review;
  - Executive PM manager/RUP reviewer–review and approve;
  - Executive IT manager–RUP reviewer–review;
  - RUP project manager and team–support.
- Project contract management plan, requirements baseline, and reports (Create, manage/integrate, review, approve, support–contract management):
  - Program delivery manager/RUP program manager–update/integrate;
  - Program business manager–support;
  - Program IT manager/RUP reviewer–support;
  - Executive business manager–review;
  - Executive PM manager/RUP reviewer–review and approve;
  - Executive IT manager–RUP reviewer–review;
  - RUP project manager and team–support.
- PIM management plan, requirements baseline, and reports (Create, manage/integrate, review, approve, support–issue management):
  - Program delivery manager/RUP program manager–update/integrate;
  - Program business manager–support;
  - Program IT manager/RUP reviewer–support;

- Executive business manager–review;
- Executive PM manager/RUP reviewer–review and approve;
- Executive IT manager–RUP reviewer–review;
- RUP project manager and team–support.

- PRM management plan, requirements baseline, and reports (Create, manage/integrate, review, approve, support–risk management):
  - Program delivery manager/RUP program manager–update/integrate;
  - Program business manager–support;
  - Program IT manager/RUP reviewer–support;
  - Executive business manager–review;
  - Executive PM manager/RUP reviewer–review and approve
  - Executive IT manager–RUP reviewer–review
  - RUP project manager and team–support
- PCM plan, requirements baseline, and reports (Create, manage/integrate, review, approve, support–communications):
  - Program delivery manager/RUP program manager–update/integrate;
  - Program business manager–support;
  - Program IT manager/RUP reviewer–support;
  - Executive business manager–review;
  - Executive PM manager/RUP reviewer–review and approve;
  - Executive IT manager–RUP reviewer–review;
  - RUP project manager and team–support.
- Project human resources management plan, requirements baseline, and reports (Create, manage/integrate, review, approve, support–human resources management):
  - Program delivery manager/RUP program manager–update/integrate;
  - Program business manager–support;
  - Program IT manager/RUP reviewer–support;
  - Executive business manager–review;
  - Executive PM manager/RUP reviewer–review and approve;
  - Executive IT manager–RUP reviewer–review;
  - RUP project manager and team–support.

7.8.2 Integration with business, RUP, and project management

Table 7.2 is an integrated representation of how the deliverable from the elaboration phase, Iterations #2 and #3, integrate with the deliverables (what), activities (how), and resources (who) from business, RUP workflow, and PMBOK–project management components. The acronyms on the matrixes are abbreviations for executive managers, program managers, and project managers' responsibilities.

Table 7.2 Integrated Business, RUP Elaboration, and PM Responsibility Matrix

	Roles						
Deliverables	*EBM*	*EPM RUP*	*EITM RUP*	*PBM*	*PDM RUP*	*PITM RUP*	*PM/Team RUP*
Business							
Project economics	RA	R	R	U	M/I	S	S
PFAs	RA	R	R	U	M/I	S	S
Project funding approvals	RA	R	R	U	M/I	S	S
Project organizational structure	RA	R	R	U	M/I	S	S
Project business architecture	RA	R	R	U	M/I	S	S
Project business support initiatives	RA	R	R	U	M/I	S	S
RUP workflows							
Business modeling artifact set	RA	R	R	S	M/I	S	U
Requirements artifact set	RA	R	R	S	M/I	S	U
Analysis/design artifact set	R	R	RA	S	M/I	S	C
Implementation artifact set	R	R	RA	S	M/I	S	C
Test artifact set	RA	R	R	S	M/I	S	C
Deployment artifact set	RA	R	R	S	M/I	S	C
Configuration and change management artifact set	R	RA	R	S	U	S	S
Project management artifact set	R	RA	R	S	U	S	S
Environment artifact set	R	RA	R	S	U	S	S
Project management							
Scope management plan	R	RA	R	S	UI	S	S
Time management plan	R	RA	R	S	UI	S	S
Cost management plan	R	RA	R	S	UI	S	S
Quality management plan	R	RA	R	S	UI	S	S
Contract management plan	R	RA	R	S	UI	S	S
Issue management plan	R	RA	R	S	UI	S	S

Table 7.2 (continued)

	Roles						
Deliverables	*EBM*	*EPM RUP*	*EITM RUP*	*PBM*	*PDM RUP*	*PITM RUP*	*PM/Team RUP*
Change management plan	R	RA	R	S	UI	S	S
Risk management plan	R	RA	R	S	UI	S	S
Communications management plan	R	RA	R	S	UI	S	S
Human Resources management plan	R	RA	R	S	UI	S	S

Note: U = update; R = review; A = approve; M/I = manage/integrate; S = support.
The shaded areas represent the major deliverables produced during this RUP phase.

7.9 Construction phase: Iterations #4, #5, and #6

A good system shortens the road to the goal.
—Orison Swett Marden

During the construction phase,[12] all remaining components and additional features of the executable prototype are further developed and tested, then migrated from the architectural baseline to a first release of the operational software product, ready for use by the business users. The project management plans, as defined in PMBOK, are updated to reflect all construction deliverables and activities for the entire project. In this phase, a decision is made as to whether or not the system can be fully operational by assessing the risks and determining the impact to the organization, business processes, technology, training, and software operational support. The focus is to gain a complete understanding of the business and technology environment without exposing the project to high risks prior to committing to the transition phase. This phase produces the first release of the operational system, Iterations #4, #5, and #6, depending on the scope, costs, time, quality, and risk factors. It also produces management, requirements, design, implementation, and deployment artifact sets, which are incrementally iterated as the project progresses throughout the RUP-PMBOK delivery life-cycle model. Iterations # 4, #5, and #6 produce one or several operational releases and supporting models and reports as the major deliverables of this phase.

The construction phase forms the basis for the following:

- The third major project baseline: Iterations #4, #5, and #6;
- A first release of the operational system to decide whether or not to commit to the transition phase;
- Updated business, RUP workflow–IT, and project management plans and deliverables;

12. Construction phase: This RUP phase builds the product and evolves the vision, the architecture, and the plans until the product—the completed vision—is ready for delivery to the user community. The construction phase concludes by delivering the initial operational capability (IOC) milestone.

- Operational system foundation for the transition phase;
- Approval and commitment to proceed with the transition phase: Iterations #7 and #8.

During this phase, the business management, RUP workflow management, and project management deliverables of Iterations #2 and #3 are updated, and the first release of the operational system and supporting models and reports are produced during Iterations #4, #5, and #6. This is the operational system to be deployed as the project progresses to production readiness or production transition.

The deliverables of this construction phase focus mainly on updates to the business management, project management, and RUP workflow artifact set developed during the elaboration phase. The RUP core workflow deliverables–implementation and testing artifact set will be defined during this phase and further refined as the project progresses through the RUP iterative PDLC. Iterations #4, #5, and #6 will include further updates to the following:

Iterations #4, #5, and #6 entail one or several operational system releases and supporting updated models, plans, and reports for business management, project management, and RUP workflow management.

- Business management deliverables:
  - BSA;
  - Project justification (benefits and priorities);
  - PFAs;
  - Project deliverables/funding approvals;
  - Program steering and working committee;
  - Project business support initiatives.
- PMBOK–project management deliverables:
  - PSM management plan, requirements baseline, and reports:
    Business requirements;
    Data requirements;
    Application requirements;
    Technology requirements;
    Scope management plan and reports.
  - PTM management plan, schedule baseline, and reports;
  - Project cost management plan, budget baseline, and reports;
  - PQM management plan, quality baseline, and reports;
  - Project change management plan, change requests, and reports;
  - Project contract management plan, contract, and reports;
  - PIM management plan, issues, and reports;
  - PRM management plan, potential risks, and reports;
  - PCM plan, progress status, and reports;
  - Project human resources–staffing plan, RAM, and reports;
  - PMO support services.

- RUP workflow management deliverables:
  - Business modeling artifact set:
    Logical business view–business model;
    Use-case view–business use–case model.
  - Requirements artifact set:
    Logical system view–systems model;
    Use-case view–system requirements use-case model.
  - Analysis and design artifact set:
    Process view–design model;
    Use-case view–system design use-case model.
  - Implementation artifact set:
    Implementation view–implementation model;
    Use-case view–system implementation use-case model.
  - Test artifact set:
    Implementation testing view–testing model;
    Use-case view–system testing use-case model.
  - Deployment artifact set:
    Deployment view–deployment model;
    Use-case view–deployment use-case model.
  - Environment artifact set: management view–project work model;
  - Project management artifact set: management view–project work model;
  - Configuration and change management artifact set: management view–project work model.

The construction phase starts as a result of completion of the business management, project management, and RUP core supporting workflow deliverables–project management, configuration and change management, and environment workflows. This phase delivers RUP implementation and testing artifact set and an operational software product. It formally concludes with the project's first release of the operational software product as a result of Iterations #4, #5, and #6. This review results in approval of the business management, RUP workflow, and project management deliverables and formally declares the first release of the operational system, as the deciding factor in determining whether or not to commit to the final transition phase.

7.9.1 Alignment with PMBOK: project management

The primary objective of this section is to demonstrate how the main deliverables of the RUP construction phase, Iterations #4, #5, and #6, align with the deliverables (what) from the PMBOK–project management component, RUP workflow components, and business management components. The roles and responsibilities from the project management processes, presented in Chapter 4, are summarized to determine the major activities (how) needed to produce the deliverables and the people resources (who) responsible to creating/update, reviewing/QA, approving/committing, and supporting the deliverables.

7.9.1.1 Deliverables (what)

This section highlights the contents of Iterations #4, #5, and #6–operational system, and supporting implementation and testing artifacts set for business management, RUP Workflows, and project management deliverables. Chapter 4 provides detailed definitions, templates, process flows, and quality measures for the creation, management, approval, and review of these deliverables. The project management deliverables are replicated in this section to provide readers with the necessary continuous and logical flow of information to enable a more analytical understanding of the deliverables produced during RUP–construction phase.

PMBOK–project management deliverables include the following:

- PSM management plan and requirements baseline updates:
  - Business requirements;
  - Data requirements;
  - Application requirements;
  - Technology requirements;
  - Scope management plan and reports updates.
- PTM management plan and schedule baseline updates;
- Project cost management plan, budget baseline, and reports updates;
- PQM management plan, quality baseline, and reports updates;
- Project change management plan, change requests, and reports updates;
- Project contract management plan, contract, and reports updates;
- PIM management plan, issues, and reports updates;
- PRM management plan, potential risks, and reports updates;
- PCM plan, progress status updates;
- Project human resources–staffing plan, RAM, and reports updates;
- PMO support services.

7.9.1.2 Activities (how)

The roles and responsibilities of the project management component outlined in Chapter 4 are classified based on the creation/updating, management/integration, review/QA, approval/commitment, and support responsibility classifications to determine the major activities required to produce the deliverables during the RUP construction phase. The integration of business management processes and RUP workflows with project management processes is highlighted to demonstrate the integrated nature of project management during the RUP-PMBOK PDLC processes. The manage and integrate responsibilities of project management are presented as similar tasks to demonstrate the integrated skills needed by the program/project manager.

- *PSM management plan, requirements baseline, and reports updates:* This plan is an integrated representation of the contents from business management–business architecture with the contents from RUP workflow management–business modeling, requirements, analysis and design, implementation, test, deployment, and project management artifact sets (Update, integrate, review, approve, support–scope management).
- *PTM management plan, schedule baseline, and reports updates:* This plan is an integrated representation of the contents from business management–project deliverables approval with the contents from RUP workflow management–project management artifact set (Update, integrate, review, approve, support–time management).
- *Project cost management plan, budget baseline, and reports updates:* This plan is an integrated representation of the contents from business management–project justifications and PFAs with the contents from RUP workflow management–project management artifact set (Update, integrate, review, approve, support–cost management).
- *PQM management plan, quality baseline, and reports updates:* This plan is an integrated representation of the contents from business management–project deliverables approval with the contents from RUP workflow management—business modeling, requirements, analysis and design, implementation, test, deployment, and project management artifact sets (Update, integrate, review, approve, support–quality management).
- *Project change management plan, change requests, and reports updates:* This is one of the supportive plans for the core project management plans—cost, time, scope, and quality. It represents the integrated management of all the changes made to the core project management deliverables, with the contents from RUP workflow management–configuration and change management artifact sets (Update, integrate, review, approve, support–change management).
- *Project contract management plan, contracts, and reports:* This is one of the supportive plans for the core project management plans—cost, time, scope, and quality. It represents the integrated management of all the contractual arrangements to support the core project management deliverables, with the contents from RUP workflow management–project management artifact set (Update, integrate, review, approve, support–contract management).
- *PIM management plan, issues, and reports updates:* This is one of the supportive plans for the core project management plans—cost, time, scope, and quality. It represents the integrated management of all issues that affect the delivery of the core project management deliverables with the contents from RUP workflow management–project management artifact set (Update, integrate, review, approve, support–issues management).

- *PRM management plan, risk, and reports updates:* This is one of the supportive plans for the core project management plans—cost, time, scope, and quality. It represents the integrated management of all risks that affect the delivery of the core project management deliverables, with the contents from RUP workflow management–project management artifact set (Update, integrate, review, approve, support–issues management).
- *PCM plan and progress status updates:* This is one of the supportive plans for the core project management plans—cost, time, scope, and quality. It represents the integrated management of all communications to support the delivery of the core project management deliverables. The contents from business management–program steering committee process form an integral part of this communications management plan with the contents from RUP workflow management–project management artifact set (Update, integrate, review, approve, support–communications).
- *PSM management plan, RAM, and reports updates:* This is one of the supportive plans for the core project management plans—cost, time, scope, and quality. It represents the integrated management of the staffing plans to support the delivery of the core project management deliverables. The contents from business management–program steering committee process also form an integral part of this communications management plan, with the contents from RUP workflow management–project management artifact set (Update, integrate, review, approve, support–human resources management).

7.9.1.3 Resources (who)

The labor resources needed to execute the activities to produce the deliverables are determined to show the type of project management resources responsible for the updating, management/integration, review/QA, and approval/commitment of the deliverables during the RUP project construction phase. It makes reference to the project managers roles, provided in Philippe Kruchten's book *The Rational Unified Process: An Introduction, Second Edition.*

- PSM management plan, requirements baseline, and reports updates (Create, manage/integrate, review, approve, support–scope management):
  - Program delivery manager/RUP program manager–update/ integrate;
  - Program business manager–support;
  - Program IT manager/RUP reviewer–support;
  - Executive business manager–review;
  - Executive PM manager/RUP reviewer–review and approve;
  - Executive IT manager/RUP reviewer–review;
  - RUP project manager and team–support.

- PTM management plan, requirements baseline, and reports (Create, manage/integrate, review, approve, support–time management):
  - Program delivery manager/RUP program manager–update/ integrate;
  - Program business manager–support;
  - Program IT manager/RUP reviewer–support;
  - Executive business manager–review;
  - Executive PM manager/RUP reviewer–review and approve;
  - Executive IT manager/RUP reviewer–review;
  - RUP project manager and team–support.
- Project cost management plan, requirements baseline, and reports (Create, manage/integrate, review, approve, support–cost management):
  - Program delivery manager/RUP program manager–update/ integrate;
  - Program business manager–support;
  - Program IT manager/RUP reviewer–support;
  - Executive business manager–review;
  - Executive PM manager/RUP reviewer–review and approve;
  - Executive IT manager/RUP reviewer–review;
  - RUP project manager and team–support.
- PQM management plan, requirements baseline, and reports (Create, manage/integrate, review, approve, support–quality management):
  - Program delivery manager/RUP program manager–update/ integrate;
  - Program business manager–support;
  - Program IT manager/RUP reviewer–support;
  - Executive business manager–review;
  - Executive PM manager/RUP reviewer–review and approve;
  - Executive IT manager/RUP reviewer–review;
  - RUP project manager and team–support.
- Project change management plan, requirements baseline, and reports (Create, manage/integrate, review, approve, support–change management):
  - Program delivery manager/RUP program manager–update/ integrate;
  - Program business manager–support;
  - Program IT manager/RUP reviewer–support;
  - Executive business manager–review;
  - Executive PM manager/RUP reviewer–review and approve;
  - Executive IT manager/RUP reviewer–review;
  - RUP project manager and team–support.
- Project contract management plan, requirements baseline, and reports (Create, manage/integrate, review, approve, support–contract management):
  - Program delivery manager/RUP program manager–update/ integrate;
  - Program business manager–support;

  - Program IT manager/RUP reviewer–support;
  - Executive business manager–review;
  - Executive PM manager/RUP reviewer–review and approve;
  - Executive IT manager/RUP reviewer–review;
  - RUP project manager and team–support.
- PIM management plan, requirements baseline, and reports (Create, manage/integrate, review, approve, support–issue management):
  - Program delivery manager/RUP program manager–update/ integrate;
  - Program business manager–support;
  - Program IT manager/RUP reviewer–support;
  - Executive business manager–review;
  - Executive PM manager/RUP reviewer–review and approve;
  - Executive IT manager/RUP reviewer–review;
  - RUP project manager and team–support.
- PRM management plan, requirements baseline, and reports (Create, manage/integrate, review, approve, support–risk management):
  - Program delivery manager/RUP program manager–update/ integrate;
  - Program business manager–support;
  - Program IT manager/RUP reviewer–support;
  - Executive business manager–review;
  - Executive PM manager/RUP reviewer–review and approve;
  - Executive IT manager/RUP reviewer–review;
  - RUP project manager and team–support.
- PCM plan, requirements baseline, and reports (Create, manage/integrate, review, approve, support–communications):
  - Program delivery manager/RUP program manager–update/ integrate;
  - Program business manager–support;
  - Program IT manager/RUP reviewer–support;
  - Executive business manager–review;
  - Executive PM manager/RUP reviewer–review and approve;
  - Executive IT manager/RUP reviewer–review;
  - RUP project manager and team–support.
- Project human resources management plan, requirements baseline, and reports (Create, manage/integrate, review, approve, support–human resources management):
  - Program delivery manager/RUP program manager–update/ integrate;
  - Program business manager–support;
  - Program IT manager/RUP reviewer–support;
  - Executive business manager–review;
  - Executive PM manager/RUP reviewer–review and approve;
  - Executive IT manager/RUP reviewer–review;
  - RUP project manager and team–support.

Table 7.3 Integrated Business, RUP Construction, and PM Responsibility Matrix

Deliverables	*Roles* EBM	EPM RUP	EITM RUP	PBM	PDM RUP	PITM RUP	PM/Team RUP
Business							
Project economics	RA	R	R	U	M/I	S	S
PFAs	RA	R	R	U	M/I	S	S
Project funding approvals	RA	R	R	U	M/I	S	S
Project organizational structure	RA	R	R	U	M/I	S	S
Project business architecture	RA	R	R	U	M/I	S	S
Project business support initiatives	RA	R	R	U	M/I	S	S
RUP workflows							
Business modeling artifact set	RA	R	R	S	M/I	S	U
Requirements artifact set	RA	R	R	S	M/I	S	U
Analysis/design artifact set	R	R	RA	S	M/I	S	C
Implementation artifact set	R	R	RA	S	M/I	S	C
Test artifact set	RA	R	R	S	M/I	S	C
Deployment artifact set	RA	R	R	S	M/I	S	C
Configuration and change management artifact set	R	RA	R	S	U	S	S
Project management artifact set	R	RA	R	S	U	S	S
Environment artifact set	R	RA	R	S	U	S	S
Project management							
Scope management plan	R	RA	R	S	U	S	S
Time management plan	R	RA	R	S	U	S	S
Cost management plan	R	RA	R	S	U	S	S
Quality management plan	R	RA	R	S	U	S	S
Contract management plan	R	RA	R	S	U	S	S
Issue management plan	R	RA	R	S	U	S	S
Change management plan	R	RA	R	S	U	S	S

Table 7.3 (continued)

	Roles						
Deliverables	*EBM*	*EPM RUP*	*EITM RUP*	*PBM*	*PDM RUP*	*PITM RUP*	*PM/Team RUP*
Risk management plan	R	RA	R	S	U	S	S
Communications management plan	R	RA	R	S	U	S	S
Human resources management plan	R	RA	R	S	U	S	S

Note: C = create; U = update; R = review; A = approve; M/I = manage/integrate; S = support. The shaded areas represent the major deliverables produced during this RUP phase.

7.9.2 Integration with business, RUP, and project management

Table 7.3 is an integrated representation of how the deliverable from the construction phase, Iterations #4, #5, and #6, integrate with the deliverables (what), activities (how), and resources (who) from business, RUP workflow, and PMBOK–project management component.

7.10 Transition phase: iterations #7 and #8

Build a system that even a fool will use, and only a fool will want to use it.
—Murphy's Law

During the transition phase,[13] the constructed operational system is deployed through the execution of parallel testing, database conversion, training, production roll-out and stakeholders' acceptance. The final product release of the operational software product is accepted and ready for production. The project management plans, as defined in PMBOK, are updated to reflect all transition deliverables and activities for the project to ensure that the actual resources expenditures versus planned expenditures are still acceptable. In this phase, a decision is made as to whether or not the system objectives were effectively and efficiently met by assessing the deployment baselines risks and determining the impact to the organization, business processes, technology, training, and software operational support. The focus is to achieve stakeholders' concurrence on the final product baseline without exposing the project to high risks prior to committing to the final production release. This phase produces the final product release milestone of the operational system, Iterations #7 and #8, depending on the scope, costs, time, quality, and risk factors. It also produces management, requirements, design, implementation, and deployment artifact sets, which are deployed to production and postimplementation or warranty support. Iterations #7 and #8

13. Transition phase: This RUP phase transitions the product to its users, which includes manufacturing, delivering, training, supporting, and maintaining the product until users are satisfied. The transition phase concludes by the delivering the product release milestone, which also concludes the cycle.

produce one or several production operational releases and supporting models and reports as the major deliverables during project completion.

The transition phase forms the basis for the following:

- The fourth major project baseline: Iterations #7 and #8;
- The final release of the operational system to decide whether or not the project objectives were met and the stakeholders agree that the deployment baselines are complete and consistent with the evaluation criteria;
- Updated business, RUP workflow–IT, and project management plans and deliverables;
- Final production release of the software product;
- Approval and commitment to meeting stakeholders' requirements for project completion.

In this phase, the business management, RUP workflow management, and project management deliverables of Iterations #4, #5, and #6 are updated and the final release of the operational system and supporting models and reports are produced during Iterations #7 and #8. This is the final release of the operational system for production as the project progresses to completion.

The deliverables of this transition phase focus mainly on updates to the business management, project management, and RUP workflow artifact set, developed during the construction phase. The RUP core workflow deliverables–deployment artifact set will be defined during this phase, and other RUP workflow artifact sets will be refined as the project progresses through to completion. Iterations #7 and #8 will include further refinements to the operational software product produced from Iterations #4, #5, and #6 and further updates to the models, plans, and reports.

Iterations #7 and #8 entail one or several production software product releases and supporting updated models, plans, and reports for business management, project management, and RUP workflow management.

- Business management deliverables:
  - BSA;
  - Project justification (benefits and priorities);
  - PFAs;
  - Project deliverables/funding approvals;
  - Program steering and working committee;
  - Project business support initiatives.
- PMBOK–project management deliverables:
  - PSM management plan, requirements baseline, and reports:
    - Business requirements;
    - Data requirements;
    - Application requirements;
    - Technology requirements;

    Scope management plan and reports.
  - PTM management plan, schedule baseline, and reports;
  - Project cost management plan, budget baseline, and reports;
  - PQM management plan, quality baseline, and reports;
  - Project change management plan, change requests, and reports;
  - Project contract management plan, contract, and reports;
  - PIM management plan, issues, and reports;
  - PRM management plan, potential risks, and reports;
  - PCM plan, progress status, and reports;
  - Project human resources–staffing plan, RAM, and reports;
  - PMO support services.
- RUP workflow management deliverables:
  - Business modeling artifact set:
    Logical business view–business model;
    Use-case view–business use-case model.
  - Requirements artifact set:
    Logical system view–systems model;
    Use-case view–system requirements use-case model.
  - Analysis and design artifact set:
    Process view–design model;
    Use-case view–system design use-case model.
  - Implementation artifact set:
    Implementation view–implementation model;
    Use-case view–system implementation use-case model.
  - Test artifact set:
    Implementation testing view–testing model;
    Use-case view–system testing use-case model.
  - Deployment artifact set:
    Deployment view–deployment model;
    Use-case view–deployment use-case model.
  - Environment artifact set: management view–project work model;
  - Project management artifact set: management view–project work model;
  - Configuration and change management artifact set: management view–project work model.

The transition phase starts as a result of the completion of Iterations #4, #5, and #6, which include business management, project management, and RUP core supporting workflow deliverables—project management, configuration and change management, and environment artifact sets. This phase delivers RUP deployment artifact set and a production operational software product. It formally concludes with the project's final release of the operational software product as a result of Iterations #7 and #8. This review results in approval of the business management, RUP workflow, and project management deliverables. It formally declares the final release of the operational system as the deciding factor in determining whether the project objectives were satisfactorily met by the stakeholders and whether or not to initiate another RUP iterative development cycle—inception, elaboration, construction and transition.

7.10.1 Alignment with PMBOK: project management

The primary objective of this section is to demonstrate how the main deliverables of the RUP transition phase, Iterations #7 and #8, align with the deliverables (what) from the PMBOK–project management component, RUP workflow components, and business management components. The roles and responsibilities from the project management processes, presented in Chapter 4, are summarized to determine the major activities (how) needed to produce the deliverables and the people resources (who) responsible for planning/managing, creating/finding, reviewing/QA, and approving/committing to the deliverables.

7.10.1.1 Deliverables (what)

This section highlights the contents of Iterations #7 and #8—production operational system, and supporting deployment artifacts set, for business management, RUP workflows, and project management deliverables. Chapter 4 provides detailed definitions, templates, process flows, and quality measures for the creation, management, approval, and review of these deliverables. The project management deliverables are replicated in this section to provide readers with the necessary continuous and logical flow of information to enable a more analytical understanding of the deliverables produced during the RUP transition phase.

Final PMBOK–project management deliverables include the following:

- PSM management plan and requirements baseline updates:
  - Business requirements;
  - Data requirements;
  - Application requirements
  - Technology requirements;
  - Scope management plan, and reports updates.
- PTM management plan and schedule baseline updates:
  - Project cost management plan, budget baseline, and reports updates;
  - PQM management plan, quality baseline, and reports updates;
  - Project change management plan, change requests, and reports updates;
  - Project contract management plan, contract, and reports updates;
  - PIM management plan, issues, and reports updates;
  - PRM management plan, potential risks, and reports updates;
  - PCM plan, progress status updates;

- Project human resources–staffing plan, RAM, and reports updates;
- PMO support services.

7.10.1.2 Activities (how)

The roles and responsibilities of the project management component outlined in Chapter 4 are classified based on the creation/updating, management/integration, review/QA, and approval/commitment, responsibility classifications to determine the major activities required to produce the deliverables during RUP transition phase. The integration of business management processes and RUP workflows with project management processes is highlighted to demonstrate the integrated nature of project management during the RUP-PMBOK PDLC processes. The manage and integrate responsibilities of project management are presented as similar tasks to demonstrate the integrated skills needed by the program/project manager.

- *PSM management plan, requirements baseline, and reports updates:* This plan is an integrated representation of the contents from business management–business architecture with the contents from RUP workflow management–business modeling, requirements, analysis and design, implementation, test, deployment, and project management artifact sets (Update, integrate, review, approve, support–scope management).
- *PTM management plan, schedule baseline, and reports updates:* This plan is an integrated representation of the contents from business management–project deliverables approval with the contents from RUP workflow management–project management artifact set (Update, integrate, review, approve, support–time management).
- *Project cost management plan, budget baseline, and reports updates:* This plan is an integrated representation of the contents from business management–project justifications and PFAs with the contents from RUP workflow management–project management artifact set (Update, integrate, review, approve, support–cost management).
- *PQM management plan, quality baseline, and reports updates:* This plan is an integrated representation of the contents from business management–project deliverables approval with the contents from RUP workflow management–business modeling, requirements, analysis and design, implementation, test, deployment, and project management artifact sets (Update, integrate, review, approve, support–quality management).
- *Project change management plan, change requests, and reports updates:* This is one of the supportive plans for the core project management plans—cost, time, scope, and quality. It represents the integrated management of all the changes made to the core project management deliverables with the contents from RUP workflow management–configuration and change management (Update, integrate, review, approve, support–change management).

- *Project contract management plan, contracts, and reports:* This is one of the supportive plans for the core project management plans—cost, time, scope, and quality. It represents the integrated management of all the contractual arrangements to support the core project management deliverables with the contents from RUP workflow management–project management artifact (Update, integrate, review, approve, support–contract management).
- *PIM management plan, issues, and reports updates:* This is one of the supportive plans for the core project management plans—cost, time, scope, and quality. It represents the integrated management of all issues that affect the delivery of the core project management deliverables with the contents from RUP workflow management–project management artifact set (Update, integrate, review, approve, support–issues management).
- *PRM management plan, risk, and reports updates:* This is one of the supportive plans for the core project management plans—cost, time, scope, and quality. It represents the integrated management of all risks that affect the delivery of the core project management deliverables with the contents from RUP workflow management–project management artifact set (Update, integrate, review, approve, support–issues management).
- *PCM plan, progress status updates:* This is one of the supportive plans for the core project management plans—cost, time, scope, and quality. It represents the integrated management of all communications to support the delivery of the core project management deliverables. The contents from business management–program steering committee process form an integral part of this communications management plan with the contents from RUP workflow management–project management artifact set (Update, integrate, review, approve, support–communications).
- *PSM plan, RAM, and reports updates:* This is one of the supportive plans for the core project management plans—cost, time, scope, and quality. It represents the integrated management of the staffing plans to support the delivery of the core project management deliverables. The contents from business management–program steering committee process also form an integral part of this communications management plan with the contents from RUP workflow management–project management artifact set (Update, integrate, review, approve, support–human resources management).

7.10.1.3 Resources (who)

The labor resources needed to execute the activities to produce the deliverables are determined to show the type of project management resources responsible for the updating, management/integration, review/QA, and approval/commitment of the deliverables during the RUP project transition phase. It makes reference to the project managers roles provided in Philippe Kruchten's book *The Rational Unified Process: An Introduction, Second Edition*.

- PSM management plan, requirements baseline, and reports updates (Create, manage/integrate, review, approve, support–scope management):
  - Program delivery manager/RUP program manager–update/integrate;
  - Program business manager–support;
  - Program IT manager/RUP reviewer–support;
  - Executive business manager–review;
  - Executive PM manager/RUP reviewer–review and approve;
  - Executive IT manager/RUP reviewer–review;
  - RUP project manager and team–support.
- PTM management plan, requirements baseline, and reports (Create, manage/integrate, review, approve, support–time management):
  - Program delivery manager/RUP program manager–update/integrate;
  - Program business manager–support;
  - Program IT manager/RUP reviewer–support;
  - Executive business manager–review;
  - Executive PM manager/RUP reviewer–review and approve;
  - Executive IT manager/RUP reviewer–review;
  - RUP project manager and team–support.
- Project cost management plan, requirements baseline, and reports (Create, manage/integrate, review, approve, support–cost management):
  - Program delivery manager/RUP program manager–update/integrate;
  - Program business manager–support;
  - Program IT manager/RUP reviewer–support;
  - Executive business manager–review;
  - Executive PM manager/RUP reviewer–review and approve;
  - Executive IT manager/RUP reviewer–review;
  - RUP project manager and team–support.
- PQM management plan, requirements baseline, and reports (Create, manage/integrate, review, approve, support–quality management):
  - Program delivery manager/RUP program manager–update/integrate;
  - Program business manager–support;
  - Program IT manager/RUP reviewer–support;
  - Executive business manager–review;
  - Executive PM manager/RUP reviewer–review and approve;
  - Executive IT manager/RUP reviewer–review;
  - RUP project manager and team–support.
- Project change management plan, requirements baseline, and reports (Create, manage/integrate, review, approve, support–change management):

- Program delivery manager/RUP program manager–update/integrate;
- Program business manager–support;
- Program IT manager/RUP reviewer–support;
- Executive business manager–review;
- Executive PM manager/RUP reviewer–review and approve;
- Executive IT manager/RUP reviewer–review;
- RUP project manager and team–support.

- Project contract management plan, requirements baseline, and reports (Create, manage/integrate, review, approve, support–contract management):
  - Program delivery manager/RUP program manager–update/integrate;
  - Program business manager–support;
  - Program IT manager/RUP reviewer–support;
  - Executive business manager–review;
  - Executive PM manager/RUP reviewer–review and approve;
  - Executive IT manager/RUP reviewer–review;
  - RUP project manager and team–support.
- PIM management plan, requirements baseline, and reports (Create, manage/integrate, review, approve, support–issue management):
  - Program delivery manager/RUP program manager–update/integrate;
  - Program business manager–support;
  - Program IT manager/RUP reviewer–support;
  - Executive business manager–review;
  - Executive PM manager/RUP reviewer–review and approve;
  - Executive IT manager/RUP reviewer–review;
  - RUP project manager and team–support.
- PRM management plan, requirements baseline, and reports (Create, manage/integrate, review, approve, support–risk management):
  - Program delivery manager/RUP program manager–update/integrate;
  - Program business manager–support;
  - Program IT manager/RUP reviewer–support;
  - Executive business manager–review;
  - Executive PM manager/RUP reviewer–review and approve;
  - Executive IT manager/RUP reviewer–review;
  - RUP project manager and team–support.
- PCM plan, requirements baseline, and reports (Create, manage/integrate, review, approve, support–communications):
  - Program delivery manager/RUP program manager–update/integrate;
  - Program business manager–support;
  - Program IT manager/RUP reviewer–support;
  - Executive business manager–review;
  - Executive PM manager/RUP reviewer–review and approve;

- Executive IT manager/RUP reviewer–review;
- RUP project manager and team–support.

- Project human resources management plan, requirements baseline, and reports (Create, manage/integrate, review, approve, support–human resources management):
  - Program delivery manager/RUP program manager–update/ integrate;
  - Program business manager–support;
  - Program IT manager/RUP reviewer–support;
  - Executive business manager–review;
  - Executive PM manager/RUP reviewer–review and approve;
  - Executive IT manager/RUP reviewer–review;
  - RUP project manager and team–support.

7.10.2 Integration with business, RUP, and project management

Table 7.4 is an integrated representation of how the deliverable from the transition phase, Iterations #7 and #8, integrate with the deliverables (what), activities (how), and resources (who) from the business, RUP workflow, and PMBOK–project management component.

7.11 PMO support services

Figure 7.1 graphically displays PMO involvement during the RUP iterative PDLC. The primary objective of this section is to demonstrate how the PMO supports the project team during each phase of the RUP PDLC. The PMO is responsible for deploying and supporting a set of consistent and IPM processes within the company, including policies, roles and responsibilities, procedures—deliverables templates, process flows, checklists, and best practices. These project management processes and program deliverables are normally stored in an IPMIR; refer to Chapter 4 for further details. This IPMIR is accessed, maintained, and reported by the IPMIS (Figure 4.10 provided a framework). This is not a one-time event, but a broad initiative that could cover a number of years to ensure consistency, completeness, integrity, and reusability of the project deliverables and processes. The program delivery manager, project managers, and supporting staff are usually part of this group.

7.11.1 Problems: real-world issues

Many companies involved in the development and deployments of projects are usually confronted with the standard project management problems of budget overrun, schedule slippage, uncontrollable scope, unsatisfactory quality, and inefficient utilization of people resources. These companies make various desperate attempts at resolving these issues, using the general

Table 7.4 Integrated Business, RUP Transition, and PM Responsibility Matrix

	Roles						
Deliverables	*EBM*	*EPM RUP*	*EITM RUP*	*PBM*	*PDM RUP*	*PITM RUP*	*PM/Team RUP*
Business							
Project economics	RA	R	R	U	M/I	S	S
PFAs	RA	R	R	U	M/I	S	S
Project funding approvals	RA	R	R	U	M/I	S	S
Project organizational structure	RA	R	R	U	M/I	S	S
Project business architecture	RA	R	R	U	M/I	S	S
Project business support initiatives	RA	R	R	U	M/I	S	S
RUP workflows							
Business modeling artifact set	RA	R	R	S	M/I	S	U
Requirements artifact set	RA	R	R	S	M/I	S	U
Analysis/design artifact set	R	R	RA	S	M/I	S	C
Implementation artifact set	R	R	RA	S	M/I	S	C
Test artifact set	RA	R	R	S	M/I	S	C
Deployment artifact set	RA	R	R	S	M/I	S	C
Configuration and change management artifact set	R	RA	R	S	U	S	S
Project management artifact set	R	RA	R	S	U	S	S
Environment artifact set	R	RA	R	S	U	S	S
Project management							
Scope management plan	R	RA	R	S	U	S	S
Time management plan	R	RA	R	S	U	S	S
Cost management plan	R	RA	R	S	U	S	S
Quality management plan	R	RA	R	S	U	S	S
Contract management plan	R	RA	R	S	U	S	S
Issue management plan	R	RA	R	S	U	S	S
Change management plan	R	RA	R	S	U	S	S

Table 7.4 (continued)

	Roles						
Deliverables	*EBM*	*EPM RUP*	*EITM RUP*	*PBM*	*PDM RUP*	*PITM RUP*	*PM/Team RUP*
Risk management plan	R	RA	R	S	U	S	S
Communications management plan	R	RA	R	S	U	S	S
Human resources management plan	R	RA	R	S	U	S	S

Note: C = create; U = update; R = review; A = approve; M/I = manage/integrate; S = support.
The shaded areas represent the major deliverables produced during this RUP phase.

management principles of leading, communicating, negotiating, problem solving, and influencing, without applying any form of standardized, consistent, or integrated processes. In most cases, these typical project management problems remain unresolved, with compounded communications, commitment, and acceptance problems. The search for the magical silver bullet solution continues with certain companies reverting to project management consulting firms to provide that silver bullet solution.

7.11.2 Recommended solutions: real-world scenarios

A major part of the solution to execute projects better, more quickly and cheaply, and with high quality should focus on the organization's ability and commitment to guide and implement common and integrated processes and practices across the entire organization. The overall strategy to achieve consistency and integration is to align and integrate the product deliverables produced from RUP workflow management components with the project deliverables produced from PMBOK project management components. If PMO is implemented properly with the right project management culture, the learning curve for the project manager and the team members as they transition from one project to another will shorten to an acceptable level for productivity gains.

Most PMOs should be responsible for deploying and supporting a set of consistent IPM processes within the company, including policies, roles and responsibilities, procedures—deliverables templates, process flows, checklists, and best practices. Some critics may argue that PMO demands additional resources, which can add overhead expenses to projects. As result of my extensive IT project management experiences, I strongly believe that if implemented properly, investments in PMO can be cost justified with productivity gains through the implementation of consistent and integrated processes and practices. This will provide the means to allow every project within the organization to be completed better, more quickly and cheaply, and with higher quality. The key guiding principle is to align or integrate RUP product deliverables with PMBOK project deliverables for each phase of the RUP PDLC.

An effective PMO can offer many potential products and services, depending on the project management needs and culture of the organization and the vision of the PMO sponsor (the person who is generally responsible for the PMO funding). Before the PMO can be successful, it must gain agreement from executive, senior management, and the project management team on its overall role and the general expectations it needs to achieve.

This section provides a framework to assist in the resolution of the typical project management problems identified above, by integrating business management and RUP workflow management with PMBOK project management process components, through the applications of standard, consistent, and integrated processes, deliverables, and practices.

7.11.3 PMO alignment with PMBOK and RUP iterative PDLC

7.11.3.1 Deliverables (what)

At the start of this chapter, the contents of RUP-PMBOK iterative development processes of business management, RUP workflow management, and PMBOK project management deliverable components are presented to highlight the overlapping deliverables. Tables 7.1 to 7.4 provide responsibility matrixes that show who does what for creation, management, review, approval, and support of these deliverables for each of the RUP phases.

The RUP-PMBOK deliverables are replicated in this section to provide readers with the necessary continuous and logical flow of information to enable a more analytical and continuous refinement in understanding the deliverables produced and the value of the PMO support services.

Business management deliverables include the following:

- BSA;
- Project justification (benefits and priorities);
- PFAs;
- Project deliverables/funding approvals;
- Program steering and working committee;
- Project business support initiatives.

PMBOK–project management deliverables include the following:

- PSM management plan, requirements baseline, and reports:
  - Business requirements;
  - Data requirements;
  - Application requirements;
  - Technology requirements;
  - Scope management plan, and reports.
- PTM management plan, schedule baseline, and reports;
- Project cost management plan, budget baseline, and reports;
- PQM management plan, quality baseline, and reports;
- Project change management plan, change requests, and reports;

- Project contract management plan, contract, and reports;
- PIM management plan, issues, and reports;
- PRM management plan, potential risks, and reports;
- PCM plan, progress status, and reports;
- Project human resources–staffing plan, RAM, and reports;
- PMO support services.

RUP workflow management deliverables include the following:

- Business modeling artifact set;
- Requirements artifact set;
- Analysis and design artifact set;
- Implementation artifact set;
- Test artifact set;
- Deployment artifact set;
- Environment artifact set;
- Project management artifact set;
- Configuration and change management artifact set.

7.11.3.2 Activities (how) and resources (who)

The major activities and resources required to create/update, manage/integrate, review/QA, approve/commit, and support the deliverables and processes to support PMO support services are shown. The deliverables are classified as product deliverables produced from execution of RUP processes and project/work deliverables produced from execution of PMBOK project management processes. PMO is responsible for the development, deployment, and support of these processes and the supporting project management tools.

- Business management processes (Create, manage, review, approve, support–business management processes):
  - Program business manager–support;
  - Program delivery manager/RUP manager–create/manage;
  - Program IT manager/RUP manager–support;
  - Executive IT manager/executive RUP manager–review and approve;
  - Executive business manager–review and approve;
  - Executive program manager/RUP (sponsor)–review and approve;
  - PMO staff–support;
  - RUP project manager/team–support.
- Business management deliverables:
  - BSA (Create, manage, review, approve, support–BSA):
    Program business manager–create;
    Program delivery manager/RUP manager–manage/integrate;
    Program IT manager/RUP manager–support;

Executive IT manager/executive RUP manager–review and approve;
Executive business manager (sponsor)–review and approve;
Executive program manager/RUP manager–review and approve;
PMO staff–support;
RUP project manager/team–support.

- Project justifications (benefits and priorities) (Create, manage, review, approve, support–project justifications):
  Program business manager–create;
  Program delivery manager/RUP manager–manage/integrate;
  Program IT manager/RUP manager–support;
  Executive IT manager/executive RUP manager–review and approve;
  Executive business manager (sponsor)–review and approve;
  Executive program manager/RUP manager–review and approve;
  PMO staff–support;
  RUP project manager/team–support.
- PFAs (Create, manage, review, approve, support–funds allocation):
  Program business manager–create;
  Program delivery manager/RUP manager–manage/integrate;
  Program IT manager/RUP manager–support;
  Executive IT manager/executive RUP manager–review and approve;
  Executive business manager (sponsor)–review and approve;
  Executive program manager/RUP manager–review and approve;
  PMO staff–support;
  RUP project manager/team–support.
- Project deliverables/funding approvals (Create, manage, review, approve, support–deliverables approvals):
  Program business manager–create;
  Program delivery manager/RUP manager–manage/integrate;
  Program IT manager/RUP manager–support;
  Executive IT manager/executive RUP manager–review and approve;
  Executive business manager (sponsor)–review and approve;
  Executive program manager/RUP manager–review and approve;
  PMO staff–support;
  RUP project manager/team–support.
- Program steering and working committee (Create, manage, review, approve, support–steering committee):
  Program business manager–create;
  Program delivery manager/RUP manager–manage/integrate;

Program IT manager/RUP manager–support;

Executive IT manager/executive RUP manager–review and approve;

Executive business manager (sponsor)–review and approve;

Executive program manager/RUP manager–review and approve;

PMO staff–support;

RUP project manager/team–support.

- Project business support initiatives (Create, manage, review, approve, support–business initiatives):

  Program business manager–create;

  Program delivery manager/RUP manager–manage/integrate;

  Program IT manager/RUP manager–support;

  Executive IT manager/executive RUP manager–review and approve;

  Executive business manager (sponsor)–review and approve;

  Executive program manager/RUP manager–review and approve;

  PMO staff–support;

  RUP project manager/team–support.

- RUP workflow management process (Create, manage, review, approve, support–workflow management process):
  - Program business manager–support;
  - Program delivery manager/RUP manager–create/manage;
  - Program IT manager/RUP manager–support;
  - Executive IT manager/executive RUP manager–review and approve;
  - Executive business manager (sponsor)–review and approve;
  - Executive program manager/RUP manager–review and approve;
  - PMO staff–support;
  - RUP project manager/team–support.
- RUP workflow management deliverables:
  - Business modeling artifact set (Create, manage, review, approve, support–business modeling set):

    Program business manager–support;

    Program delivery manager/RUP manager–manage/integrate;

    Program IT manager/RUP manager–support;

    Executive IT manager/executive RUP manager–review and approve;

    Executive business manager (sponsor)–review and approve;

    Executive program manager/RUP manager–review and approve;

    PMO staff–support;

    RUP project manager/team–create/update.
  - Requirements artifact set (Create, manage, review, approve, support–requirements set):

Program business manager–support;

Program delivery manager/RUP manager–manage/integrate;

Program IT manager/RUP manager–support;

Executive IT manager/executive RUP manager–review and approve;

Executive business manager (sponsor)–review and approve;

Executive program manager/RUP manager–review and approve;

PMO staff–support;

RUP project manager/team–create/update.

- Analysis and design artifact set (Create, manage, review, approve, support–analysis and design set):

  Program business manager–support;

  Program delivery manager/RUP manager–manage/integrate;

  Program IT manager/RUP manager–support;

  Executive IT manager/executive RUP manager–review and approve;

  Executive business manager–review and approve;

  Executive program manager/RUP manager–review and approve;

  PMO staff–support;

  RUP project manager/team–create/update.

- Implementation artifact set (Create, manage, review, approve, support–implementation set):

  Program business manager–support;

  Program delivery manager/RUP manager–manage/integrate;

  Program IT manager/RUP manager–support;

  Executive IT manager/RUP manager (sponsor)–review and approve;

  Executive business manager–review and approve;

  Executive program manager/RUP manager–review and approve;

  PMO staff–support;

  RUP project manager/team–create/update.

- Test artifact set (Create, manage, review, approve, support–testing set):

  Program business manager–support;

  Program delivery manager/RUP manager–manage/integrate;

  Program IT manager/RUP manager–support;

  Executive IT manager/executive RUP manager–review and approve;

  Executive business manager (sponsor)–review and approve;

  Executive program manager/RUP manager–review and approve;

  PMO staff–support;

  RUP project manager/team–create/update.

- Deployment artifact set (Create, manage, review, approve, support–deployment set):
  - Program business manager–support;
  - Program delivery manager/RUP manager–manage/integrate;
  - Program IT manager/RUP manager–support;
  - Executive IT manager/executive RUP manager–review and approve;
  - Executive business manager–review and approve;
  - Executive program manager/RUP manager–review and approve;
  - PMO staff–support;
  - RUP project manager/team–create/update.
- Environment artifact set (Create, manage, review, approve, support–environment set):
  - Program business manager–support;
  - Program delivery manager/RUP–create/manage/integrate;
  - Program IT manager/RUP manager–support;
  - Executive IT manager/executive RUP manager–review and approve;
  - Executive business manager–review and approve;
  - Executive program manager/RUP (sponsor)–review and approve;
  - PMO staff–support;
  - RUP project manager/team–support.
- Project management artifact set (Create, manage, review, approve, support–project management set):
  - Program business manager–support;
  - Program delivery manager/RUP–create/manage/integrate;
  - Program IT manager/RUP manager–support;
  - Executive IT manager/executive RUP manager–review and approve;
  - Executive business manager–review and approve;
  - Executive program manager (sponsor)–review and approve;
  - PMO staff–support;
  - RUP project manager/team–support.
- Configuration and change management artifact set (Create, manage, review, approve, support–configuration and change management set):
  - Program business manager–support;
  - Program delivery manager/RUP–create/manage/integrate;
  - Program IT manager/RUP manager–support;
  - Executive IT manager/executive RUP manager–review and approve;
  - Executive business manager–review and approve;

Executive program manager/RUP (sponsor)–review and approve;
PMO staff–support;
RUP project manager/team–support.

- Project management processes (Create, manage, review, approve, support–project management processes):
  - Program business manager–support;
  - Program delivery manager/RUP–create/manage/integrate;
  - Program IT manager/RUP manager–support;
  - Executive IT manager/executive RUP manager–review and approve;
  - Executive business manager–review and approve;
  - Executive program manager/RUP (sponsor)–review and approve;
  - PMO staff–support;
  - RUP project manager/team–support.
- Project management deliverables:
  - PSM management plan, requirements baseline, and reports updates (Create, manage/integrate, review, approve–scope management):

    Program business manager–support;
    Program delivery manager/RUP–create/manage/integrate;
    Program IT manager/RUP manager–support;
    Executive IT manager/executive RUP manager–review and approve;
    Executive business manager–review and approve;
    Executive program manager/RUP (sponsor)–review and approve;
    PMO staff–support;
    RUP project manager/team–support.
  - PTM management plan, requirements baseline, and reports (Create, manage/integrate, review, approve–time management):

    Program business manager–support;
    Program delivery manager/RUP–create/manage/integrate;
    Program IT manager/RUP manager–support;
    Executive IT manager/executive RUP manager–review and approve;
    Executive business manager–review and approve;
    Executive program manager/RUP (sponsor)–review and approve;
    PMO staff–support;
    RUP project manager/team–support.
  - Project cost management plan, requirements baseline, and reports (Create, manage/integrate, review, approve–cost management):

    Program business manager–support;
    Program delivery manager/RUP–create/manage/integrate;
    Program IT manager/RUP manager–support;

Executive IT manager/executive RUP manager–review and approve;

Executive business manager–review and approve;

Executive program manager/RUP (sponsor)–review and approve;

PMO staff–support;

RUP project manager/team–support.

- PQM management plan, requirements baseline, and reports (Create, manage/integrate, review, approve–quality management):

  Program business manager–support;

  Program delivery manager/RUP–create/manage/integrate;

  Program IT manager/RUP manager–support;

  Executive IT manager/executive RUP manager–review and approve;

  Executive business manager–review and approve;

  Executive program manager/RUP (sponsor)–review and approve;

  PMO staff–support;

  RUP project manager/team–support.

- Project change management plan, requirements baseline, and reports (Create, manage/integrate, review, approve–change management):

  Program business manager–support;

  Program delivery manager/RUP–create/manage/integrate;

  Program IT manager/RUP manager–support;

  Executive IT manager/executive RUP manager–review and approve;

  Executive business manager–review and approve;

  Executive program manager/RUP (sponsor)–review and approve;

  PMO staff–support;

  RUP project manager/team–support.

- Project contract management plan, requirements baseline, and reports (Create, manage/integrate, review, approve–contract management):

  Program business manager–support;

  Program delivery manager/RUP–create/manage/integrate;

  Program IT manager/RUP manager–support;

  Executive IT manager/executive RUP manager–review and approve;

  Executive business manager–review and approve;

  Executive program manager/RUP (sponsor)–review and approve;

  PMO staff–support;

  RUP project manager/team–support.

- PIM management plan, requirements baseline, and reports (Create, manage/integrate, review, approve–issue management):

  Program business manager–support;

  Program delivery manager/RUP–create/manage/integrate;

  Program IT manager/RUP manager–support;

  Executive IT manager/executive RUP manager–review and approve;

  Executive business manager–review and approve;

  Executive program manager/RUP (sponsor)–review and approve;

  PMO staff–support;

  RUP project manager/team–support.

- PRM management plan, requirements baseline, and reports (Create, manage/integrate, review, approve–risk management):

  Program business manager–support;

  Program delivery manager/RUP–create/manage/integrate;

  Program IT manager/RUP manager–support;

  Executive IT manager/executive RUP manager–review and approve;

  Executive business manager–review and approve;

  Executive program manager/RUP (sponsor)–review and approve;

  PMO staff–support;

  RUP project manager/team–support.

- PCM plan, requirements baseline, and reports (Create, manage/integrate, review, approve–communications):

  Program business manager–support;

  Program delivery manager/RUP–create/manage/integrate;

  Program IT manager/RUP manager–support;

  Executive IT manager/executive RUP manager–review and approve;

  Executive business manager–review and approve;

  Executive program manager/RUP (sponsor)–review and approve;

  PMO staff–support;

  RUP project manager/team–support.

- Project human resources management plan, requirements baseline, and reports (Create, manage/integrate, review, approve–human resources management):

  Program business manager–support;

  Program delivery manager/RUP–create/manage/integrate;

  Program IT manager/RUP manager–support;

  Executive IT manager/executive RUP manager–review and approve;

  Executive business manager–review and approve;

Executive program manager/RUP (sponsor)–review and approve;
PMO staff–support;
RUP project manager/team–support.

7.11.4 Integration with business, RUP, and project management

Table 7.5 is an integrated representation to show how PMO support services, integrate with the deliverables (what), processes (how), and resources (who) from business, RUP workflow, and PMBOK–project management components.

7.12 Summary

The PMI-PMBOK alignment with Rational Corporation RUP presented in this chapter describes how the components of PMI-PMBOK project management and RUP workflow are aligned based on the phases of the RUP. for each phase, the alignment with PMBOK project management and RUP workflow is demonstrated, based on the deliverables (what), activities (how), and key responsibilities of the resources (who) needed to meet the milestone. Real-world problems and recommended solutions for PMO support services are also discussed to provide readers with scenarios of situations for proactive actions. The business management, project management, and RUP workflow management deliverables (what), activities (how), and roles and responsibilities (who does what) matrixes provide excellent representations of how these components are horizontally integrated during the RUP iterative PDLC.

IT organizations who have standardized on RUP technologies and PMI-PMBOK project management process for managing and delivering multiple IT projects must understand how PMBOK is aligned with RUP. They should consider the following recommendations to guide them towards successful management and delivery of multiple IT projects during the RUP iterative PDLC:

- Construct Iteration #1 to consist of key PMBOK project management and RUP workflow deliverables, similar to the model presented in Section 7.7. This Iteration #1 will form the baseline to better manage the scope deliverables, effort, costs, schedule, and quality objectives of the development projects during the project inception phase.
- Construct Iterations #2 and #3 similar to the PMBOK-RUP evolutionary prototype model presented in Section 7.8. This evolutionary prototype will form the baseline to ensure a smooth transition from Iteration #1 to a working automated solution. This prototype will produce a concrete solution that business users can easily understand and refine as the project progresses. It will ensure that business users and management

Table 7.5 PMO–RUP Support Services Responsibility Matrix

Deliverables	Roles EBM	EPM RUP (PMO)	EITM RUP	PBM	PDM RUP (PMO)	PITM RUP	PM/Team RUP (PMO)	PMO Support Staff
Business management processes	RA*	RA	RA	S	C/M/I	S	S	S
Project economics	RA*	RA	RA	C	M/I	S	S	S
PFAs	RA*	RA	RA	C	M/I	S	S	S
Project funding approvals	RA	RA	RA	C	M/I	S	S	S
Project organizational structure	RA	RA	RA	C	M/I	S	S	S
Project business architecture	RA	RA	RA	C	M/I	S	S	S
Project business support initiatives	RA	RA	RA	C	M/I	S	S	S
RUP workflow processes	RA	RA*	RA	S	C/M/I	S	S	S
Business modeling artifact set	RA*	RA	RA	S	M/I	S	U	S
Requirements artifact set	RA*	RA	RA	S	M/I	S	U	S
Analysis/design artifact set	RA	RA	RA*	S	M/I	S	C	S
Implementation artifact set	RA	RA	RA*	S	M/I	S	C	S
Test artifact set	RA*	RA	RA	S	M/I	S	C	S
Deployment artifact set	RA*	RA	RA	S	M/I	S	C	S
Configuration and change management artifact set	RA	RA*	RA	S	C	S	S	S
Project management artifact set	RA	RA*	RA	S	C	S	S	S
Environment artifact set	RA	RA*	RA	S	C	S	S	S
Project management process	RA	RA*	RA	S	C/M/I	S	S	S
Cost management plan	RA	RA*	RA	S	C	S	S	S
Time management plan	RA	RA*	RA	S	C	S	S	S
Scope management plan	RA	RA*	RA	S	C	S	S	S
Quality management plan	RA	RA*	RA	S	C	S	S	S
Change management plan	RA	RA*	RA	S	C	S	S	S

Table 7.5 (continued)

	Roles							
Deliverables	*EBM*	*EPM RUP (PMO)*	*EITM RUP*	*PBM*	*PDM RUP (PMO)*	*PITM RUP*	*PM/Team RUP (PMO)*	*PMO Support Staff*
Contract management plan	RA	RA*	RA	S	C	S	S	S
Issue management plan	RA	RA*	RA	S	C	S	S	S
Risk management plan	RA	RA*	RA	S	C	S	S	S
Communications management plan	RA	RA*	RA	S	C	S	S	S
Human resources management plan	RA	RA*	RA	S	C	S	S	S

Note: C = create; U = update; R = review; A = approve; M/I = manage/integrate; S = support.
The shaded areas represent the major deliverables produced during the RUP phases; *sponsor (main approval).

commit to the requirements or scope of the project. The major root causes of project failures usually occur during the analysis phase.

- Construct iterations #4, #5, and #6 similar to the PMBOK-RUP model presented in Section 7.9. These iterations will ensure a smooth transition from the evolutionary prototype to a working automated solution. This development will ensure that business and IT staff commit to the RUP design by providing a concrete solution that business users can easily understand and refine as the project progresses.
- Construct Iterations #7 and #8 similar to the PMBOK-RUP model presented in Section 7.10. These iterations will ensure commitment to and acceptance of the final product by providing concrete solutions that business and IT staff can understand and refine prior to production migration. One of the critical root causes of most project failures usually occurs during the transition-deployment-conversion phase.
- Establish a PMO function that provides the services similar to those discussed in Section 7.11. This will ensure completeness, consistency, integrity, and reusability of project deliverables to improve project team productivity.

7.13 Questions to think about: management perspectives

1. Think about how your organization identifies, prioritizes, and manages multiple IT projects using RUP and PMI-PMBOK project management processes. How does your organization control project deliverables? What are the key rationales to obtaining business users' commitments for project requirements? What are the major

components of RUP-PMBOK alignment? How do these components relate to your project environment? What is the perception of senior management to aligning RUP with PMBOK to manage project deliverables?

2. Think about how your organization uses RUP during project inception. What are the three components of Iteration #1? How do these three major deliverables relate to your project definition approach? How does PMBOK align with RUP during project inception?
3. Think about how your organization uses RUP during project elaboration. What are the major deliverables of RUP–elaboration phase? How do these deliverables relate to your organization? How does PMBOK align with RUP during project elaboration?
4. Think about how your organization uses RUP during project construction. What are the major deliverables of RUP–construction phase? How do these deliverables relate to your organization? How does PMBOK align with RUP during project construction?
5. Think about how your organization uses RUP during project transition. What are the major deliverables of RUP–transition phase? How do these deliverables relate to your organization? How does PMBOK align with RUP during project transition?
6. Think about how your organization ensures consistency, completeness, and integration of project deliverables using RUP. What are the services of the PMO on projects that use RUP? How do these services relate to your organization? What is the perception of senior management of the value of PMO on projects that use RUP? What are the major causes of PMO failures? What is the key recommended solution? How does PMO align with RUP PDLC? What are the major responsibilities of the program delivery manager?

Selected bibliography

CSC—Computer Sciences Corporation, *Catalyst Methodology*, CA CSC, 1999.

Fowler, M., and K. Scott, *UML Distilled*, 2nd ed., Reading, MA: Addison-Wesley, 1999.

Jacobson, I., G. Booch, and J. Rumbaugh, *The Unified Software Development Process*, Reading, MA: Addison-Wesley, 1999.

Kerzner, H., *Project Management: A Systems Approach to Planning, Scheduling, and Controlling*, 7th ed., New York: John Wiley & Sons, 2001.

Kruchten, P., *The Rational Unified Process: An Introduction*, 2nd ed., Reading, MA: Addison-Wesley, 2000.

Lewis, J. P., *Project Planning: Scheduling and Control*, 3rd ed., New York: McGraw-Hill, 2001.

Muller, R. J., *Productive Objects: An Applied Software Project Management Framework*, San Francisco, CA: Morgan Kaufmann, 1998.

PMI, *Project Management Institute: A Guide to Project Management Body of Knowledge*, 2000 Edition, Newton Square, PA: PMI, 2000.

Rosenberg, D., and K. Scott, *Use Case Driven Object Modeling with UML*, Reading, MA: Addison-Wesley, 1999.

Appendixes

Integrated IT Project Management Presentation and Microsoft Project Schedules

The integrated IT project management presentations provide some highlights that I presented at:

- Association of Information Technology Professionals (AITP)—2003 National Collegiate Conference, Purdue University;
- IFIP 2002 World Computer Congress—Montreal, Canada;
- 2001 International Project Management (PMI) Conference, Trinidad.

These presentations created much interest among the audiences, which resulted in the compilation of this book. The presentation materials highlighted the model-centric approach to integrated IT project management (IPM-IT) presented in this book. Highlights of the presentations can be viewed at http://www.ICCP.org and http://www.kenbainey.ca.

Microsoft Project schedules based on real-world examples are also provided to demonstrate implementation of this model-based work breakdown structure (WBS) and to show the practical applicability of using this WBS as the foundation structure during the management and delivery of IT projects. The Microsoft Project schedules are summarized at the deliverables level to show how the major project deliverables relate to the model-based WBS.

Appendix A: Detailed Microsoft Project Schedule—Deliverables Level

Microsoft Project - BSP.mpp

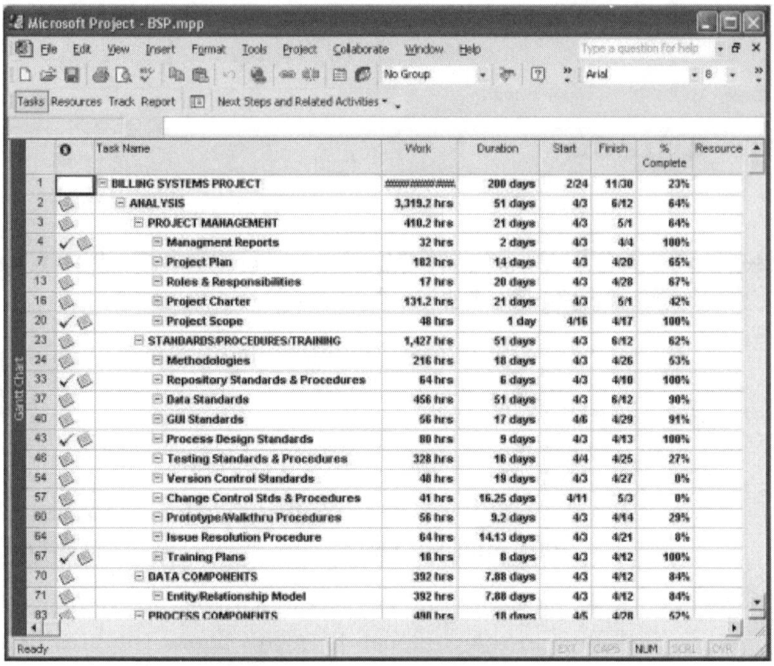

	Indicators	Task Name	Work	Duration	Start	Finish	% Complete	Resource
1		BILLING SYSTEMS PROJECT	#########	200 days	2/24	11/30	23%	
2		ANALYSIS	3,319.2 hrs	51 days	4/3	6/12	64%	
3		PROJECT MANAGEMENT	410.2 hrs	21 days	4/3	5/1	64%	
4	✓	Managment Reports	32 hrs	2 days	4/3	4/4	100%	
7		Project Plan	182 hrs	14 days	4/3	4/20	65%	
13		Roles & Responsibilities	17 hrs	20 days	4/3	4/28	67%	
16		Project Charter	131.2 hrs	21 days	4/3	5/1	42%	
20	✓	Project Scope	48 hrs	1 day	4/16	4/17	100%	
23		STANDARDS/PROCEDURES/TRAINING	1,427 hrs	51 days	4/3	6/12	62%	
24		Methodologies	216 hrs	18 days	4/3	4/26	53%	
33	✓	Repository Standards & Procedures	64 hrs	6 days	4/3	4/10	100%	
37		Data Standards	456 hrs	51 days	4/3	6/12	90%	
40		GUI Standards	56 hrs	17 days	4/6	4/29	91%	
43	✓	Process Design Standards	80 hrs	9 days	4/3	4/13	100%	
46		Testing Standards & Procedures	328 hrs	16 days	4/4	4/25	27%	
54		Version Control Standards	48 hrs	19 days	4/3	4/27	0%	
57		Change Control Stds & Procedures	41 hrs	16.25 days	4/11	5/3	0%	
60		Prototype/Walkthru Procedures	56 hrs	9.2 days	4/3	4/14	29%	
64		Issue Resolution Procedure	64 hrs	14.13 days	4/3	4/21	8%	
67	✓	Training Plans	18 hrs	8 days	4/3	4/12	100%	
70		DATA COMPONENTS	392 hrs	7.88 days	4/3	4/12	84%	
71		Entity/Relationship Model	392 hrs	7.88 days	4/3	4/12	84%	
83		PROCESS COMPONENTS	490 hrs	18 days	4/5	4/28	52%	

Figure A.1 Detailed Microsoft Project schedule—deliverables level: analysis, part 1.

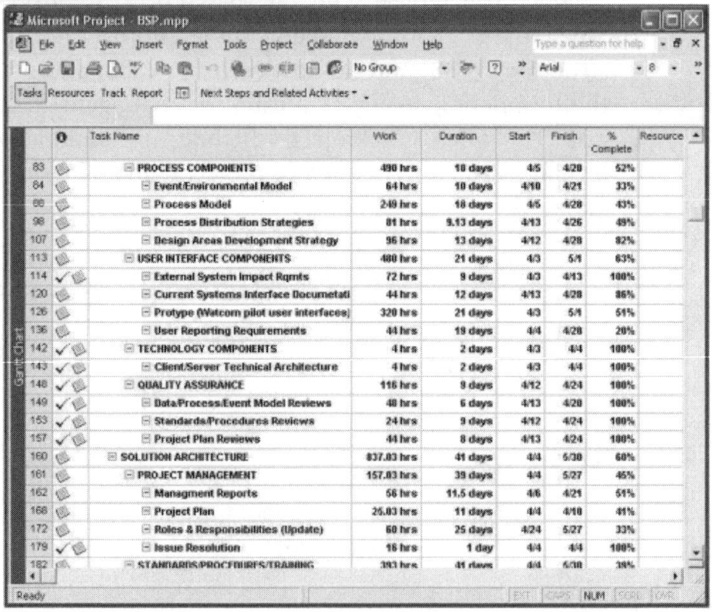

Microsoft Project - BSP.mpp

	Indicators	Task Name	Work	Duration	Start	Finish	% Complete	Resource
83		PROCESS COMPONENTS	490 hrs	18 days	4/5	4/28	52%	
84		Event/Environmental Model	64 hrs	10 days	4/10	4/21	33%	
88		Process Model	249 hrs	18 days	4/5	4/28	43%	
98		Process Distribution Strategies	81 hrs	9.13 days	4/13	4/26	49%	
107		Design Areas Development Strategy	96 hrs	13 days	4/12	4/28	82%	
113		USER INTERFACE COMPONENTS	480 hrs	21 days	4/3	5/1	63%	
114	✓	External System Impact Rqmts	72 hrs	9 days	4/3	4/13	100%	
120		Current Systems Interface Documetati	44 hrs	12 days	4/13	4/28	86%	
126		Protype (Watcom pilot user interfaces)	320 hrs	21 days	4/3	5/1	51%	
136		User Reporting Requirements	44 hrs	19 days	4/4	4/28	20%	
142	✓	TECHNOLOGY COMPONENTS	4 hrs	2 days	4/3	4/4	100%	
143	✓	Client/Server Technical Architecture	4 hrs	2 days	4/3	4/4	100%	
148	✓	QUALITY ASSURANCE	116 hrs	9 days	4/12	4/24	100%	
149	✓	Data/Process/Event Model Reviews	48 hrs	6 days	4/13	4/20	100%	
153	✓	Standards/Procedures Reviews	24 hrs	9 days	4/12	4/24	100%	
157	✓	Project Plan Reviews	44 hrs	8 days	4/13	4/24	100%	
160		SOLUTION ARCHITECTURE	837.03 hrs	41 days	4/4	5/30	60%	
161		PROJECT MANAGEMENT	157.03 hrs	39 days	4/4	5/27	45%	
162		Managment Reports	56 hrs	11.5 days	4/6	4/21	51%	
168		Project Plan	25.03 hrs	11 days	4/4	4/18	41%	
172		Roles & Responsibilities (Update)	60 hrs	25 days	4/24	5/27	33%	
179	✓	Issue Resolution	16 hrs	1 day	4/4	4/4	100%	
182		STANDARDS/PROCEDURES/TRAINING	393 hrs	41 days	4/4	5/30	39%	

Figure A.2 Detailed Microsoft Project schedule—deliverables level: analysis, part 2.

	Indicators	Task Name	Work	Duration	Start	Finish	% Complete	Resource
182		STANDARDS/PROCEDURES/TRAINING	393 hrs	41 days	4/4	5/30	39%	
183		Library Management Stds	88 hrs	15 days	4/16	5/5	6%	
189		SA Repository Stds/Procedures	20 hrs	2 days	5/3	5/4	0%	
192		Reusable Objects Components Stds	48 hrs	22 days	4/4	5/3	67%	
196		Conversion Strategies	40 hrs	4 days	4/20	4/25	94%	
199		Development/Production Tools	69 hrs	41 days	4/4	5/30	36%	
207		Training (Project Team)	128 hrs	41 days	4/4	5/30	43%	
216	✓	DATA COMPONENTS	81 hrs	6 days	4/4	4/11	100%	
217	✓	Data Architecture	81 hrs	6 days	4/4	4/11	100%	
223		PROCESS COMPONENTS	96 hrs	34 days	4/4	5/20	93%	
224		Application Architecture	96 hrs	34 days	4/4	5/20	93%	
236	✓	USER INTERFACE COMPONENTS	16 hrs	3 days	4/4	4/6	100%	
237	✓	User Interface Architecture	16 hrs	3 days	4/4	4/6	100%	
240		TECHNOLOGY COMPONENTS	72 hrs	35.25 days	4/4	5/23	74%	
241		Technology Architecture	72 hrs	35.25 days	4/4	5/23	74%	
247	✓	QUALITY ASSURANCE	22 hrs	1 day	4/4	4/4	100%	
248	✓	Data/Applications/Technology Architec	8 hrs	1 day	4/4	4/4	100%	
253	✓	Standards/Procedures (Reviews)	6 hrs	1 day	4/4	4/4	100%	
257	✓	Project Plan (Reviews)	8 hrs	1 day	4/4	4/4	100%	
260		PROTOTYPE/DESIGN	4,367 hrs	118 days	4/19	9/29	21%	
261		PROJECT MANAGEMENT	153 hrs	34 days	4/19	6/5	0%	
262		System Migration Marketing	33 hrs	34 days	4/19	6/5	0%	
266		Managment Reports	32 hrs	2.38 days	4/19	4/21	0%	

Figure A.3 Detailed Microsoft Project schedule—deliverables level: architecture.

	Indicators	Task Name	Work	Duration	Start	Finish	% Complete	Resource
260		PROTOTYPE/DESIGN	4,367 hrs	118 days	4/19	9/29	21%	
261		PROJECT MANAGEMENT	153 hrs	34 days	4/19	6/5	0%	
262		System Migration Marketing	33 hrs	34 days	4/19	6/5	0%	
266		Managment Reports	32 hrs	2.38 days	4/19	4/21	0%	
269		Project Plan	24 hrs	5.25 days	4/19	4/26	0%	
272		Issue Resolution Procedures	64 hrs	13 days	4/19	5/5	0%	
275		STANDARDS/PROCEDURES/TRAINING	1,178 hrs	45 days	4/26	6/27	24%	
276		Conversion/Transition	328 hrs	18 days	5/2	5/25	66%	
288		Test Plan	133 hrs	11 days	4/26	5/10	0%	
296		Construction Plan (Developers)	61 hrs	21 days	5/30	6/27	0%	
315		Training Plan (Users/Developers)	656 hrs	5 days	5/31	6/6	13%	
335		DATA COMPONENTS	800 hrs	22 days	4/19	5/18	56%	
336		Refine Entity Model	424 hrs	19 days	4/19	5/15	71%	
349		Physical Data Base Design	288 hrs	10 days	5/5	5/18	49%	
360		Database Access/Security/Integrity Rqr	88 hrs	3 days	5/16	5/18	0%	
367		PROCESS COMPONENTS	1,496 hrs	118 days	4/19	9/29	8%	
368		Preprocessor (Design, construct, test,	400 hrs	60 days	5/15	8/4	0%	
376		Billing Preprocessor	400 hrs	75 days	6/19	9/29	0%	
381		Physical Program Structure	584 hrs	43 days	4/26	6/23	23%	
403		Distributed Program Structures	112 hrs	30 days	4/19	5/30	0%	
414		USER INTERACE COMPONENTS	512 hrs	22 days	5/2	5/31	0%	
415		Prototype (Oracle Tested User Interfac	512 hrs	22 days	5/2	5/31	0%	
423		TECHNOLOGY COMPONENTS	104 hrs	13.5 days	5/11	5/30	26%	

Figure A.4 Detailed Microsoft Project schedule—deliverables level: prototype/design.

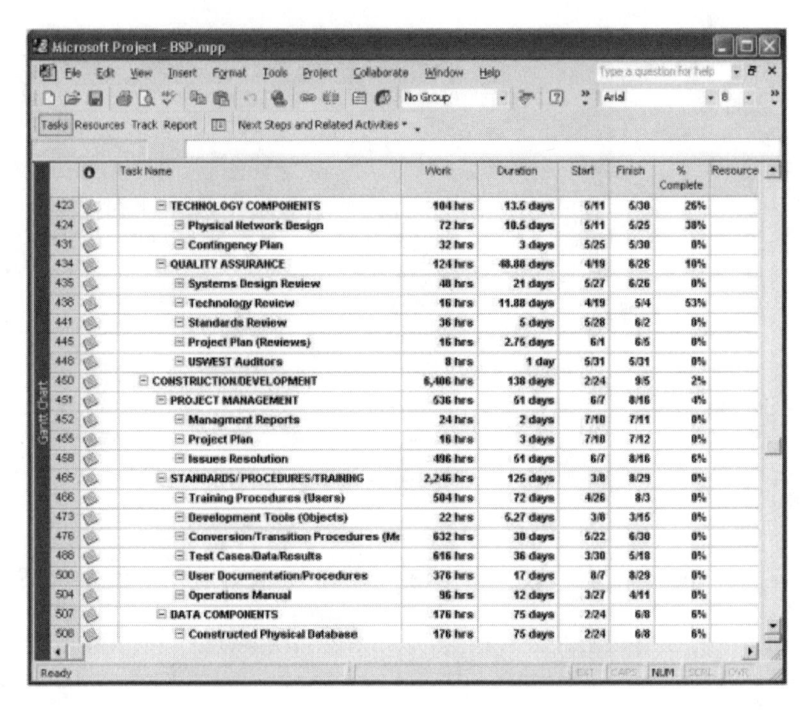

	Task Name	Work	Duration	Start	Finish	% Complete	Resource
423	TECHNOLOGY COMPONENTS	104 hrs	13.5 days	5/11	5/30	26%	
424	Physical Network Design	72 hrs	10.5 days	5/11	5/25	38%	
431	Contingency Plan	32 hrs	3 days	5/25	5/30	0%	
434	QUALITY ASSURANCE	124 hrs	40.88 days	4/19	6/26	10%	
435	Systems Design Review	48 hrs	21 days	5/27	6/26	0%	
438	Technology Review	16 hrs	11.88 days	4/19	5/4	53%	
441	Standards Review	36 hrs	5 days	5/28	6/2	0%	
445	Project Plan (Reviews)	16 hrs	2.75 days	6/1	6/5	0%	
448	USWEST Auditors	8 hrs	1 day	5/31	5/31	0%	
450	CONSTRUCTION/DEVELOPMENT	6,406 hrs	138 days	2/24	9/5	2%	
451	PROJECT MANAGEMENT	536 hrs	61 days	6/7	8/16	4%	
452	Managment Reports	24 hrs	2 days	7/10	7/11	0%	
455	Project Plan	16 hrs	3 days	7/10	7/12	0%	
458	Issues Resolution	496 hrs	61 days	6/7	8/16	6%	
465	STANDARDS/ PROCEDURES/TRAINING	2,246 hrs	125 days	3/8	8/29	0%	
466	Training Procedures (Users)	504 hrs	72 days	4/26	8/3	0%	
473	Development Tools (Objects)	22 hrs	5.27 days	3/8	3/15	0%	
476	Conversion/Transition Procedures (M	632 hrs	30 days	5/22	6/30	0%	
488	Test Cases/Data/Results	616 hrs	36 days	3/30	5/18	0%	
500	User Documentation/Procedures	376 hrs	17 days	8/7	8/29	0%	
504	Operations Manual	96 hrs	12 days	3/27	4/11	0%	
507	DATA COMPONENTS	176 hrs	75 days	2/24	6/8	6%	
508	Constructed Physical Database	176 hrs	75 days	2/24	6/8	6%	

Figure A.5 Detailed Microsoft Project schedule—deliverables level: construction, part 1.

	Task Name	Work	Duration	Start	Finish	% Complete	Resource
507	DATA COMPONENTS	176 hrs	75 days	2/24	6/8	6%	
508	Constructed Physical Database	176 hrs	75 days	2/24	6/8	6%	
516	PROCESS COMPONENTS	2,136 hrs	61 days	6/12	9/4	0%	
517	Constructed Modules/Programs	2,072 hrs	60 days	6/12	9/1	0%	
532	Bridges (Conversion Routines)	56 hrs	7 days	7/20	7/28	0%	
537	Build/Install	8 hrs	3 days	8/30	9/4	0%	
541	USER INTERFACE COMPONENTS	72 hrs	2 days	6/17	6/20	0%	
542	Constructed User Interface Objects & I	72 hrs	2 days	6/17	6/20	0%	
546	TECHNOLOGY COMPONENTS	224 hrs	45 days	6/7	8/8	30%	
547	Constructed Network	224 hrs	45 days	6/7	8/8	30%	
553	QUALITY ASSURANCE(SYSTEM TESTS)	1,016 hrs	27 days	7/31	9/5	0%	
554	Verified Systems Construction	960 hrs	27 days	7/31	9/5	0%	
565	Verified Network Installation	0 hrs	2 days	8/9	8/11	0%	
568	Verified Users/Operations Documentat	56 hrs	1 day	9/4	9/4	0%	
575	IMPLEMENTATION-TRANSITION/CONVERSION	2,008 hrs	174 days	4/3	11/30	0%	
576	PROJECT MANAGEMENT	104 hrs	10.2 days	4/3	4/17	0%	
577	Managment Reports	24 hrs	10.08 days	4/3	4/17	0%	
580	Project Plan	24 hrs	9.88 days	4/3	4/14	0%	
583	Issue Resolution	32 hrs	5.25 days	4/3	4/10	0%	
586	User Marketing	24 hrs	6 days	4/3	4/10	0%	
589	STANDARDS/PROCEDURES/TRAINING	112 hrs	2 days	8/4	8/7	0%	
590	Training Implementation	112 hrs	2 days	8/4	8/7	0%	
598	CONVERSION/TRANSITION IMPLEMENTATIO	56 hrs	5 days	7/3	7/7	0%	

Figure A.6 Detailed Microsoft Project schedule—deliverables level: construction, part 2.

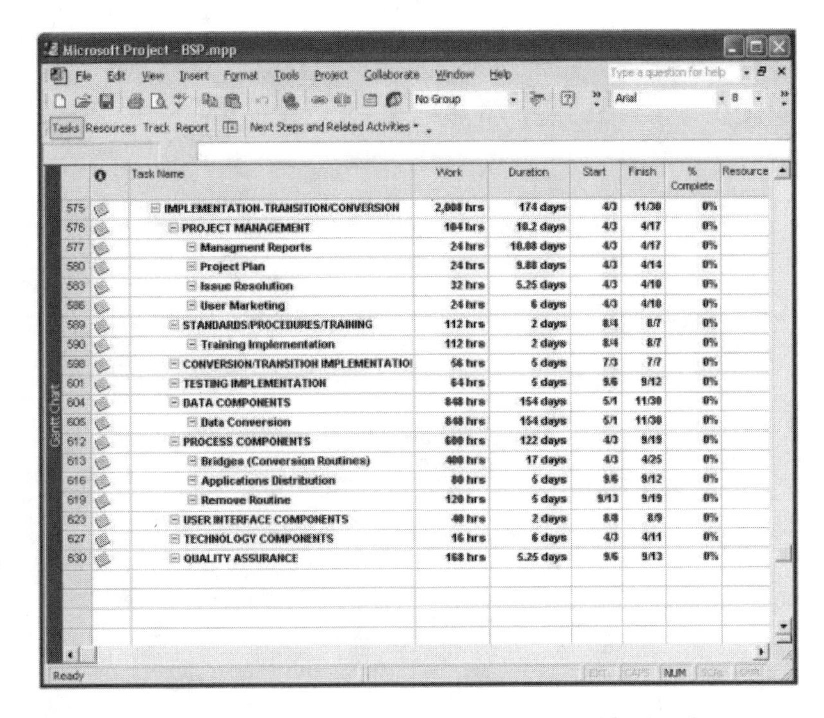

	Task Name	Work	Duration	Start	Finish	% Complete	Resource
575	IMPLEMENTATION-TRANSITION/CONVERSION	2,008 hrs	174 days	4/3	11/30	0%	
576	PROJECT MANAGEMENT	104 hrs	10.2 days	4/3	4/17	0%	
577	Managment Reports	24 hrs	10.08 days	4/3	4/17	0%	
580	Project Plan	24 hrs	9.88 days	4/3	4/14	0%	
583	Issue Resolution	32 hrs	5.25 days	4/3	4/10	0%	
586	User Marketing	24 hrs	6 days	4/3	4/10	0%	
589	STANDARDS/PROCEDURES/TRAINING	112 hrs	2 days	8/4	8/7	0%	
590	Training Implementation	112 hrs	2 days	8/4	8/7	0%	
598	CONVERSION/TRANSITION IMPLEMENTATIO	56 hrs	5 days	7/3	7/7	0%	
601	TESTING IMPLEMENTATION	64 hrs	5 days	9/6	9/12	0%	
604	DATA COMPONENTS	848 hrs	154 days	5/1	11/30	0%	
605	Data Conversion	848 hrs	154 days	5/1	11/30	0%	
612	PROCESS COMPONENTS	600 hrs	122 days	4/3	9/19	0%	
613	Bridges (Conversion Routines)	400 hrs	17 days	4/3	4/25	0%	
616	Applications Distribution	80 hrs	5 days	9/6	9/12	0%	
619	Remove Routine	120 hrs	5 days	9/13	9/19	0%	
623	USER INTERFACE COMPONENTS	40 hrs	2 days	8/8	8/9	0%	
627	TECHNOLOGY COMPONENTS	16 hrs	6 days	4/3	4/11	0%	
630	QUALITY ASSURANCE	168 hrs	5.25 days	9/6	9/13	0%	

Figure A.7 Detailed Microsoft Project schedule—deliverables level: implementation.

Appendix B: IPM-IT Detailed Microsoft Project Schedule—Deliverables Level

	Task Name	Work	Duration	Start	Finish	% Complete	Resource
1	BILLING SYSTEMS PROJECT	##########	192 days	3/8	11/30	21%	
2	DEFINITION PHASE-PDP	3,319.2 hrs	51 days	4/3	6/12	64%	
3	PROJECT MANAGEMENT	418.2 hrs	21 days	4/3	5/1	64%	
4	Managment Reports	32 hrs	2 days	4/3	4/4	100%	
7	Project Plan	182 hrs	14 days	4/3	4/20	65%	
13	Roles & Responsibilities	17 hrs	20 days	4/3	4/28	67%	
16	Project Charter	131.2 hrs	21 days	4/3	5/1	42%	
20	Project Scope	48 hrs	1 day	4/16	4/17	100%	
23	STANDARDS/PROCEDURES/TRAINING	1,427 hrs	51 days	4/3	6/12	63%	
24	Methodologies	216 hrs	18 days	4/3	4/26	53%	
33	Repository Standards & Procedures	64 hrs	6 days	4/3	4/10	100%	
37	Data Standards	456 hrs	51 days	4/3	6/12	90%	
40	GUI Standards	56 hrs	17 days	4/6	4/29	91%	
43	Process Design Standards	80 hrs	9 days	4/3	4/13	100%	
46	Testing Standards & Procedures	328 hrs	16 days	4/4	4/25	27%	
54	Version Control Standards	48 hrs	19 days	4/3	4/27	0%	
57	Change Control Stds & Procedures	41 hrs	16.25 days	4/11	5/3	0%	
60	Prototype/Walkthru Procedures	56 hrs	9.2 days	4/3	4/14	29%	
64	Issue Resolution Procedure	64 hrs	14.13 days	4/3	4/21	10%	
67	Training Plans	18 hrs	8 days	4/3	4/12	100%	
70	DATA COMPONENTS	392 hrs	7.88 days	4/3	4/12	85%	
71	Entity/Relationship Model	392 hrs	7.88 days	4/3	4/12	85%	
83	PROCESS COMPONENTS	490 hrs	18 days	4/5	4/28	53%	

Figure B.1 IPM-IT detailed Microsoft Project schedule—deliverables level: PDP definition, part 1.

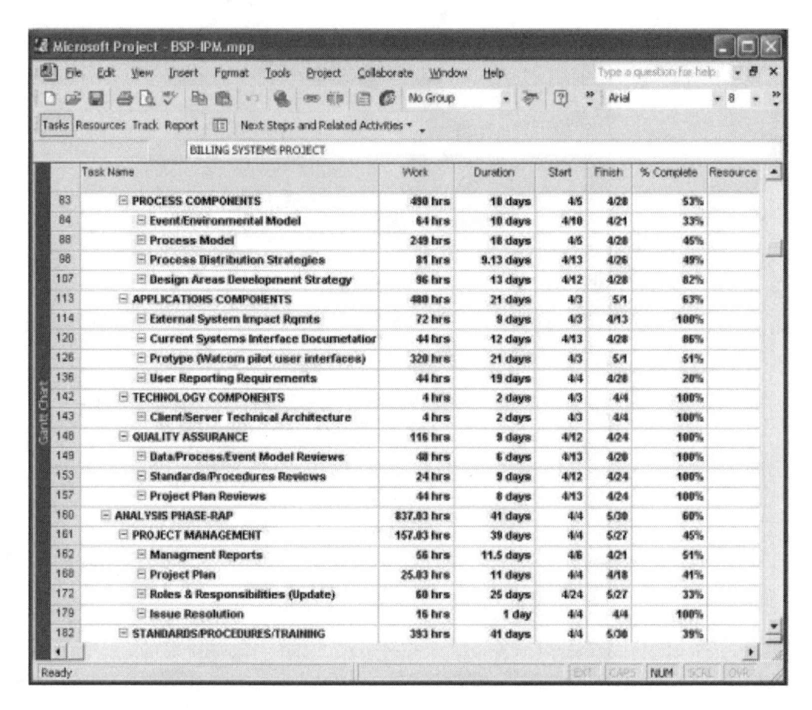

Microsoft Project - BSP-IPM.mpp

BILLING SYSTEMS PROJECT

	Task Name	Work	Duration	Start	Finish	% Complete	Resource
83	PROCESS COMPONENTS	490 hrs	18 days	4/5	4/28	53%	
84	Event/Environmental Model	64 hrs	10 days	4/10	4/21	33%	
88	Process Model	249 hrs	18 days	4/5	4/28	45%	
98	Process Distribution Strategies	81 hrs	9.13 days	4/13	4/26	49%	
107	Design Areas Development Strategy	96 hrs	13 days	4/12	4/28	82%	
113	APPLICATIONS COMPONENTS	480 hrs	21 days	4/3	5/1	63%	
114	External System Impact Rqmts	72 hrs	9 days	4/3	4/13	100%	
120	Current Systems Interface Documetatior	44 hrs	12 days	4/13	4/28	86%	
126	Protype (Watcom pilot user interfaces)	320 hrs	21 days	4/3	5/1	51%	
136	User Reporting Requirements	44 hrs	19 days	4/4	4/28	20%	
142	TECHNOLOGY COMPONENTS	4 hrs	2 days	4/3	4/4	100%	
143	Client/Server Technical Architecture	4 hrs	2 days	4/3	4/4	100%	
148	QUALITY ASSURANCE	116 hrs	9 days	4/12	4/24	100%	
149	Data/Process/Event Model Reviews	48 hrs	6 days	4/13	4/20	100%	
153	Standards/Procedures Reviews	24 hrs	9 days	4/12	4/24	100%	
157	Project Plan Reviews	44 hrs	8 days	4/13	4/24	100%	
160	ANALYSIS PHASE-RAP	837.03 hrs	41 days	4/4	5/30	60%	
161	PROJECT MANAGEMENT	157.03 hrs	39 days	4/4	5/27	45%	
162	Managment Reports	56 hrs	11.5 days	4/6	4/21	51%	
168	Project Plan	25.03 hrs	11 days	4/4	4/18	41%	
172	Roles & Responsibilities (Update)	60 hrs	25 days	4/24	5/27	33%	
179	Issue Resolution	16 hrs	1 day	4/4	4/4	100%	
182	STANDARDS/PROCEDURES/TRAINING	393 hrs	41 days	4/4	5/30	39%	

Figure B.2 IPM-IT detailed Microsoft Project schedule—deliverables level: PDP definition, part 2.

Microsoft Project - BSP-IPM.mpp

BILLING SYSTEMS PROJECT

	Task Name	Work	Duration	Start	Finish	% Complete	Resource
160	ANALYSIS PHASE-RAP	837.03 hrs	41 days	4/4	5/30	60%	
161	PROJECT MANAGEMENT	157.03 hrs	39 days	4/4	5/27	45%	
162	Managment Reports	56 hrs	11.5 days	4/6	4/21	51%	
168	Project Plan	25.03 hrs	11 days	4/4	4/18	41%	
172	Roles & Responsibilities (Update)	60 hrs	25 days	4/24	5/27	33%	
179	Issue Resolution	16 hrs	1 day	4/4	4/4	100%	
182	STANDARDS/PROCEDURES/TRAINING	393 hrs	41 days	4/4	5/30	39%	
183	Library Management Stds	88 hrs	15 days	4/16	5/5	6%	
189	SA Repository Stds/Procedures	20 hrs	2 days	5/3	5/4	0%	
192	Reusable Objects Components Stds	48 hrs	22 days	4/4	5/3	67%	
196	Conversion Strategies	40 hrs	4 days	4/20	4/25	94%	
199	Development/Production Tools	69 hrs	41 days	4/4	5/30	36%	
207	Training (Project Team)	128 hrs	41 days	4/4	5/30	43%	
216	DATA COMPONENTS	81 hrs	6 days	4/4	4/11	100%	
217	Data Architecture	81 hrs	6 days	4/4	4/11	100%	
223	PROCESS COMPONENTS	96 hrs	34 days	4/4	5/20	93%	
224	Application Architecture	96 hrs	34 days	4/4	5/20	93%	
236	USER INTERFACE COMPONENTS	16 hrs	3 days	4/4	4/6	100%	
237	User Interface Architecture	16 hrs	3 days	4/4	4/6	100%	
240	TECHNOLOGY COMPONENTS	72 hrs	35.25 days	4/4	5/23	74%	
241	Technology Architecture	72 hrs	35.25 days	4/4	5/23	74%	
247	QUALITY ASSURANCE	22 hrs	1 day	4/4	4/4	100%	
248	Data/Applications/Technology Architectu	8 hrs	1 day	4/4	4/4	100%	

Figure B.3 IPM-IT detailed Microsoft Project schedule—deliverables level: RAP analysis.

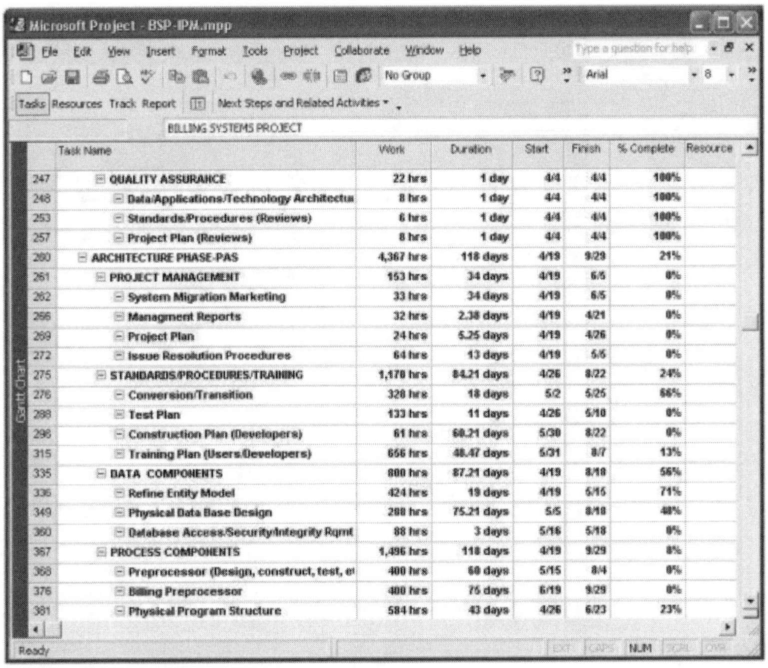

Microsoft Project - BSP-IPM.mpp

BILLING SYSTEMS PROJECT

	Task Name	Work	Duration	Start	Finish	% Complete	Resource
247	QUALITY ASSURANCE	22 hrs	1 day	4/4	4/4	100%	
248	Data/Applications/Technology Architectu	8 hrs	1 day	4/4	4/4	100%	
253	Standards/Procedures (Reviews)	6 hrs	1 day	4/4	4/4	100%	
257	Project Plan (Reviews)	8 hrs	1 day	4/4	4/4	100%	
260	ARCHITECTURE PHASE-PAS	4,367 hrs	118 days	4/19	9/29	21%	
261	PROJECT MANAGEMENT	153 hrs	34 days	4/19	6/5	0%	
262	System Migration Marketing	33 hrs	34 days	4/19	6/5	0%	
266	Managment Reports	32 hrs	2.38 days	4/19	4/21	0%	
269	Project Plan	24 hrs	5.25 days	4/19	4/26	0%	
272	Issue Resolution Procedures	64 hrs	13 days	4/19	5/6	0%	
275	STANDARDS/PROCEDURES/TRAINING	1,178 hrs	84.21 days	4/26	8/22	24%	
276	Conversion/Transition	328 hrs	18 days	5/2	5/25	66%	
288	Test Plan	133 hrs	11 days	4/26	5/10	0%	
296	Construction Plan (Developers)	61 hrs	60.21 days	5/30	8/22	0%	
315	Training Plan (Users/Developers)	656 hrs	48.47 days	5/31	8/7	13%	
335	DATA COMPONENTS	800 hrs	87.21 days	4/19	8/18	56%	
336	Refine Entity Model	424 hrs	19 days	4/19	5/16	71%	
349	Physical Data Base Design	288 hrs	75.21 days	5/5	8/18	48%	
360	Database Access/Security/Integrity Rqmt	88 hrs	3 days	5/16	5/18	0%	
367	PROCESS COMPONENTS	1,496 hrs	118 days	4/19	9/29	8%	
368	Preprocessor (Design, construct, test, e	400 hrs	60 days	5/15	8/4	0%	
376	Billing Preprocessor	400 hrs	75 days	6/19	9/29	0%	
381	Physical Program Structure	584 hrs	43 days	4/26	6/23	23%	

Ready NUM

Figure B.4 IPM-IT detailed Microsoft Project schedule—deliverables level: PAS architecture.

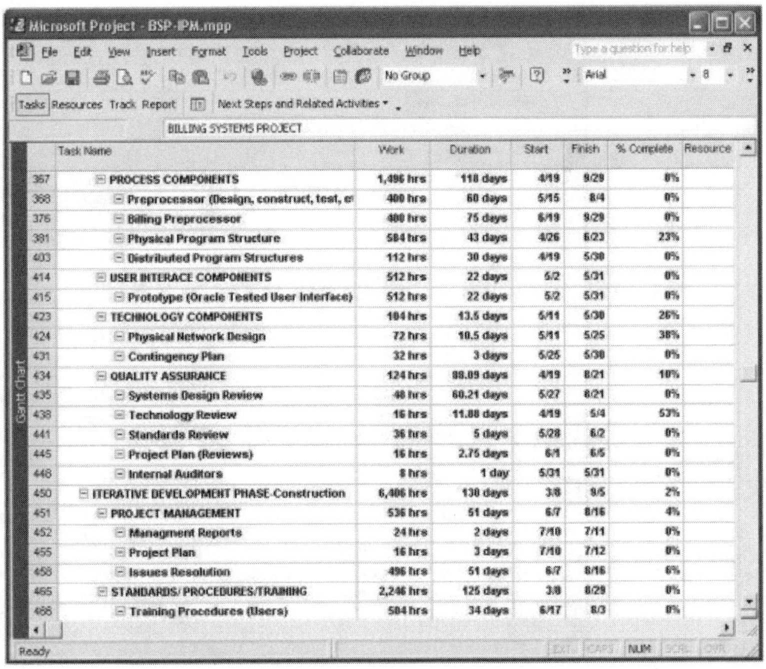

Microsoft Project - BSP-IPM.mpp

BILLING SYSTEMS PROJECT

	Task Name	Work	Duration	Start	Finish	% Complete	Resource
367	PROCESS COMPONENTS	1,496 hrs	118 days	4/19	9/29	8%	
368	Preprocessor (Design, construct, test, e	400 hrs	60 days	5/15	8/4	0%	
376	Billing Preprocessor	400 hrs	75 days	6/19	9/29	0%	
381	Physical Program Structure	584 hrs	43 days	4/26	6/23	23%	
403	Distributed Program Structures	112 hrs	30 days	4/19	5/30	0%	
414	USER INTERACE COMPONENTS	512 hrs	22 days	5/2	5/31	0%	
415	Prototype (Oracle Tested User Interface)	512 hrs	22 days	5/2	5/31	0%	
423	TECHNOLOGY COMPONENTS	104 hrs	13.5 days	5/11	5/30	26%	
424	Physical Network Design	72 hrs	10.5 days	5/11	5/25	38%	
431	Contingency Plan	32 hrs	3 days	5/25	5/30	0%	
434	QUALITY ASSURANCE	124 hrs	88.09 days	4/19	8/21	10%	
435	Systems Design Review	48 hrs	60.21 days	5/27	8/21	0%	
438	Technology Review	16 hrs	11.88 days	4/19	5/4	53%	
441	Standards Review	36 hrs	5 days	5/28	6/2	0%	
445	Project Plan (Reviews)	16 hrs	2.75 days	6/1	6/5	0%	
448	Internal Auditors	8 hrs	1 day	5/31	5/31	0%	
450	ITERATIVE DEVELOPMENT PHASE-Construction	6,406 hrs	130 days	3/8	9/5	2%	
451	PROJECT MANAGEMENT	536 hrs	51 days	6/7	8/16	4%	
452	Managment Reports	24 hrs	2 days	7/10	7/11	0%	
455	Project Plan	16 hrs	3 days	7/10	7/12	0%	
458	Issues Resolution	496 hrs	51 days	6/7	8/16	6%	
465	STANDARDS/ PROCEDURES/TRAINING	2,246 hrs	125 days	3/8	8/29	0%	
466	Training Procedures (Users)	504 hrs	34 days	6/17	8/3	0%	

Ready NUM

Figure B.5 IPM-IT detailed Microsoft Project schedule—deliverables level: IDP construction, part 1.

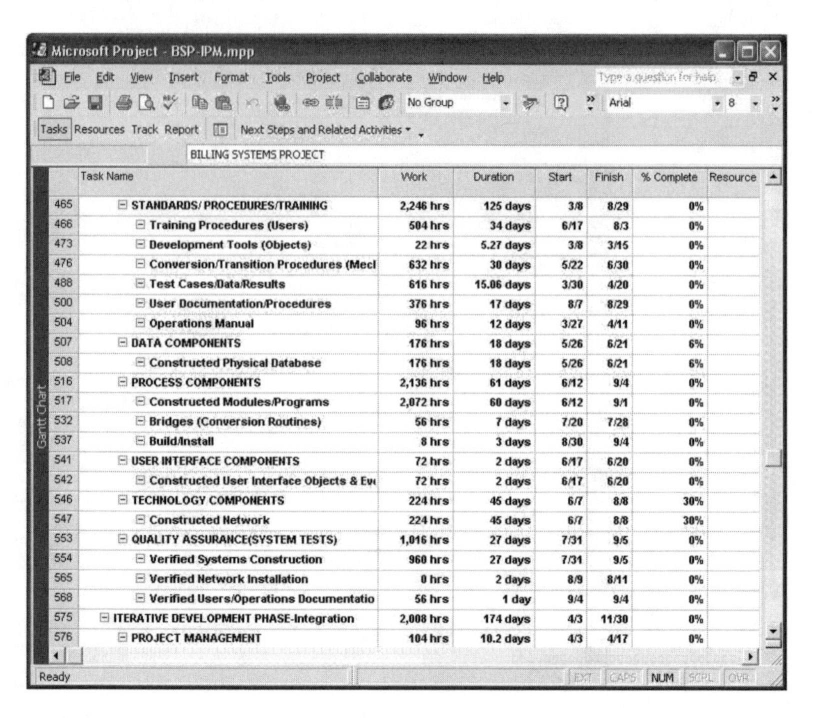

Microsoft Project - BSP-IPM.mpp

BILLING SYSTEMS PROJECT

	Task Name	Work	Duration	Start	Finish	% Complete	Resource
465	STANDARDS/ PROCEDURES/TRAINING	2,246 hrs	125 days	3/8	8/29	0%	
466	Training Procedures (Users)	504 hrs	34 days	6/17	8/3	0%	
473	Development Tools (Objects)	22 hrs	5.27 days	3/8	3/15	0%	
476	Conversion/Transition Procedures (Mecl	632 hrs	30 days	5/22	6/30	0%	
488	Test Cases/Data/Results	616 hrs	15.06 days	3/30	4/20	0%	
500	User Documentation/Procedures	376 hrs	17 days	8/7	8/29	0%	
504	Operations Manual	96 hrs	12 days	3/27	4/11	0%	
507	DATA COMPONENTS	176 hrs	18 days	5/26	6/21	6%	
508	Constructed Physical Database	176 hrs	18 days	5/26	6/21	6%	
516	PROCESS COMPONENTS	2,136 hrs	61 days	6/12	9/4	0%	
517	Constructed Modules/Programs	2,072 hrs	60 days	6/12	9/1	0%	
532	Bridges (Conversion Routines)	56 hrs	7 days	7/20	7/28	0%	
537	Build/Install	8 hrs	3 days	8/30	9/4	0%	
541	USER INTERFACE COMPONENTS	72 hrs	2 days	6/17	6/20	0%	
542	Constructed User Interface Objects & Ev	72 hrs	2 days	6/17	6/20	0%	
546	TECHNOLOGY COMPONENTS	224 hrs	45 days	6/7	8/8	30%	
547	Constructed Network	224 hrs	45 days	6/7	8/8	30%	
553	QUALITY ASSURANCE(SYSTEM TESTS)	1,016 hrs	27 days	7/31	9/5	0%	
554	Verified Systems Construction	960 hrs	27 days	7/31	9/5	0%	
565	Verified Network Installation	0 hrs	2 days	8/9	8/11	0%	
568	Verified Users/Operations Documentatio	56 hrs	1 day	9/4	9/4	0%	
575	ITERATIVE DEVELOPMENT PHASE-Integration	2,008 hrs	174 days	4/3	11/30	0%	
576	PROJECT MANAGEMENT	104 hrs	10.2 days	4/3	4/17	0%	

Figure B.6 IPM-IT detailed Microsoft Project schedule—deliverables level: IDP construction, part 2.

Microsoft Project - BSP-IPM.mpp

BILLING SYSTEMS PROJECT

	Task Name	Work	Duration	Start	Finish	% Complete	Resource
575	ITERATIVE DEVELOPMENT PHASE-Integration	2,008 hrs	174 days	4/3	11/30	0%	
576	PROJECT MANAGEMENT	104 hrs	10.2 days	4/3	4/17	0%	
577	Managment Reports	24 hrs	10.08 days	4/3	4/17	0%	
580	Project Plan	24 hrs	9.88 days	4/3	4/14	0%	
583	Issue Resolution	32 hrs	5.13 days	4/3	4/10	0%	
586	User Marketing	24 hrs	6 days	4/3	4/10	0%	
589	STANDARDS/PROCEDURES/TRAINING	112 hrs	2 days	8/4	8/7	0%	
590	Integration Implementation	112 hrs	2 days	8/4	8/7	0%	
598	INTEGRATION/CONVERSION IMPLEMENTATION	56 hrs	5 days	7/3	7/7	0%	
601	INTEGRATION TESTING IMPLEMENTATION	64 hrs	5 days	9/6	9/12	0%	
604	DATA COMPONENTS	848 hrs	154 days	5/1	11/30	0%	
605	Data Integration/Conversion	848 hrs	154 days	5/1	11/30	0%	
612	PROCESS COMPONENTS	600 hrs	122 days	4/3	9/19	0%	
613	Bridges (Conversion Routines)	400 hrs	17 days	4/3	4/25	0%	
616	Applications Integration	80 hrs	5 days	9/6	9/12	0%	
619	Integration Performance	120 hrs	5 days	9/13	9/19	0%	
623	USER INTERFACE COMPONENTS	40 hrs	2 days	8/8	8/9	0%	
627	TECHNOLOGY COMPONENTS	16 hrs	6 days	4/3	4/11	0%	
630	QUALITY ASSURANCE	168 hrs	5.25 days	9/6	9/13	0%	
638	ITERATIVE DEVELOPMENT PHASE-Deployment	2,008 hrs	174 days	4/3	11/30	0%	
639	PROJECT MANAGEMENT	104 hrs	10.88 days	4/3	4/17	0%	
640	Managment Reports	24 hrs	10.88 days	4/3	4/17	0%	
643	Project Plan	24 hrs	5.88 days	4/3	4/10	0%	

Figure B.7 IPM-IT detailed Microsoft Project schedule—deliverables level: IDP integration.

Microsoft Project - BSP-IPM.mpp

BILLING SYSTEMS PROJECT

	Task Name	Work	Duration	Start	Finish	% Complete	Resource
638	⊟ ITERATIVE DEVELOPMENT PHASE-Deployment	2,008 hrs	174 days	4/3	11/30	0%	
639	⊟ PROJECT MANAGEMENT	104 hrs	10.88 days	4/3	4/17	0%	
640	⊟ Managment Reports	24 hrs	10.88 days	4/3	4/17	0%	
643	⊟ Project Plan	24 hrs	5.88 days	4/3	4/10	0%	
646	⊟ Issue Resolution	32 hrs	6.88 days	4/3	4/11	0%	
649	⊟ User Marketing	24 hrs	2 days	4/10	4/11	0%	
652	⊟ STANDARDS/PROCEDURES/TRAINING	112 hrs	7.13 days	4/3	4/12	0%	
653	⊟ Deployment/Training Implementation	112 hrs	7.13 days	4/3	4/12	0%	
661	⊟ DEPLOYMENT IMPLEMENTATION	56 hrs	5 days	4/10	4/14	0%	
664	⊟ ACCEPTANCE TESTING IMPLEMENTATION	64 hrs	8.13 days	4/10	4/20	0%	
667	⊟ DATA COMPONENTS	848 hrs	174 days	4/3	11/30	0%	
668	⊟ Data Base Deployment	848 hrs	174 days	4/3	11/30	0%	
675	⊟ PROCESS COMPONENTS	600 hrs	17 days	4/3	4/25	0%	
676	⊟ Bridges Deployment	400 hrs	17 days	4/3	4/25	0%	
679	⊟ Applications Deployment	80 hrs	5.13 days	4/10	4/17	0%	
682	⊟ System Production Performance	120 hrs	5 days	4/17	4/24	0%	
686	⊟ USER INTERFACE COMPONENTS	40 hrs	2 days	4/12	4/14	0%	
690	⊟ TECHNOLOGY COMPONENTS	16 hrs	6.13 days	4/3	4/11	0%	
693	⊟ QUALITY ASSURANCE	168 hrs	13.2 days	4/3	4/20	0%	

Figure B.8 IPM-IT detailed Microsoft Project schedule—deliverables level: IDP deployment.

Appendix C: RUP Detailed Microsoft Project Schedule—RUP Deliverables Level

Microsoft Project - BSP-RUP.mpp

BILLING SYSTEMS PROJECT

	Task Name	Work	Duration	Start	Finish	% Complete	Resource
1	⊟ BILLING SYSTEMS PROJECT	###########	387.82 days	4/3	9/25	15%	
2	⊟ INCEPTION PHASE-Iteration #1	3,327.2 hrs	51 days	4/3	6/12	64%	
3	⊟ PROJECT MANAGEMENT	410.2 hrs	21 days	4/3	5/1	64%	
4	⊟ Project Management Artifact Set	32 hrs	2 days	4/3	4/4	100%	
7	⊟ Environment Artifact Set	182 hrs	14 days	4/3	4/20	65%	
13	⊞ Configuration & Change Artifact Set	196.2 hrs	21 days	4/3	5/1	55%	
23	⊟ STANDARDS/PROCEDURES/TRAINING	1,427 hrs	51 days	4/3	6/12	63%	
24	⊟ Methodologies	216 hrs	18 days	4/3	4/26	53%	
33	⊟ Repository Standards & Procedures	64 hrs	6 days	4/3	4/10	100%	
37	⊟ Data Standards	456 hrs	51 days	4/3	6/12	90%	
40	⊟ GUI Standards	56 hrs	17 days	4/6	4/29	91%	
43	⊟ Process Design Standards	80 hrs	9 days	4/3	4/13	100%	
46	⊟ Testing Standards & Procedures	328 hrs	16 days	4/4	4/25	27%	
54	⊟ Version Control Standards	48 hrs	19 days	4/3	4/27	0%	
57	⊟ Change Control Stds & Procedures	41 hrs	23 days	4/3	5/3	0%	
60	⊟ Prototype/Walkthru Procedures	56 hrs	9.2 days	4/3	4/14	29%	
64	⊟ Issue Resolution Procedure	64 hrs	5.5 days	4/3	4/10	9%	
67	⊟ Training Plans	18 hrs	8 days	4/3	4/12	100%	
70	⊟ ANALYSIS MODEL	400 hrs	6 days	4/3	4/10	82%	
71	⊞ Business Modeling Artifact Set	144 hrs	2 days	4/3	4/4	100%	
77	⊞ Requirements Artifact Set	256 hrs	6 days	4/3	4/10	73%	SI,UI,TT,C
83	⊟ DESIGN MODEL	490 hrs	18 days	4/5	4/28	52%	
84	⊞ Analysis & Design Artifact Set	490 hrs	18 days	4/5	4/28	52%	

Figure C.1 RUP detailed Microsoft Project schedule—RUP deliverables level: inception, Iteration #1.

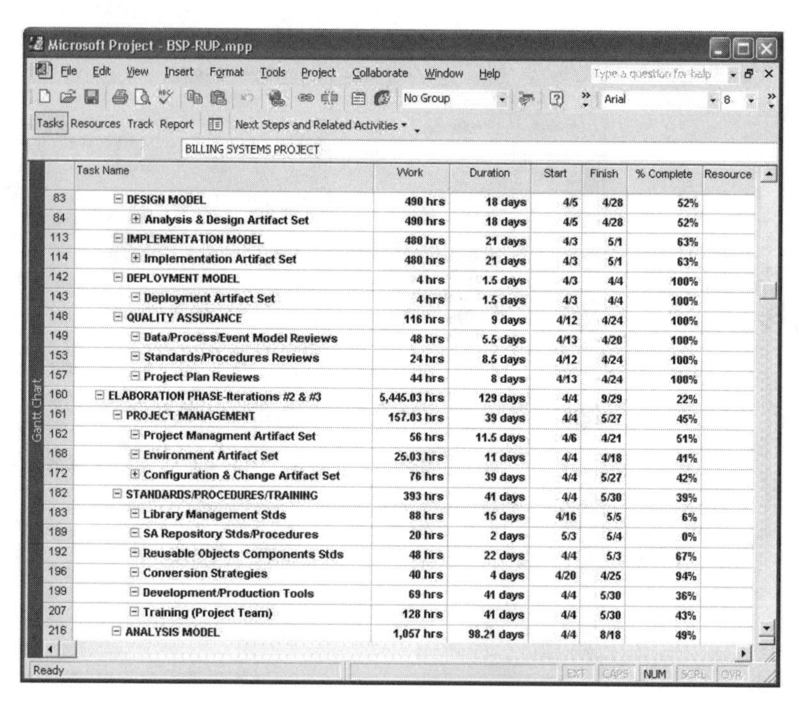

	Task Name	Work	Duration	Start	Finish	% Complete	Resource
83	⊟ DESIGN MODEL	490 hrs	18 days	4/5	4/28	52%	
84	⊞ Analysis & Design Artifact Set	490 hrs	18 days	4/5	4/28	52%	
113	⊟ IMPLEMENTATION MODEL	480 hrs	21 days	4/3	5/1	63%	
114	⊞ Implementation Artifact Set	480 hrs	21 days	4/3	5/1	63%	
142	⊟ DEPLOYMENT MODEL	4 hrs	1.5 days	4/3	4/4	100%	
143	⊟ Deployment Artifact Set	4 hrs	1.5 days	4/3	4/4	100%	
148	⊟ QUALITY ASSURANCE	116 hrs	9 days	4/12	4/24	100%	
149	⊟ Data/Process/Event Model Reviews	48 hrs	5.5 days	4/13	4/20	100%	
153	⊟ Standards/Procedures Reviews	24 hrs	8.5 days	4/12	4/24	100%	
157	⊟ Project Plan Reviews	44 hrs	8 days	4/13	4/24	100%	
160	⊟ ELABORATION PHASE-Iterations #2 & #3	5,445.03 hrs	129 days	4/4	9/29	22%	
161	⊟ PROJECT MANAGEMENT	157.03 hrs	39 days	4/4	5/27	45%	
162	⊟ Project Managment Artifact Set	56 hrs	11.5 days	4/6	4/21	51%	
168	⊟ Environment Artifact Set	25.03 hrs	11 days	4/4	4/18	41%	
172	⊞ Configuration & Change Artifact Set	76 hrs	39 days	4/4	5/27	42%	
182	⊟ STANDARDS/PROCEDURES/TRAINING	393 hrs	41 days	4/4	5/30	39%	
183	⊟ Library Management Stds	88 hrs	15 days	4/16	5/5	6%	
189	⊟ SA Repository Stds/Procedures	20 hrs	2 days	5/3	5/4	0%	
192	⊟ Reusable Objects Components Stds	48 hrs	22 days	4/4	5/3	67%	
196	⊟ Conversion Strategies	40 hrs	4 days	4/20	4/25	94%	
199	⊟ Development/Production Tools	69 hrs	41 days	4/4	5/30	36%	
207	⊟ Training (Project Team)	128 hrs	41 days	4/4	5/30	43%	
216	⊟ ANALYSIS MODEL	1,057 hrs	98.21 days	4/4	8/18	49%	

Figure C.2 RUP detailed Microsoft Project schedule—RUP deliverables level: elaboration, Iterations #2 and #3.

Microsoft Project - BSP-RUP.mpp

File Edit View Insert Format Tools Project Collaborate Window Help

Tasks Resources Track Report Next Steps and Related Activities

BILLING SYSTEMS PROJECT

	Task Name	Work	Duration	Start	Finish	% Complete	Resource
216	⊟ ANALYSIS MODEL	1,057 hrs	98.21 days	4/4	8/18	49%	
217	⊞ Business Modelling Artifact Set	457 hrs	30 days	4/4	5/15	72%	
233	⊞ Requirements Artifact Set	600 hrs	98.21 days	4/4	8/18	31%	
264	⊟ DESIGN MODEL	3,728 hrs	129 days	4/4	9/29	7%	
265	⊞ Analysis & Design Artifact Set	3,728 hrs	129 days	4/4	9/29	7%	
349	⊟ IMPLEMENTATION MODEL	16 hrs	2.5 days	4/4	4/6	100%	
350	⊟ Implementation Artifact Set	16 hrs	2.5 days	4/4	4/6	100%	
353	⊟ DEPLOYMENT MODEL	72 hrs	35.25 days	4/4	5/23	74%	
354	⊟ Deployment Artifact Set	72 hrs	35.25 days	4/4	5/23	74%	
360	⊟ QUALITY ASSURANCE	22 hrs	1 day	4/4	4/4	100%	
361	⊟ Data/Applications/Technology Architectu	8 hrs	0.5 days	4/4	4/4	100%	
366	⊟ Standards/Procedures (Reviews)	6 hrs	0.5 days	4/4	4/4	100%	
370	⊟ Project Plan (Reviews)	8 hrs	1 day	4/4	4/4	100%	
373	⊟ CONSTRUCTION PHASE- Iterations # 4, #5 & #6	9,207 hrs	191.82 days	4/26	1/18	11%	
374	⊟ PROJECT MANAGEMENT	153 hrs	83 days	5/3	8/28	0%	
375	⊟ Project Management Artifact Set	33 hrs	58.38 days	6/2	8/23	0%	
379	⊞ Environment Artifact Set	56 hrs	5.25 days	8/21	8/28	0%	
385	⊟ Configuration & Change Artifact Set	64 hrs	79.75 days	5/3	8/22	0%	
388	⊟ STANDARDS/PROCEDURES/TRAINING	1,178 hrs	103 days	5/2	9/21	24%	
389	⊟ Conversion/Transition	328 hrs	81.4 days	5/2	8/23	66%	
401	⊟ Test Plan	133 hrs	79 days	5/3	8/21	0%	
409	⊟ Construction Plan (Developers)	61 hrs	83 days	5/30	9/21	0%	
428	⊟ Training Plan (Users/Developers)	656 hrs	63 days	5/31	8/25	13%	

Ready NUM

Figure C.3 RUP detailed Microsoft Project schedule—RUP deliverables level: construction, Iterations #4, #5, and #6.

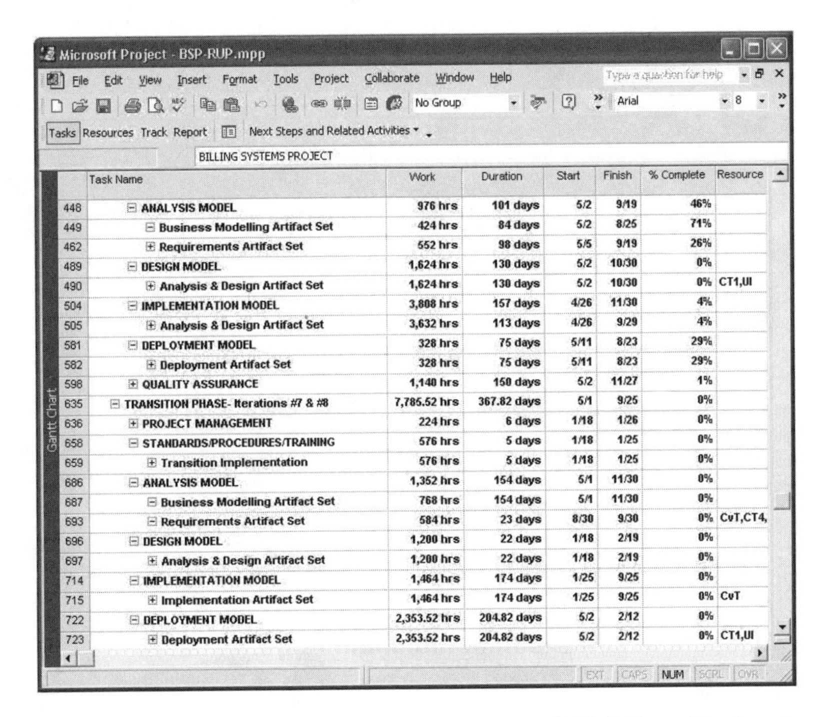

Microsoft Project - BSP-RUP.mpp

BILLING SYSTEMS PROJECT

	Task Name	Work	Duration	Start	Finish	% Complete	Resource
448	ANALYSIS MODEL	976 hrs	101 days	5/2	9/19	46%	
449	Business Modelling Artifact Set	424 hrs	84 days	5/2	8/25	71%	
462	Requirements Artifact Set	552 hrs	98 days	5/5	9/19	26%	
489	DESIGN MODEL	1,624 hrs	130 days	5/2	10/30	0%	
490	Analysis & Design Artifact Set	1,624 hrs	130 days	5/2	10/30	0%	CT1,UI
504	IMPLEMENTATION MODEL	3,808 hrs	157 days	4/26	11/30	4%	
505	Analysis & Design Artifact Set	3,632 hrs	113 days	4/26	9/29	4%	
581	DEPLOYMENT MODEL	328 hrs	75 days	5/11	8/23	29%	
582	Deployment Artifact Set	328 hrs	75 days	5/11	8/23	29%	
598	QUALITY ASSURANCE	1,140 hrs	150 days	5/2	11/27	1%	
635	TRANSITION PHASE- Iterations #7 & #8	7,785.52 hrs	367.82 days	5/1	9/25	0%	
636	PROJECT MANAGEMENT	224 hrs	6 days	1/18	1/26	0%	
658	STANDARDS/PROCEDURES/TRAINING	576 hrs	5 days	1/18	1/25	0%	
659	Transition Implementation	576 hrs	5 days	1/18	1/25	0%	
686	ANALYSIS MODEL	1,352 hrs	154 days	5/1	11/30	0%	
687	Business Modelling Artifact Set	768 hrs	154 days	5/1	11/30	0%	
693	Requirements Artifact Set	584 hrs	23 days	8/30	9/30	0%	CvT,CT4,
696	DESIGN MODEL	1,200 hrs	22 days	1/18	2/19	0%	
697	Analysis & Design Artifact Set	1,200 hrs	22 days	1/18	2/19	0%	
714	IMPLEMENTATION MODEL	1,464 hrs	174 days	1/25	9/25	0%	
715	Implementation Artifact Set	1,464 hrs	174 days	1/25	9/25	0%	CvT
722	DEPLOYMENT MODEL	2,353.52 hrs	204.82 days	5/2	2/12	0%	
723	Deployment Artifact Set	2,353.52 hrs	204.82 days	5/2	2/12	0%	CT1,UI

Figure C.4 RUP detailed Microsoft Project schedule—RUP deliverables level: transition, Iterations #7 and #8.

About the Author

Kenneth R. Bainey is a senior information technology (IT) professional with 28 years of industry experience, including 10 years as a consultant, performing leadership roles in project management and the implementation of numerous business application systems in Canada and the United States. As a senior project manager, senior systems/data architect, and IT management consultant, he has acquired excellent specialized skills in project management, IT management, and business management.

His senior-level responsibilities include 10 years in telecommunications, 5 years in oil and gas, and 10 years in transportation and utilities, where he has performed various program/project management responsibilities. He performed the roles of IT program/project manager, enterprise/systems architect, and IT management consultant on many projects and implemented this model-centric approach to project management at various large corporations in Canada and the United States.

His professional qualifications include the PMP professional designation in project management from PMI-USA; the CCP professional designation in information technology management from ICCP-USA, and the CIM professional designation in business management from the Canadian Institute of Management. He is a graduate of the University of Alberta, Canada, where he majored in mathematics and computing science; he has also completed many graduate-level courses in computing science, business administration, and project management.

Mr. Bainey is currently the director and chief information officer of information and communications technology with the Ministry of Transportation in the government of Alberta, Canada. He is also a member of the Certification Council for ICCP.

Index

M

Q

Recent Titles in the Artech House Effective Project Management Library

Robert K. Wysocki, Series Editor

Integrated IT Project Management: A Model-Centric Approach, Kenneth R. Bainey

The Project Management Communications Toolkit, Carl Pritchard

Project Management Process Improvement, Robert K. Wysocki

For further information on these and other Artech House titles, including previously considered out-of-print books now available through our In-Print-Forever® (IPF®) program, contact:

Artech House
685 Canton Street
Norwood, MA 02062
Phone: 781-769-9750
Fax: 781-769-6334
e-mail: artech@artechhouse.com

Artech House
46 Gillingham Street
London SW1V 1AH UK
Phone: +44 (0)20 7596-8750
Fax: +44 (0)20 7630-0166
e-mail: artech-uk@artechhouse.com

Find us on the World Wide Web at:
www.artechhouse.com